AF772332

Bernd J. Krämer, Wolfgang A. Halang (Eds.)

Contributions to Ubiquitous Computing

Studies in Computational Intelligence, Volume 42

Editor-in-chief
Prof. Janusz Kacprzyk
Systems Research Institute
Polish Academy of Sciences
ul. Newelska 6
01-447 Warsaw
Poland
E-mail: kacprzyk@ibspan.waw.pl

Further volumes of this series can be found on our homepage: springer.com

Vol. 26. Nadia Nedjah, Luiza de Macedo Mourelle (Eds.)
Swarm Intelligent Systems, 2006
ISBN 3-540-33868-3

Vol. 27. Vassilis G. Kaburlasos
Towards a Unified Modeling and Knowledge-Representation based on Lattice Theory, 2006
ISBN 3-540-34169-2

Vol. 28. Brahim Chaib-draa, Jörg P. Müller (Eds.)
Multiagent based Supply Chain Management, 2006
ISBN 3-540-33875-6

Vol. 29. Sai Sumathi, S.N. Sivanandam
Introduction to Data Mining and its Application, 2006
ISBN 3-540-34689-9

Vol. 30. Yukio Ohsawa, Shusaku Tsumoto (Eds.)
Chance Discoveries in Real World Decision Making, 2006
ISBN 3-540-34352-0

Vol. 31. Ajith Abraham, Crina Grosan, Vitorino Ramos (Eds.)
Stigmergic Optimization, 2006
ISBN 3-540-34689-9

Vol. 32. Akira Hirose
Complex-Valued Neural Networks, 2006
ISBN 3-540-33456-4

Vol. 33. Martin Pelikan, Kumara Sastry, Erick Cantú-Paz (Eds.)
Scalable Optimization via Probabilistic Modeling, 2006
ISBN 3-540-34953-7

Vol. 34. Ajith Abraham, Crina Grosan, Vitorino Ramos (Eds.)
Swarm Intelligence in Data Mining, 2006
ISBN 3-540-34955-3

Vol. 35. Ke Chen, Lipo Wang (Eds.)
Trends in Neural Computation, 2007
ISBN 3-540-36121-9

Vol. 36. Ildar Batyrshin, Janusz Kacprzyk, Leonid Sheremetor, Lotfi A. Zadeh (Eds.)
Preception-based Data Mining and Decision Making in Economics and Finance, 2006
ISBN 3-540-36244-4

Vol. 37. Jie Lu, Da Ruan, Guangquan Zhang (Eds.)
E-Service Intelligence, 2007
ISBN 3-540-37015-3

Vol. 38. Art Lew, Holger Mauch
Dynamic Programming, 2007
ISBN 3-540-37013-7

Vol. 39. Gregory Levitin (Ed.)
Computational Intelligence in Reliability Engineering, 2007
ISBN 3-540-37367-5

Vol. 40. Gregory Levitin (Ed.)
Computational Intelligence in Reliability Engineering, 2007
ISBN 3-540-37371-3

Vol. 41. Mukesh Khare, S.M. Shiva Nagendra
Artificial Neural Networks in Vehicular Pollution Modelling, 2007
ISBN 3-540-37417-5

Vol. 42. Bernd J. Krämer, Wolfgang A. Halang (Eds.)
Contributions to Ubiquitous Computing, 2007
ISBN 3-540-44909-4

Bernd J. Krämer
Wolfgang A. Halang
(Eds.)

Contributions to Ubiquitous Computing

With 131 Figures and 12 Tables

Springer

Prof. Dr. Bernd J. Krämer
FernUniversität in Hagen
Lehrstuhl für Datenverarbeitungstechnik
58084 Hagen
Germany
E-mail: bernd.kraemer@fernuni-hagen.de

Prof. Dr. Wolfgang A. Halang
FernUniversität in Hagen
Lehrstuhl für Informationstechnik
58084 Hagen
Germany
E-mail: wolfgang.halang@FernUni-Hagen.de

Library of Congress Control Number: 2006933175

ISSN print edition: 1860-949X
ISSN electronic edition: 1860-9503
ISBN-10 3-540-44909-4 Springer Berlin Heidelberg New York
ISBN-13 978-3-540-44909-6 Springer Berlin Heidelberg New York

This work is subject to copyright. All rights are reserved, whether the whole or part of the material is concerned, specifically the rights of translation, reprinting, reuse of illustrations, recitation, broadcasting, reproduction on microfilm or in any other way, and storage in data banks. Duplication of this publication or parts thereof is permitted only under the provisions of the German Copyright Law of September 9, 1965, in its current version, and permission for use must always be obtained from Springer-Verlag. Violations are liable to prosecution under the German Copyright Law.

Springer is a part of Springer Science+Business Media
springer.com
© Springer-Verlag Berlin Heidelberg 2007

The use of general descriptive names, registered names, trademarks, etc. in this publication does not imply, even in the absence of a specific statement, that such names are exempt from the relevant protective laws and regulations and therefore free for general use.

Cover design: deblik, Berlin
Typesetting by the authors and SPi using a Springer LaTeX macro package
Printed on acid-free paper SPIN: 11679684 89/SPi 5 4 3 2 1 0

Contents

Survey of this Book

Bernd J. Krämer

Faculty of Electrical and Computer Engineering
FernUniversität in Hagen, 58084 Hagen, Germany
bernd.kraemer@fernuni-hagen.de

Information and communication technologies (ICT) continuously sketch new horizons and create new computing visions. Recent examples include the disappearing computer, autonomic computing and organic computing. According to Weiser, the disappearing computer describes a new era of computing, which he called *ubiquitous computing*, and which follows the first wave of mainframe computing and the second wave of personal computing [7]. This era is characterised by the fact that computing and communication technology, though being integrated in almost every object we deal with, will disappear into the background, and make computing an integral but invisible part of people's lives — just like electricity. Ubiquitous computing will keep people continuously connected to other persons as well as to Internet applications and services. Weiser considers the Internet age, which enabled the massive interconnection of personal, business and government information and on-line services, just a transitional phase leading us from the PC era with its one-to-one relationship between users and PCs to ubiquitous computing when a multitude of computers will serve one person any time and everywhere. While Weiser's notion of ubiquitous computing refers to an idealised human-centric and unobtrusive long-term vision, industry has coined the term *pervasive computing* for the same idea. This term emphasises the existence of omnipresent ICT that aims to turn web-based business processes to practical use in the context of mobile commerce scenarios on a rather short to mid-size range [6]. The former IBM chairman Gerstner stressed this view in his 1998 shareholders speech [4] by claiming that "the networked world, which will soon allow perhaps a billion people to interact with a million e-businesses will extend further to interconnect perhaps a trillion intelligent devices."

Autonomic computing deals with the increasing difficulty of managing distributed computing systems, which become more and more interconnected and diverse, and software applications that go beyond corporate boundaries into the Internet. Autonomic computing anticipates systems consisting of myriads of interacting autonomous components that are able to manage themselves according to pre-defined goals and policies. Self-management capabilities include automated configuration of components, automated recognition of opportunities for performance improvement (self-optimisation), automatic detection, diagnosis and repair of local hardware and software problems, and automatic defense against attacks [5]. The objective is to minimise the degree of external intervention, e.g., by human system managers, and, at the same time, to preserve the architectural properties imposed by its specification [3]. The concept of self-configuration refers to a process in which an application's internal structure changes depending on its environment. Though self-organisation is a relatively new area of research, prototype implementations of this concept became reality in the implementation of distributed sensor networks. As a collection of mobile sensor nodes is deployed, the nodes are able to create an impromptu network, which can reconfigure itself automatically in case sensor nodes move, and which dynamically adapts to device failure or degradation. In some cases, such a network can even adapt to new task and network requirements.

Integration, deployment and management of complex computing systems begin to overwhelm the capabilities of software developers and system administrators. The only viable long-term solution is to create computer systems that manage themselves, perhaps utilising mechanisms taken from biological systems, in accordance with high-level guidance from humans.

Organic computing investigates design and implementation of self-organising systems that are self-configuring, self-optimising, self-healing, self-protecting, self-describing, context-aware, and anticipatory. Thus, organic computing includes the autonomic computing targets of IBM's initiative. Organic computing emphasises biological and organic-inspired systems, as well as aspects of self-organisation and emergence. Meeting the grand challenge of organic computing requires scientific and technological advances in a wide variety of fields.

Future Applications

Implementations of these visions are gradually becoming reality. In its remote space exploration projects NASA envisions, for example, to employ swarms of light-weight autonomous spacecrafts equipped with all kinds of miniaturised instruments, and controlled by software agents that mimic insect-type social behaviour to prospect the frontier of our solar system [2]. These components will produce huge volumes of sensor data to be processed with limited computing and communication resources. Different types of agents will posses different types of knowledge and expertise and, thus, play a particular rôle in a swarm. Some are supposed to take autonomous decisions of what data need to be observed, maintained and in what fashion, while others will take responsibility for navigating or for operating the spacecraft.

Current research on smart factories anticipates autonomous robots capable of forming ad hoc federations through spontaneous networking to master a given task. Component overload or failure is detected automatically and tasks are re-allocated. Smart home initiatives integrate home automation products, including appliances, monitoring and security systems, controllers and all kinds of sensors, entertainment equipment and ICT to enhance safety, comfort, and convenience of the home-owners. Leading consulting companies predict that the warehouse of the future will be pervasive and personalised. Billions of sensors will be attached to products, shelves, containers, shopping carts and electronic shopping lists organising themselves in flexible networks, e.g., to calculate the content of a shopping cart and to direct customers to missing articles contained on their shopping list or to special offers. Products can be monitored throughout the whole supply chain, e.g., to ensure that frozen food is kept sufficiently cool.

Novel approaches to container security in cargo shipping propose to use so-called motes, i.e., tiny battery-controlled computers that are equipped with different types of sensors and include software for sensing, alerting, spontaneous networking and mobile communication. The alerting function continuously monitors the sensors for reportable events, such as suspected intrusion, unsafe cargo, shock, high temperature, or out-of-bounds movement. It either allows automatic setting of alert thresholds or users to set alert thresholds. The alerting function provides the signal processing necessary to increase detection probability and to reduce rates of false alarms.

Commonalities and Base Technologies

All these computing paradigms and the applications exploiting them share properties such as distribution, openness, transparency, heterogeneity, mobility, continuous change, and local autonomy. Distribution refers to the fact that applications consist of components running on different computers concurrently. Openness is concerned with extensions and improvements of a system based on standard interfaces and plugging facilities. Transparency lets users perceive a distributed system as a whole rather than collection of cooperating components. Transparency has different dimensions including: *access transparency* by which information objects can be accessed using identical operations like browsing web pages; *location transparency*, by which users need not be aware of where information resources, services or software agents are physically located; *failure transparency*, which hides failures from users and allows applications to continue their execution despite the failure of particular system components; and *migration transparency*, which allows logical or physical movement of services, data resources or agents without affecting system operations. Both hardware and software may be heterogeneous as an application may include sensors, computing and communications devices from different vendors, and the software components may be implemented in different programming languages and run under disparate operating systems including, for instance, tinyOS for motes and similar devices, Symbian for mobile phones, Palm OS, and different PC operating system for notebook PCs and stationary servers. Mobility may be an issue both for computing devices and application components, i.e., information resources and services.

Change comes with several connotations: application software is likely to evolve to accommodate new user needs, or it may automatically adapt to particular system parameters. For mobile users in ubiquitous systems change is concerned with sensing, fusing and distributing location information and providing location- and context-awareness. One strand of current research on ubiquitous mobile computing aims to seamlessly combine today's wireless networking technologies including GSM[1], HSCSD[2], GPRS[3], UMTS[4], CDMA[5], IEEE 802.11[6], and Bluetooth[7]. Each local node in a distributed system is highly autonomous as it may follow entirely different operating or even decision-making policies. The autonomy of a node must, however, be balanced with the need for coherence or co-operation with other nodes in the system.

The applications envisioned above also rely on common base technologies including portable computing devices, spontaneous and wireless networking, or middleware that abstracts from the differences of various operating and communication systems involved. They also share system-level engineering requirements including security and privacy, location-awareness, trust, dependability, and decentralised management capabilities. Since a ubiquitous computing world deals with a range of ownership issues and a variety of personal and corporate policies about information and service use, interacting parties must be readily authenticated and authorised, and interfering policies must be co-ordinated. E-enterprise, e-government and similar visions — be they mobile or stationary — will have no economic chance in the long run without these requirements being satisfied. On the other hand, these objectives are often in conflict with ease of use. ActiveX controls, Javascript and other mechanisms, for instance, which were designed

[1] Global System for Mobile Communications, a set of standardised specifications and interfaces necessary for a functioning mobile radio network.

[2] High Speed, Circuit Switched Data, a mobile data communication connection that bundles several GSM data channels to a virtual one.

[3] General Packet Radio Service, a service for packet-based shared data communication.

[4] Universal Mobile Telecommunications System, the third-generation wireless communication standard.

[5] Code Division Multiple Access, a technique to broadcast differently encoded multiple parallel bit streams over the same frequency.

[6] A family of wireless LAN standards.

[7] Another wireless communication standard designed for short distance communication (10 to 100m).

for enabling components in networked environments to interact smoothly and automatically in activating remote functions, have been recognised as security threats.

In autonomic computing, local management capabilities are needed to manage relationships between components including needs and preferences expressed, e.g., in terms of goals, policies, constraints, or rules; they autonomously monitor and enforce agreements, and negotiate with other au tonomous components to obtain particular services. A user's environment is dynamic, adaptive, and interactive. It contains sensors, wireless devices, and personal and service agents operating autonomously with the aid of context. By context, we mean information about physical characteristics (such as location and network elements), the system (such as applications running and services available), and the user (such as privacy and presence). The environment becomes context aware when it can capture, interpret, and reason about this information.

Application Scenario

As technology advances, users desire ever more convenient computing options that will enhance their productivity while fitting seamlessly into their busy lives. Business processes have become distributed, they extend organisational borders, and users began to exploit mobile communication features to access or update corporate data, e.g., in field sales, maintenance or market research. As we travel, we want to have access to applications that we currently work on, or that are of particular relevance in the context of the actual travel or meeting purpose, irrespective of administrative network boundaries we may cross on our route. We also want to easily access services that specifically support our actual activities independent of time and location.

Assume the following scenario: The calendar on our PDA indicates a management group meeting of project P in some location L for tomorrow, and we just want to carry our PDA and mobile telephone while traveling to that location. Yet, besides personal data such as agenda, address book, telephone numbers and MP3-encoded music, we want all documents, applications and services relevant to project P be accessible, as we need them before, during, or directly after the meeting, e.g., while traveling home. Ambient communication connections, constraints in storage capacity, and computing capabilities of our electronic equipment should be taken into account without human interference, e.g., by automatic pre-fetching of data, code migration, best service selection, and other automatic facilities. To make this happen without being bothered with the task to load all documents and applications needed on our PDA and to manually remove and save those we may not need or to set or adapt communication parameters of the actual location ourselves, we want our documents and applications to configure and manage themselves, and the services to be orchestrated automatically.

Transferring the concept of self-organisation to our travel scenario, we want our equipment to show the following features. Assume that one of our frequently used applications is a text processing system like Word and some Word documents are to be at our hands during the meeting. Then, to be able to work on these documents, our PDA should have a small text processing system compatible with Word and running equipped with a spelling checker.

Self-managing systems are a topic of research in distributed processing and network management aiming at the (partial) automation of tasks typically carried out by human administrators. Self-managing systems are able to detect failures or performance degradations, trigger diagnostic procedures, and carry out corrective actions on their own. Self-management is particularly important for mobile systems as they include a large number of heterogeneous elements and changes occur frequently.

Contents of this Book

This book takes a holistic view on mobile and distributed computing systems. All system layers ranging from hardware over vertical and horizontal infrastructure services and novel middleware solutions to various types of application software are touched. The book's 15 chapters are organised in three parts. The five chapters of Part I address core properties of ubiquitous applications including mobility, self-healing and self-organisation of both technical and social-technical systems whose constituent parts or members are autonomous and geographically dispersed. The chapters of Part II present software solutions for business and engineering problems that occur in a specific application domain, like industrial automation or semantic web applications, or provide support across an enterprise or organisation, like embedded systems design or electronic payment. Ubiquitous applications depends and reliable, secure and effective communication channels. The risks for society will be unbearably high if ubiquitous systems will be as vulnerable as today's computer networks. Part III introduces a predictive approach towards distributed systems management and presents novel solutions to cope with denial of service attacks against mobile agent systems, Internet congestions and data encryption, including both software- and hardware-based methods. Seen from an engineering perspective, the book introduces novel modeling and design techniques and deals with service-oriented computing and component- and agent-based software development methods.

Mobile applications are typically hosted in resource-constrained environments and may have to dynamically re-organise to accommodate changes of user needs, connectivity problems, and changes in execution context and physical environment. The first chapter of this book presents SATIN, a middleware system for mobile applications that builds on a light-weight component model incorporating code mobility primitives to assist application developers in building adaptive mobile systems. Concepts and implementation of the SATIN middleware are discussed and evaluated by means of a case study, in which existing open source software is mapped into SATIN components through a simple adaptation process. In a second example the dynamic update of components on mobile hosts is built on top of the SATIN middleware and tested extensively.

Service-Oriented Computing (SOC) is a new computing paradigm that utilises services as lightweight constructs to support the development of rapid, low-cost, and easy composition of distributed applications. Services constitute a new approach to programming, based on the idea of creating applications by discovering resources and applications available in the network and integrating them to accomplish specific tasks. The visionary promise of SOC is a world of co-operating services, where application components are assembled with minimum effort into a network of services that span organisations and are independent of programming languages, operating systems, or computing platforms. Services-Oriented Architecture (SOA) is a standardised design and co-ordination framework for the dynamic assembly and linkage of software components exposed as services. The flexibility and dynamism of SOA have, however, a price, because bindings between service provider and client may fail. Chapter 2 studies this problem and self-healing solutions including the identification of faults in service binding and the activation of appropriate recovery actions. The authors of this chapter propose an extension to current service composition techniques that allows them to provide better monitoring and reconfiguration solutions than have been proposed in the literature.

Chapter 3 introduces a new self-configuring wireless ad hoc network with multirooted tree topology. The network was designed for scalability and robustness, and is to perform in noisy environments. It is made up of two broad node types. The generic sensor node is equipped with a short-range transceiver, whereas the border nodes are equipped with an additional network interface to the external world. Border nodes act as gateways between the sensor cluster and the cyberspace. All traffic is assumed to originate and terminate at the sensors. A new cost metric is defined to determine optimum routes. This novel cost function takes into account several metrics including link capacity, link reliability, number of hops to a gateway, and power constraints of the node. The proposed cost function is configurable by means of weights assigned to each contributing metric. Simulations revealed that the proposed architecture adapted well to changes in the state of the links/nodes and always converged to a valid solution.

Chapters 4 and 5 address the field of e-learning in higher education, which many experts consider a growth area of considerable importance. Chapter 4 reflects on the aspect of self-organisation in the context of geographically dispersed students of e-learning courses, who share common learning interests and, therefore, want to form learning groups governed by a high degree of overlap of personal interests. As direct social contacts between these students are limited or non-existent, this chapter proposes an agent-based architecture, including agents that maintain and update the learning preferences and interests of individual students, and agents that dynamically re-organise learner groups with the objective to achieve a stable system of groups with high interest cohesion after a short period of observation and re-organisation. Chapter 17 presents the core ideas, a prototype implementation, and evaluation results from experiments gained with students of Shanghai Jiaotong University's e-learning branch.

Chapter 5 views e-learning as a business process supporting mobility of students across different educational institutions. This idea varies the concept of a virtual university as it was pursued in different projects in the sector "Flexible Universities" of the 5th Framework Programme of the European Union. Web services are proposed here as a mechanism to integrate educational services of different providers in such a way that a student can realise a personalised curriculum involving courses and credits from different universities. Service component architecture is a relatively new engineering model in SOC that lifts the concept of services to components. The e-learning scenario treated in this chapter illustrates the contributions of web services and service components to distributed computing.

Chapter 6 shifts the focus to a quite different type of applications, namely, industrial process control and plant automation in distributed configurations. To enable the design of optimum maintenance schedules for plant components, and to improve the performance of overall plant operation, it is proposed to attach a novel asset management component, which compiles relevant plant and device data from the plants components and transmits this information via the Internet to dedicated management clients. To keep interventions in the existing hardware at a minimum and, thus, limit undesired effects on plants, a wireless communication system is used. After introducing the base technology, a few novel scenarios for on-line and wireless asset management are specified.

Chapter 7 remains in the realm of industrial automation systems and extends on the issue of asset maintenance. But emphasis is put here on automated acquisition and management of knowledge about properties and the current status of hardware devices and software components of automation systems with the particular objective to allow for proactive maintenance. This approach aims to commission maintenance tasks prior to potential asset failure or degradation, while keeping costs under control and avoiding unnecessary performance constraints. The prototype infrastructure proposed in this chapter builds on distributed agent technology, which is also used in Chapter 4 to provide self-organisation support to e-learners, while Chapter 12 copes with denial-of-service attacks run against mobile agents.

The semantic web, a vision first stated by Tim Berners-Lee and others [1], foresees enriched data resources on the world-wide web that are not only useful for humans but are also machine-processable, and which specify what users can do with the data found and how data can be converted in other forms to ease comparison and integration of data from heterogeneous sources following their own naming and structuring conventions. Chapter 8 focuses on the problem of knowledge discovery, organisation, and presentation in semantic web applications. It surveys a range of mining techniques and, in particular, on extensions that rely on formal concept analysis, a mathematically well-defined theory of data analysis, knowledge representation, and information management, that identifies conceptual structures among data sets.

Completely anonymous electronic cash systems are subject to mis-use by malicious users. To solve this problem, Chapter 9 proposes a new fair off-line electronic cash system, where fair means that anonymity can be uncovered under certain circumstances. All users of this system are organised in groups, which are associated with an efficient group signature scheme. This allows system providers to revoke anonymity of electronic coins in certain suspicious cases, and to prevent the system from being mis-used by criminals for blackmailing or money laundry. Three parties are involved in a normal transaction: customer, merchant, and a bank. A trusted third party, with help of the bank, is involved to revoke anonymity. This cash-like system includes four kinds of protocols: a registration protocol connecting the trusted third party and the bank, a withdrawal protocol to control the communication between customer and bank, a payment protocol used between customers and merchants, and a deposit protocol between the bank and a merchant. In addition, the bank deploys a tracing protocol, which plays the rôle of the trusted third party and serves to discover the identity of a suspect. The system presented is capable to trace coins and owners, it is untraceable by unauthorised entities, resistant to double-spending, and un-linkable, i.e., merchant and bank cannot decide to whom different e-coins belong. Compared to previous e-cash systems, the approach is computationally more efficient, as well.

The systematic development of dependable distributed software systems needs proper modeling tools that allow application developers to provide explicit abstractions of different views on a system including data, interaction, configuration, process or state transition views. With the advent of the diagrammatic modeling language UML, which is targeted to large and complex software (-intensive) systems, multiview specification and modeling are becoming daily practice in software engineering. Though the Object Management Group claims UML to be normative, the semantics of UML is just an informally defined, popular set of conventions leaving the modeling approach vague and incomplete. In consequence, objective assessments of a particular UML model's meaning are impossible, and debates about its UML compliance or questions about the interoperability of UML tools must remain controversial. An important application domain that has been neglected by UML designers is the whole area of industrial real-time automation systems, which has been governed for more than three decades by high-level programming languages for multitasking real-time systems like PEARL. Chapter 10 builds on the long-term experience in the process automation domain, and presents well-defined UML extensions that allow for dealing with hardware and software configurations, dynamic allocation, and real-time features. The adequacy of the approach is illustrated with an automotive on-board system.

Traditionally, network and systems management are largely manually controlled processes. System operators are typically supported by management software providing monitoring and measurement functions to support trouble-shooting and diagnostic tasks. But the ever-increasing number of managed components in the vision of the disappearing computer, the growing heterogeneity of hardware and software components involved, and the interference of different management policies let the complexity of the systems management task grow beyond human capabilities. Chapter 11 promises to bridge this chasm between current practice and future needs in the management of distributed systems by applying a probabilistic model that allows the authors to deduce predictive knowledge about the state of managed components, and to propose pro-active measures that are likely to prevent failures, degradation or faults. The approach is able to cope with incomplete information about the current configuration of a distributed system and the properties of managed components. It builds on pre-defined or empirically acquired knowledge about dependencies between managed components.

Chapter 12 studies security problems in the context of mobile agents acting in malicious environments. A specific type of denial-of-service attack is considered which focuses on malicious hosts that either delete foreign agents or prevent them from continuing their routes as they migrate from host to host. Since the prevention of such attacks is generally not possible, this chapter presents a detection method for a posteriori identification of such malicious hosts. The solution is based on a new protocol that combines the application of well-known cryptographic primitives and of a set of behaviour rules. The protocol takes into account given visiting order constraints for routes, which arise due to dependencies of computation results on an agent's journey.

Chapter 13 takes on the problem of Internet traffic congestion caused by the never-seen growth of multimedia and voice communication over the standard Internet protocol. As packet buffering in routers cannot be scaled by increasing buffer sizes in simple ways, various active buffer management mechanisms pro-actively drop packets before buffers overflow. Current solutions like the random early detection (RED) algorithm are, however, instable as they rely on heuristic parameter settings, which attempt to approximate chaotic traffic behaviour. After discussing the strengths and weaknesses of RED algorithms, this chapter introduces a mathematical model based on chaos theory that reflects core properties of TCP-RED, and makes them amenable to analytical treatment. Further, adaptive control schemes able to stabilise the average buffer size across a wide area network are proposed. It is shown that they are easy to implement and work even in the absence of exact knowledge about a network configuration.

Networked PCs are notoriously insecure. The risks for society will be unbearable high if ubiquitous systems will be as vulnerable as today's PCs. Chapter 14 provides a solution to communication security based on chaos theory. On the side of the message sender, signals are modulated over a chaotic signal, while the message receiver retrieves the information communicated by synchronising and demodulating the chaos signal. The effectiveness of the approach is discussed along an illustrative example.

Chapter 15 presents a solution to security requirements in multimedia communication by means of a hardware chip that implements a novel stream encryption scheme. A central objective of this research is to replace traditional encryption schemes by an algorithm that exploits cryptological concepts based on chaos theory, and to map the algorithm into efficient circuit structures that allow for parallel computation. The proposed solution also has the advantage that it obviates the need for floating point operations. Empirical results acquired with a prototype implemented on a field programmable gate array are reported. The chip can be embedded into devices such as mobile telephones, handheld computers, or Personal Data Assistants to en- and decrypt data sent over communication links of any kind and, thus, accomplish security of critical mobile applications such as bank transactions.

Acknowledgement

This book comprises the scientific outcome of the binational Sino-German networking project *Self-organising Networks and Software Applications* (see also `http://www.fernuni-hagen.de/IQN/`), which was financially supported by the German Academic Exchange Service (DAAD) within its programme *International Quality Networks* (IQN). This innovative programme particularly aimed at the integration of highly qualified foreign scientists into research and teaching activities at German universities.

This IQN project formed a partnership between FernUniversität, Institut für Automation und Kommunikation e.V. Magdeburg (ifak), and Forschungsinstitut für Telekommunikation e.V. Dortmund (FTK) on the German side, and East China Institute of Computer Technology (ECI), Jiaotong University and Tongji University, all Shanghai, Lanzhou University, and City University of Hong Kong on the Chinese side to prepare young Chinese scientists for excellence in research, competence in learning for the future, and engineering excellence in the topic areas nanotechnology, embedded systems and distributed software.

Hence, we are indebted to many people for their valuable contributions towards carrying through this project. First and foremost, we are grateful to Katja Jaeckel, Beate Pasch, Eva Rode-Luettges, Marina Steinmann, and Susanne Wohlleben of DAAD who made the IQN project run smoothly with their continuous advice and administrative support. We also express our sincere appreciation to Dagmar Hacker-Thünken, Hans-Hermann Kriegeskorte, and Volker Winkler of FernUniversität who were responsible for the financial statements of the project, supported the presentation of outstanding project results at CeBIT Fair, Hanover, 2003, and designed the project's web-site, respectively. Thanks are also due to our colleague Luise Unger who, when chairing the Committee for Fostering Women in Science, was co-responsible for supporting the second three months' research stay of Fan Yang of Shanghai Jiaotong University at FernUniversität in Hagen, which took place after the IQN project had already ended.

We are, of course, most grateful to the authors of this book's chapters for their insightful work and for managing to take time to provide us with such excellent material. Renate Zielinski invested much time in streamlining the references of each chapter and Jutta Düring was extremely supportive in copy-editing the draft of this book. We especially thank the editor, Prof. Janusz Kacprzyk, and the publisher, Dr. Thomas Ditzinger, of Springer-Verlag's series *Studies in Computational Intelligence* for their patience and helpful and considerate collaboration.

Finally, I should like to thank my wife Mija, who frequently tried to convince me to spend some time with her in the Southern Californian sun rather than working on that book, for her continuous support and unconditional love.

References

1. Berners-Lee T, Hendler J, Lassila O (2001) The Semantic Web. Scientific American
2. Curtis S, Mica J, Nuth J, Marr G, Rilee M, Bhat M (2000) ANTS (Autonomous Nano-Technology Swarm): An Artificial Intelligence Approach to Asteroid Belt Resource Exploration. International Astronautical Federation, 51st Congress
3. Georgiadis I, Magee J, Kramer J (2002) Self-organizing Software Architecture for Distributed Systems
4. Gerstner L (1998) `http://www.ibm.com/lvg/annual98.phtml`, last visited: February 23, 2005
5. Kephart J, Ches D (2003) The Vision of Autonomic Computing, IEEE Computer
6. Mattern F (2004) Ubiquitous Computing: Schlaue Alltagsgegenstände – Die Vision von der Informatisierung des Alltags (in German). *Bulletin SEV/VSE* 19: 9–13
7. Weiser M (1991) "The Computer for the Twenty-First Century". Scientific American, pp. 94–10

Part I

Supporting Mobility, Self-Healing and Self-Organisation

1

Building Adaptable Mobile Middleware Services Using Logical Mobility Techniques

Stefanos Zachariadis, Manish Lad, Cecilia Mascolo, and Wolfgang Emmerich

Department of Computer Science
University College London, Gower Street, London WC1E 6BT, Great Britain
{s.zachariadis|m.lad|c.mascolo|w.emmerich}@cs.ucl.ac.uk

Summary. In recent years, various mobile middleware systems have been developed that build on top of the protocol stacks provided by the underlying network operating systems, to offer a set of higher-level services. Common services provided include remote procedure calls, data sharing, or service discovery. With the number of mobile devices, such as PDAs, mobile telephones, or smart phones increasing, as well as the impeding flourishing of embedded and wearable computing, the rôle of middleware in offering services that hide the complexity in the underlying environment is at the same time important and complex. The mobile environment is becoming highly dynamic, with very heterogeneous devices, in terms of both hardware and software, forming short lived ad hoc networks. Here, it is argued that the provision of static services cannot tackle the peculiarities of this environment. Logical mobility is presented as a technique for dynamic adaptation, a specification of a framework that offers its flexible use is detailed, and a number of dynamic middleware services is presented, which can intelligently be acquired by a device at run-time.

Key words: Distributed objects, components, pervasive computing, model validation and analysis, process models.

1.1 Introduction

The further decentralisation of computing into increasingly inexpensive mobile devices, such as personal digital assistants, mobile telephones, sensors and laptop computers, together with the advances in wireless networking (UMTS and other standards for wireless personal area networks including IEEE 802.11, 802.15.4, or 802.15.1), create a distributed computing environment that is becoming highly dynamic and heterogeneous.

Developing applications in this environment is very difficult. Building directly on the operating system of a device and only having a networking protocol stack available to program against is particularly tedious and error-prone, as it forces the developer to directly handle the complexities of the

S. Zachariadis et al.: *Building Adaptable Mobile Middleware Services Using Logical Mobility Techniques*, Studies in Computational Intelligence (SCI) **42**, 3–26 (2007)
www.springerlink.com © Springer-Verlag Berlin Heidelberg 2007

environment, such as intermittent connectivity and heterogeneity. As such, multiple mobile-computing middleware systems have been developed, that build on top of the operating system, hiding some of the complexities of the environment. Networked middleware systems build on the protocol stack to offer higher-level services to applications. These services range from remote procedure calls to data sharing and service discovery. This has led to the development of various incompatible middleware systems, which are able to provide a fixed set of services.

We argue that this approach does not adequately tackle the dynamicity of a mobile environment, because the provision of fixed and static services cannot scale. The constant introduction of new devices into networks, which may offer different services, ranging from temperature sensing to printing, over different hardware, requires middleware systems to *adapt* or *mutate* to provide the functionality needed to allow applications to communicate with them. Pre-installing all the services that may be required before a device is brought to the market is not feasible, because mobile devices have strict memory limitations. Moreover, we argue that it is also very difficult to predict the services that may be needed throughout a device's lifetime.

In this chapter, we argue for offering *logical mobility*, or code relocation primitives, as major constituents in mobile middleware systems. The primitives can be used to build dynamic middleware services. In this context, a dynamic service is a service that can dynamically be deployed into a running system, or a service that can be changed at run-time. We discuss the utility of logical mobility for mobile systems, and present the process algebra specification for a platform that allows systems to dynamically send, receive, and deploy code. We then briefly present SATIN, a component-based mobile-computing middleware system, that encapsulates this platform, offering flexible use of mobile computing paradigms to applications. We also show how this was used to build dynamic middleware services, ranging from advertising and discovery to ubiquitous security.

The chapter continues as follows: Section 1.2 defines logical mobility and outlines its advantages for physically mobile systems. Section 1.3 details our logical mobility platform. Section 1.4 briefly outlines SATIN, and shows how the platform of Section 1.3 was instantiated. It concludes by presenting, as case studies, a number of mobile-computing middleware services built and deployed. Section 1.5 gives a brief outline of related work, while Section 1.6 discusses and concludes the chapter.

1.2 Defining Logical Mobility

Logical mobility refers to the ability to change the configuration of the software of a distributed system, by *transferring* logical units between nodes. Logical mobility has been argued [19] to have great potential for engineering mobile systems, a potential that has not yet been realised. This section claims that logical mobility can be used as a technique that can adapt services, because it can *encapsulate functionality* which can dynamically be deployed onto a running system. Logical mobility is usually offered using *code mobility* techniques. In the following paragraphs, we define what code mobility is, outline its relationship to logical mobility, and describe its usage paradigms.

1.2.1 Introduction to Code Mobility

Code mobility has been defined [2] as '*the capability to dynamically change the bindings between code fragments and the location where they are executed*'. More informally, code mobility can be defined as the ability to move code between nodes in a network. It has been argued [5] that code mobility is a technology that can be used to engineer configurable, scalable, and customisable large-scale distributed systems, by allowing code to migrate and bind to different nodes of a running system. Mobile code systems usually define a *code unit* as a conceptual or realised abstraction that encapsulates a form of code. A code unit is the *minimum unit of transfer* or *unit of mobility*.

Whereas code mobility specifically refers to the transfer of code between nodes, logical mobility builds on this notion and refers to the reconfiguration of systems by moving any part of the logical layer between nodes. Logical mobility is usually offered using code mobility techniques to transfer information, including binary code, compiled for a specific architecture, interpreted textual scripts, bytecode compiled for a virtual platform, such as the Java Virtual Machine (JVM), but also application data such as profiles. In this context, data are defined to be anything that cannot be directly executed by the underlying platform.

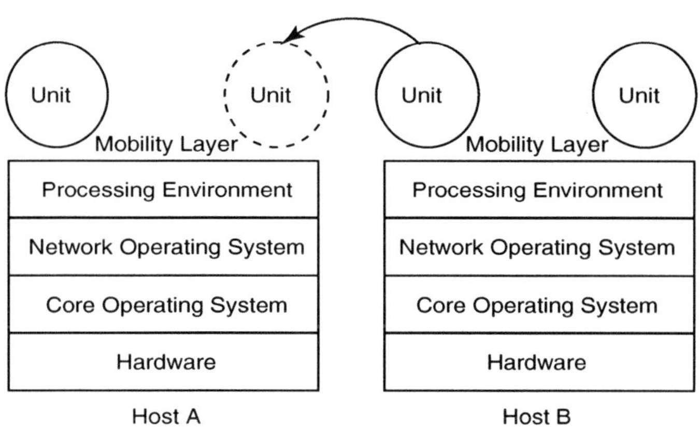

Fig. 1.1: A mobile code system, showing the transfer of a unit from Host B to Host A

Fig. 1.1, adapted from [14], shows an outline of two systems using mobile code. The core operating system is built on the hardware layer, and provides abstractions to access the hardware and basic services such as memory management. The networking aspects of the operating system are built on top of this; they provide basic networking services, such as a TCP/IP protocol stack. Layered on the network operating system is the Processing Environment (PE). The processing environment is a *container* which allows *code units* to run; it provides a set of primitives (the extent of which varies between platforms) to allow for code migration, or even for access to local resources. Any co-ordination between various units as well as between the units and the rest of the system takes place at this layer. A processing environment usually acts as a sandbox, restricting the access of a unit to protect from malicious code. Fig. 1.1 shows a unit transferred from one node to another.

There are two manifestations of code mobility: *weak* mobility, where a code unit transferred cannot include execution state information, and *strong* mobility, where this is possible. Strong mobility allows for a process or thread to suspend execution, move to another host, and resume execution from the exact point at which it stopped on the originating host. At a conceptual level, the migration of a thread or process using a strong mobility mechanism can be completely hidden from the application programmer. A weak mobility mechanism can approximate the operation of a strong mobility one, provided that the application programmer is aware of the migration process; the programmer would need to explicitly save any data that are needed to resume execution at the recipient host before the transfer takes place, and to use the data to resume appropriately after the transfer process is completed. By definition, weak mobility platforms consume less computational resources than strong mobility ones.

1.2.2 Paradigms of Code Mobility

Usage of code mobility has been classified [5] into a set of paradigms: code on demand, remote evaluation, client-server interactions, and mobile agents. These paradigms mainly differ on the party that initiates transfer of code, and on the party that actually transfers it. We extend these concepts to apply to *logical* mobility. As such, *client-server* (CS), a popular paradigm in traditional distributed systems, suggests the execution of some code in a computing device (the server) triggered by a request of another device (the client). The most common example of this paradigm are remote procedure calls (RPCs). *Remote evaluation* (REV) suggests that a host sends a particular unit to another host, to be executed there. This paradigm is employed by Distributed.NET [21] and other similar distributed computing environments, which work using the divide-and-conquer paradigm to break large computational challenges into smaller, more manageable problems, and distribute those to machines around the world. The results are then sent back to the server orchestrating the problem, which can re-compose the answer to the original challenge. The *code on demand* (COD) paradigm enables a host to request a particular unit from another machine. Following the request, the code is transferred to the requesting host and can, then, be executed there. This is an example of dynamic code update, whereby a host or application can update its libraries and available code-base at run-time. A *mobile agent* (MA), is an autonomous unit of code. It is injected into a network by a host, to perform some tasks on behalf of a user or an application. The agent can autonomously migrate from one execution environment to another.

1.2.3 Logical over Physical Mobility

In previous work [23], we identified a number of examples showing that all paradigms of logical mobility can bring tangible benefits to mobile computing. The examples included dynamic updating in the face of limited resources, offering location-based services, active networking, exploiting distributed computational resources, and limiting connectivity costs. Those benefits can be summarised in the following interrelated points:

- Logical mobility allows applications to update their code-base, hence acquiring new functionality.
- Logical mobility may permit interoperability with remote applications and environments, which have not been envisioned at design time.
- Logical mobility potentially achieves the efficient use of peer resources, as computationally expensive calculations can be off-loaded to the environment.
- Logical mobility facilitates the efficient use of local resources, as infrequently used functionality can be removed to free some of the limited memory that mobile devices are equipped with. The functionality may potentially be retrieved later when needed.
- Logical mobility primitives can be used to encapsulate, request and transfer functionality between nodes; hence, it is a tool that can be used to create adaptable systems.
- By allowing functionality to be retrieved locally, logical mobility allows for autonomous operation instead of relying on an externally provided service.

1.3 Conceptualising the Use of Logical Mobility over Physical Mobility

Having outlined the benefits of logical mobility for adaptable mobile computing systems, this section presents a conceptual object-oriented framework for logical mobility targeting mobile systems. The system is general enough as to not be tied to any particular application or paradigm. The framework presented is built on the notion of *weak mobility*, because its implementation requires less resources than a strong mobility one.

This section does not discuss in great detail the client-server paradigm defined above. Client-server interactions are not directly beneficial for purposes of mobile adaptation, as they do not involve the transfer of functionality. This section will describe a conceptual platform for logical mobility, which will offer symmetric operation (the ability to both send and receive execution units), and will be engineered for the intricacies of physical mobility, taking into account, in particular, heterogeneity, security, and limited resources. The platform describes is composed of various conceptual layers. The remainder of the section starts by describing a container that is used to encapsulate aspects of logical mobility; it then builds on that to define all the layers and aspects of the framework.

1.3.1 The Logical Mobility Unit

Fig. 1.2 presents a conceptual encapsulation of logical mobility as an extension of the UML [13] meta-model version 2.0 compliant with the Meta Object Facility [12]. It builds upon and extends the concepts of Classifier, Class, InstanceSpecification, and DataType. The diagram defines three aspects of logical mobility: Classes, ObjectSpecifications, and DataTypes; DataTypes are defined as bitstreams that are not directly executable by the underlying architecture, and ObjectSpecifications are defined as specialisations of InstanceSpecifications that reify classes. As such, the framework specifically addresses the transfer of classes, instances, and data as aspects of logical mobility. The *Logical Mobility Entity* (LME) is defined as an abstract generalisation of a Class, Instance, or Data. Consequently, an instantiation of an LME represents an aspect of the logical layer of a system.

The *Logical Mobility Unit* (LMU) is defined as the minimum unit of transfer in this framework. An LMU is a container, that can encapsulate various constructs and representations of code and data. As such, an LMU is, in part, a composition of an arbitrary number of LMEs. This allows an LMU to contain anything from a single class to a collection of classes, instances, and data. The LMU provides operations that permit inspection of contents. This allows a recipient to inspect an LMU before using it.

The LMU can potentially encapsulate a *Handler* class. The Handler can be instantiated, and the resulting object used by the recipient to deploy and manipulate the contents of the LMU. This can allow sender-customised deployment and binding. The Handler concept and name is taken from [15]. Handlers and deployment in general are further discussed in the next section.

An LMU also encapsulates a set of *attributes*, called the *properties* of the LMU. An attribute is a tuple containing a *key* and a *value*, and the properties of the LMU map each key to its associated value. As such, a reference to an

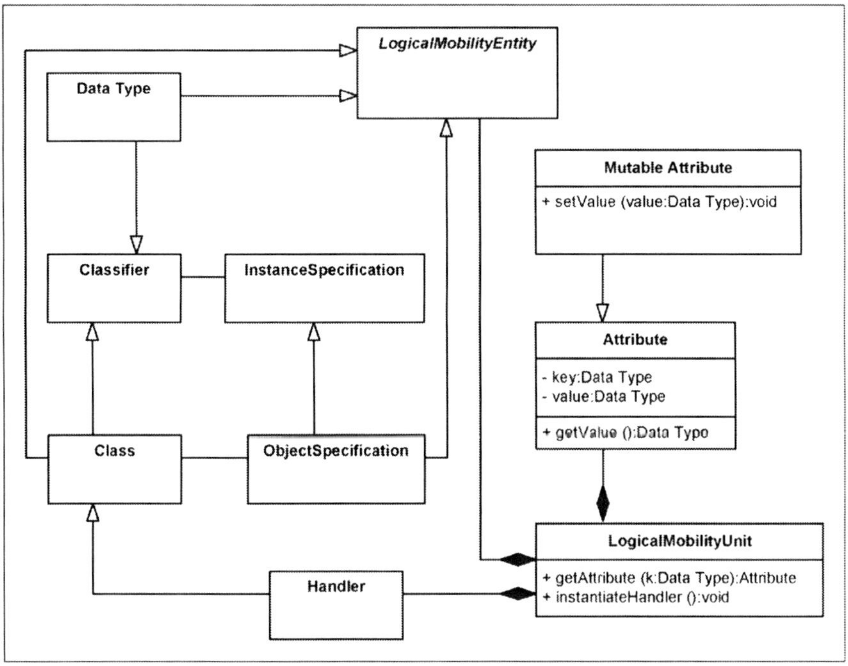

Fig. 1.2: The Logical Mobility Unit

attribute encapsulated in the LMU can be obtained by identifying its key. Attributes represent the meta-data of the LMU. Attributes can be either mutable or immutable. The number and type of attributes is not fixed. The properties are used to describe the LMU they are associated with. For example, logical (software) or physical (hardware) dependencies, digital signatures and even end-user textual descriptions can be expressed as attributes. As such, they can be used to express the heterogeneity of the target environment. An LMU that contains Java classes, for instance, may specify that it requires a Java Virtual Machine that implements version 2 of the appropriate specification as an attribute. An ontology for attribute keys and values is not defined at this stage. Mutable attributes are useful, because they allow for storing the state of the Logical Mobility Entities separately to their logic. This allows users, in principle, to update the logic of a logical mobility entity, while maintaining its state. This is useful in many scenarios; for example, in self-updating mobile agents.

The LMU and its contents can be serialised and deserialised. As such, the use of logical mobility techniques is equivalent to composing the LMU, serialising it, transferring it, deserialising it, and deploying it, as well as triggering this sequence of operations. The next section describes a framework that allows this.

1.3.2 A Framework for Logical Mobility

Fig. 1.3 outlines a framework for the use of logical mobility techniques by mobile systems, as a collection of conceptual layers, built on top of a network operating system. The following paragraphs describe each layer in detail. The operations of the framework are modeled as a collection of interacting concurrent processes, using the Finite State Processes (FSP) process algebra [8], with each layer represented by a process. A process algebra was chosen over alternatives such as axiomatic and denotational models, because the algebra allows for a clear definition of the various states of each process and of the system as a whole. FSP was chosen in particular for reasons of familiarity, ease of use, and tool support. The process algebra allows us to express safety and liveness properties on the framework, and verify that it operates correctly, while allowing callers to use any logical mobility paradigm.

The process algebra specification for the platform reads as:

```
//Trust & Security Layer
TRUSTANDSEC = ( inspect -> INSPECTION
              |examine -> EXAMINATION ),
EXAMINATION = ( trusted -> TRUSTANDSEC
              |mistrusted -> TRUSTANDSEC ),
INSPECTION = ( accepted -> TRUSTANDSEC
             |rejected-> TRUSTANDSEC ).

//Serialisation & Deserialisation Engine
SERDESERENGINE = ( deserialise -> DESERIALISING
                 |serialise -> SERIALISING ),
DESERIALISING = ( deserialised -> CONFLICTCHECK
                |deserFailed -> SERDESERENGINE ),
CONFLICTCHECK = ( conflict -> RESOLVECONFLICT
                |noConflict -> deserSuccess -> SERDESERENGINE ),
RESOLVECONFLICT = ( deserReject -> deserFailed -> SERDESERENGINE
                  |conflictResolved -> deserSuccess -> SERDESERENGINE ),
SERIALISING = ( serSuccess -> SERDESERENGINE
              |serFailed -> SERDESERENGINE ).

//Communications Layer
CONTROLLER = ( controllerStart -> ON ),
ON = ( sendRequest -> ON
     |receiveRequest -> REQUESTRECEIVED
     |controllerStop -> CONTROLLER ),
REQUESTRECEIVED = ( acceptRequest -> ON
                  |rejectRequest -> ON).

SENDERRECEIVERCONTROL = ( srStart -> ON ),
ON = ( receiveLMU -> ON
     |sendLMU -> ON
     |srStop -> SENDERRECEIVERCONTROL ).
RECEIVER = ( receiveLMU -> deserialise -> DESERIALISATION
           |srStop -> RECEIVER ),
DESERIALISATION = ( deserSuccess -> inspect -> INSPECTION
                  |deserFailed -> RECEIVER ),
INSPECTION = ( accepted -> deployLMU -> deployed -> RECEIVER
             |rejected -> RECEIVER ).
SENDER = ( sendLMU -> examine -> EXAMINATION
         |srStop -> SENDER ),
EXAMINATION = ( trusted -> serialise -> SERIALISING
              |mistrusted -> SENDER ),
SERIALISING = ( serSuccess -> lmuSend -> SENDER
              |serFailed -> SENDER ).
```

```
||SENDERRECEIVER = ( SENDERRECEIVERCONTROL || SENDER || RECEIVER ).

||API = ( SENDERRECEIVERCONTROL || SENDER || RECEIVER
        || TRUSTANDSEC || SERDESERENGINE ||CONTROLLER).

//APPLICATION
APPLICATION = ( deployLMU -> DEPLOYLMU ),
DEPLOYLMU = ( lmuPartialAccept -> deployed -> APPLICATION
            |lmuAccept -> deployed -> APPLICATION
            |lmuInstantiateHandler -> deployed -> APPLICATION
            |lmureject -> deployed -> APPLICATION ).

//Safety and Liveness Properties
property SERIALISETRUSTED = ( trusted -> serialise -> SERIALISETRUSTED ).
property SENDSERIALISED = ( serSuccess -> lmuSend  -> SENDSERIALISED ).
property INSPECTDESERIALISED = ( deserSuccess -> inspect ->INSPECTDESERIALISED).
property DEPLOYACCEPTED = ( accepted -> deployLMU -> DEPLOYACCEPTED ).
progress SENDANDRECEIVELMUS = { sendLMU, receiveLMU }
progress SENDANDRECEIVEREQUESTS = { sendRequest, sendRequest }

//Framework
||FRAMEWORK = ( SENDERRECEIVERCONTROL || SENDER || RECEIVER
            || TRUSTANDSEC || SERDESERENGINE || CONTROLLER
            || APPLICATION || SERIALISETRUSTED || SENDSERIALISED
            || INSPECTDESERIALISED || DEPLOYACCEPTED ). )
```

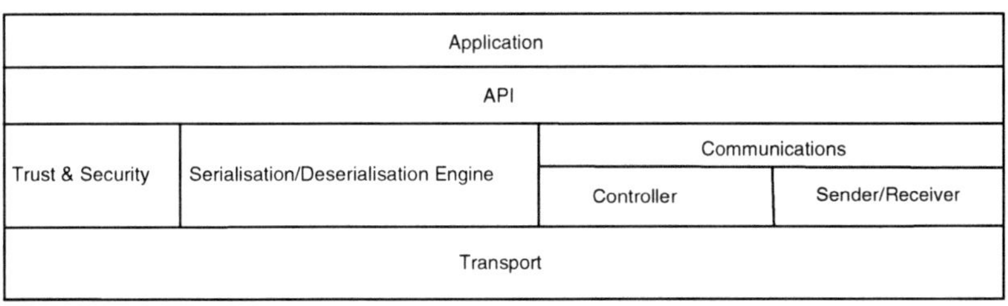

Fig. 1.3: A Logical Mobility Framework

The specification is discussed in the following paragraphs.

The Trust & Security Layer

There are many aspects to consider when trying to establish a secure mobile computing environment that uses logical mobility. In particular, *privacy of communications*, *system integrity* against malicious LMUs, and *trust* should be calculated and maintained between the various nodes.

System integrity and trust are the responsibility of the *trust & security* layer, of which there can be various realisations that employ different trust models, digital signatures and trusted third party-based verifications, heuristic virus scanning, or even logic-based techniques such as proof carrying code [10] to offer various levels of security.

It is represented as the TRUSTANDSEC process, which may either *inspect* an incoming LMU (thus trying to maintain system integrity), or *examine* the host to which an LMU is to be sent (thus implementing a trust mechanism). The result of the inspection (represented by INSPECTION) is either *accepted*, which denotes that the LMU is not malicious and that it behaves as advertised, or *rejected* otherwise. The result of the examination (represented by EXAMINATION) is either *trusted*, or *mistrusted*. The exact semantics of accepted, rejected, trusted, and mistrusted depend on the particular realisation.

The Serialisation & Deserialisation Engine

The *serialisation & deserialisation* engine is responsible for converting an LMU into a bitstream and vice versa. Different implementations may use different encodings to write and read the stream. When deserialising a bitstream, the engine is responsible to instantiate an LMU with the contents of the bitstream into a processing environment where it can be inspected by the trust & security layer. Deserialisation may fail if an element in the LMU has references which cannot be restored in the recipient node or if, because of a failure in the transport layer, the bitstream was not successfully received. Upon deserialisation, the elements of the LMU are checked for conflicts with elements already in the system. Essentially, conflicts may occur if elements of the LMU define themselves using names that are already in use in the recipient node. If a conflict is detected, implementations of the serialisation & deserialisation engine may either reject the LMU, or try to resolve the conflict, by loading it, for example, into a private namespace. The layer is represented by process SERDESERENGINE. The process can either *deserialise* an incoming LMU or *serialise* an outgoing one. In the former scenario, represented by DESERIALISING, the process can either fail (*deserFailed*), if, for example, the incoming bitstream was invalid (due to a transport layer failure) and could not be read or references contained in the LMU could not be restored, or succeed (*deserialised*). In the latter case, the serialisation & deserialisation engine checks whether the contents of the LMU conflict with the running system (CONFLICTCHECK). If a conflict is not detected (*noConflict*), then the deserialisation process is successfully completed (*deserSuccess*). If a conflict is detected, then the engine may try to resolve it (RESOLVECONFLICT), by, for example, loading the contents in a private namespace. If the conflict resolution process is successful (*conflictResolved*), then the deserialisation process is successfully completed (*deserSuccess*). Otherwise, the LMU is rejected (*deserReject*) and the deserialisation process fails (*deserFailed*). When serialising an outgoing LMU (SERIALISING), the process may either successfully complete (*serSuccess*) or fail (*serFailed*). A reason for failure is, for instance, that the contents of the LMU contained non-serialisable references, such as a reference to a hardware resource that does not exist on the recipient host.

The Communications Layer

The communications layer builds on the basic primitives provided by the transport layer for sending and receiving LMUs. In particular, it is composed of two different modules, the *controller* and the *sender-receiver*, both of which are described below. The *controller* implements an application layer Client/Server protocol that allows hosts to *request* the composition and transfer of a particular LMU from a remote host. The protocol encapsulating the request is considered to be specific to the implementation, but it is expected that the request message will be based on LMU attributes. Thus, the controller allows remote hosts to *pull* logical mobility units. Note that the protocol implemented by the controller is *asynchronous*; a request is *non-blocking*, and an LMU requested may be retrieved at a later stage. Moreover, a request that has been denied simply results in the requested LMU not being sent — no other information is generated and the requesting host is not notified of the failure.

This chapter considers potential failure to be typical of the dynamicity of a mobile distributed system — as such, failure is not an exception, rather it is a frequent event that the application programmer (or a middleware system built around this framework) must be aware of.

In the process algebra specification, the controller, represented by process `CONTROLLER`, is not active initially — this means that no requests can be sent or received. The controller can be activated (*controllerStart*), and an active controller (`ON`) can receive a request (*receiveRequest*), send a request (*sendRequest*), or be deactivated (*controllerStop*). Notice that this implies that application programmers using this framework should be made explicitly aware of the fact that their requests may fail, and that no reply is expected when sending a request, as explained above. When a request is received (`REQUESTRECEIVED`), it can be either rejected (*rejectRequest*) or accepted (*acceptRequest*). An accepted request implies that an LMU will be composed and sent (*sendLMU* in the sender-receiver — see below). Note that failures in the transport layer may result in a request not being successfully sent or received. Realisations of the framework may notify the caller about these failures.

The Sender-Receiver. Using the infrastructure provided by the serialisation & deserialisation engine and the trust & security layer, the *sender-receiver* allows for sending and receiving LMUs. In the process algebra specification, the Sender-Receiver is modeled as process `SENDERRECEIVER` in form of a concurrent composition of processes `SENDER`, `RECEIVER`, and `SENDERRECEIVERCONTROL`. The latter is responsible for enabling and disabling the functionality of sending and receiving LMUs, while `SENDER` and `RECEIVER` are responsible for sending and receiving LMUs, respectively. The sender-receiver is inactive initially, meaning that no LMUs can be sent or received. It can be activated (*srStart*), and an active sender-receiver (`ON`) can receive an LMU (*receiveLMU*), send an LMU (*sendLMU*), or be deactivated (*srStop*). When receiving an LMU, the serialisation & deserialisation engine is used to *deserialise* the incoming bitstream into an LMU. The deserialisation process (`DESERIALISATION`) can either succeed (*deserSuccess*) or fail (*deserFailed*). In the former scenario, the sender-receiver uses the trust & security layer to *inspect* the deserialised LMU (`INSPECTION`) for malicious elements. This can either result in rejecting the LMU (*rejected*), or accepting it (*accepted*) and passing it to the application for deployment (*deployLMU*).

When sending an LMU (*sendLMU*), the recipient host is first examined (*examine*) by the trust & security layer, to see whether the local host trusts it to send it information. The result of this process (`EXAMINATION`) is that the host is either *trusted* or *mistrusted*. If the host is trusted, then the serialisation & deserialisation engine attempts to *serialise* the LMU. As mentioned above, the serialisation process (`SERIALISING`) can either result in success (*serSuccess*), allowing the LMU to be sent (*lmuSend*), or in failure (*serFailed*).

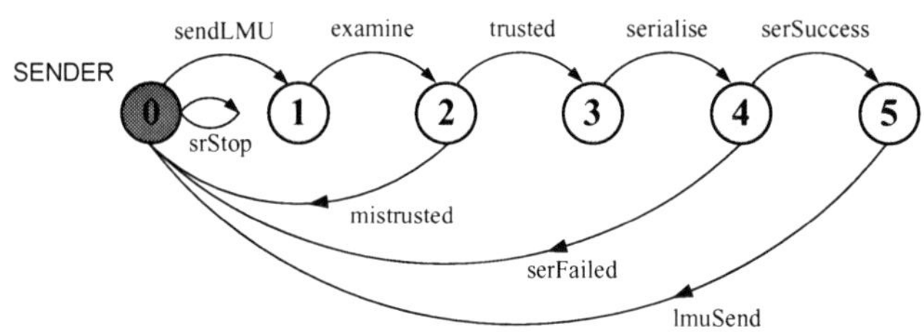

Fig. 1.4: A state machine representing the Sender

Note that failures in the transport layer may result in the LMU not being successfully sent or received. In this case, the deserialisation process in the receiving host, as performed by the serialisation and deserialisation engine, will fail. `SENDER` and `RECEIVER` are visualised as state machines in Fig. 1.4 and 1.5, respectively.

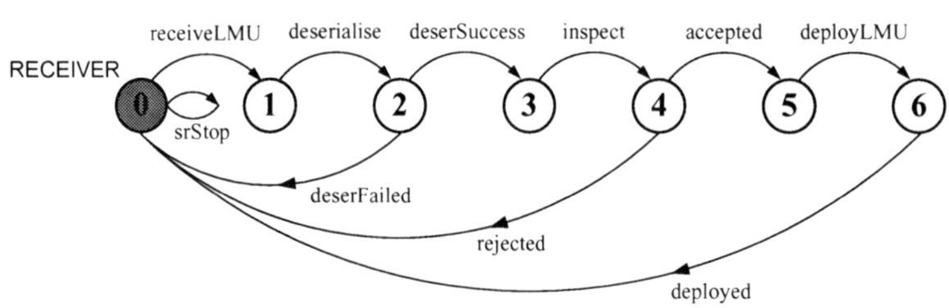

Fig. 1.5: A state machine representing the Receiver

Note that LMUs can be sent independently of the controller, i.e., an LMU can be sent without the host having requested it. The recipient does, however, have the option of rejecting an incoming LMU. As such, when modeled as concurrent processes, the *sendLMU* action is available independently of whether a request has been received and accepted. This allows for the operation of the Mobile Agent and Remote Evaluation paradigms, as will be shown in Section 1.3.3.

Both the *controller* and the *sender-receiver* can be realised as concurrent threads which can be started and stopped by the Application Programmer Interface. By allowing this to happen, we allow implementations to stop monitoring for requests, conserving resources, such as battery, and catering for the eventuality of network disconnection.

The Application Programmers Interface

The Application Programmers Interface (API) builds on the functionality provided by the lower layers, and provides primitives that an application can use to create and send an LMU, to request an LMU to be received, as well as to start and stop the controller and the sender-receiver. Algebraically, the functionality exposed by the API can be represented as a *concurrent composition* of the above.

The Application Layer

Applications built using this framework are part of the *application layer*. When the communication layer receives an LMU which is successfully deserialised and inspected, it passes it on to an application for deployment. Note that, in this context, an application may represent any software abstraction that uses this framework for logical mobility; as such, applications can range from end-user applications to aspects of the system.

In the specification, an application deploying an LMU is modeled by process `APPLICATION`. When an LMU is passed to an application for deployment (*deployLMU*), it is inspected (shown in `DEPLOYLMU`). The results of the inspection can be the following:

Partial Acceptance Some aspects of the LMU are accepted, while others are rejected. This is represented by *lmuPartialAccept*.

Full Acceptance All the contents of the LMU are accepted by the application. This is represented by *lmuAccept*.

Instantiation of the Handler The application may not know how to deploy the LMU received. If the LMU contains a *Handler*, then the latter can be instantiated to take care of the deployment. This is represented by *lmuInstantiateHandler*.

Rejection The LMU may also be rejected by the application. There can be many reasons for this — the application may, for example, have no need for the contents of the LMU. This is represented by *lmuReject*.

SERIALISETRUSTED, SENDSERIALISED, INSPECTDESERIALISED, and DEPLOYACCEPTED are the *safety properties* for this framework. In particular, the safety and liveness properties for the platform state the following:

SERIALISETRUSTED is a safety property that ensures that an LMU will only be serialised to be sent (action *serialise*) if the recipient host is trusted (action *trusted*) to receive it.

SENDSERIALISED is a safety property that ensures that an LMU will only be sent (action *lmuSend*) if it has been successfully serialised (action *serSuccess*).

INSPECTDESERIALISED is a safety property that ensures that an LMU will only be inspected for security reasons (action *inspect*) if it has been successfully deserialised (action *deserSuccess*).

DEPLOYACCEPTED is a safety property that ensures that an LMU will only be deployed (action *deployLMU*) if it is accepted (action *accepted*) by the inspection process.

SENDANDRECEIVELMUS is a liveness property that ensures that, given an infinite length of time, the framework will be able to send and receive an infinite number of LMUs (actions *sendLMU* and *receiveLMU*), avoiding deadlocks. In other words, that both sending and receiving will happen an infinite number of times.

SENDANDRECEIVEREQUESTS is a liveness property that ensures that, given an infinite length of time, the framework will be able to send and receive an infinite number of requests for LMUs (actions *sendRequest* and *receiveRequest*) avoiding deadlocks.

The framework was found to satisfy all safety and liveness properties, by using the FSP model checking tool, LTSA [8]. The next section describes the framework and shows how it can be used to offer all the paradigms discussed in Section 1.2.2.

1.3.3 Transferring Logical Mobility Units

The following paragraphs demonstrate the generality and applicability of this framework, by showing how it can be used by applications to employ the logical mobility paradigms outlined in Section 1.2.2. In particular, this section shows how Code On Demand, Remote Evaluation, and Mobile Agents can be mapped onto a sequence of actions on the model of the framework. To illustrate this, two instances of the framework, A and B, are composed. The composition is as follows:

```
||TWOINSTANCES = ( a:FRAMEWORK || b:FRAMEWORK )
                  /{a.sendRequest/b.receiveRequest, a.receiveLMU/b.lmuSend,
                    b.sendRequest/a.receiveRequest, b.receiveLMU/a.lmuSend}.
```

The / notation re-names actions. This results, for example, in b.receiveRequest to be re-named to a.sendRequest. This results in modeling that when A sends a request, B receives it.

Code On Demand

The use of Code on Demand is equivalent to sending a request via the controller and receiving the code requested by the sender-receiver. In the following trace, node A requests and receives an LMU from node B:

Step Number	Action	Description
0	`a.srStart`	starts the sender-receiver on node A
1	`a.controllerStart`	starts the controller on node A
2	`b.srStart`	starts the sender-receiver on node B
3	`b.controllerStart`	starts the controller on node B
4	`a.sendRequest`	A sends a request for the code required
5	`b.acceptRequest`	B accepts the request
6	`b.sendLMU`	B packs and tries to send the LMU
7	`b.examine`	B inspects the target node (A) to see whether it is trusted
8	`b.trusted`	B finds that A is trusted
9	`b.serialise`	B tries to serialise the LMU
10	`b.serSuccess`	B successfully serialises the LMU
11	`a.receiveLMU`	B sends the LMU / A receives it
12	`a.deserialise`	A deserialises the LMU
13	`a.deserialised`	the LMU is deserialised and checked for conflicts
14	`a.conflict`	a conflict is detected
15	`a.conflictResolved`	the conflict is resolved
16	`a.deserSuccess`	deserialisation process is successfully completed
17	`a.inspect`	LMU is inspected for security
18	`a.accepted`	it is accepted into the system
19	`a.deployLMU`	LMU is passed on to the application for deployment
20	`a.lmuAccept`	the application fully accepts it
21	`a.deployed`	LMU is successfully deployed on A

Remote Evaluation

The use of Remote Evaluation is equivalent to sending the LMU via the sender-receiver. The recipient host may decline the LMU. In the following trace, node A sends an LMU to node B.

Step Number	Action	Description
0	`a.srStart`	starts the sender-receiver on node A
1	`b.srStart`	starts the sender-receiver on node B
2	`a.sendLMU`	A packs and tries to send the LMU to B
3	`a.examine`	A inspects the target node (B) to see whether it is trusted
4	`a.trusted`	A finds that B is trusted
5	`a.serialise`	A tries to serialise the LMU
6	`a.serSuccess`	A successfully serialises the LMU
7	`b.receiveLMU`	A sends the LMU / B receives it
8	`b.deserialise`	B deserialises the LMU
9	`b.deserialised`	the LMU is deserialised and checked for conflicts
10	`b.noConflict`	no conflict was found
11	`b.deserSuccess`	deserialisation process is successfully completed
12	`b.inspect`	LMU is inspected for security
13	`b.accepted`	it is accepted into the system
14	`b.deployLMU`	LMU is passed on to the application for deployment
15	`b.lmuPartialAccept`	the application partially accepts it (i.e., parts of the LMU are discarded)
16	`b.deployed`	the LMU is successfully deployed on B

Note that this framework does not directly address the issue that an application using Remote Evaluation may request a reply based on the execution of the LMU sent. The request may be stored in the properties of the LMU. The reply sent is considered to be an application level issue.

Mobile Agents

The use of Mobile Agents is equivalent to sending an LMU with a Handler, responsible to activate a thread representing the agent on the recipient host. In the following trace, node A sends an agent to node B.

Step Number	Action	Description
0	`a.srStart`	starts the sender-receiver on node A
1	`b.srStart`	starts the sender-receiver on node B
2	`a.sendLMU`	A packs and tries to send the LMU to B
3	`a.examine`	A inspects the target node (B) to see whether it is trusted
4	`a.trusted`	A finds that B is trusted
5	`a.serialise`	A tries to serialise the LMU
6	`a.serSuccess`	A successfully serialises the LMU
7	`b.receiveLMU`	A sends the LMU / B receives it
8	`b.deserialise`	B deserialises the LMU
9	`b.deserialised`	the LMU is deserialised and checked for conflicts
10	`b.noConflict`	no conflict was found
11	`b.deserSuccess`	deserialisation process is successfully completed
12	`b.inspect`	LMU is inspected for security
13	`b.accepted`	it is accepted into the system
14	`b.deployLMU`	LMU is passed on to the application for deployment
15	`b.lmuInstantiateHandler`	application instantiates the handler of the LMU
16	`b.deployed`	the LMU is successfully deployed on B
17		handler starts a thread representing the agent

The agent can then use the API of the framework to migrate itself to another host.

It is important to note that the framework offers the ability to reject an incoming LMU at many different stages. As such, an LMU can be rejected if deserialisation fails, if it is malicious, or if it creates an unresolvable conflict in the system. Moreover, finer-grained control is given to the application programmer, who may inspect the contents of the LMU before accepting or rejecting it, partially or in full. Similarly, the process of sending an LMU can fail on two stages: if the target host is not trusted, or if the serialisation process fails. Finally, realisations of the serialisation & deserialisation engine may decline serialising an LMU if it contains data that should not be shared (for legal reasons, for example) or cannot be shared (as the data can contain non-serialisable elements).

1.4 Logical Mobility for Mobile Computing Middleware

This section starts by outlining a middleware system that we have developed, which makes use of the logical mobility platform described above. It then proceeds by describing a number of services that were built using it.

1.4.1 The SATIN Component Metamodel and Middleware System

In order to engineer a mobile computing system to take advantage of the logical mobility platform described above, we designed the SATIN component metamodel, which we instantiated as the SATIN middleware system. The metamodel is described in [24] in detail, and is outlined in the following paragraph.

SATIN is a local component model, where components reside in the same address space and are interconnected using local references. A SATIN component encapsulates functionality, from user interfaces and libraries, to protocol stacks. A component offers functionality through well-defined interfaces, called *facets*. Components can have arbitrary metadata attached; SATIN allows for dynamically querying the local system as to what components are available, and provides support for late binding of components. Moreover, the SATIN metamodel allows instances to send and receive components dynamically. This is abstracted in the metamodel by a *Deployer* component. Finally, SATIN provides an event mechanism, that allows listeners to be notified of changes in component availability.

The platform described above was implemented in MiToolkit [7], and was used by SATIN as a library to send and receive Java-based LMUs. The implementations of SATIN and MiToolkit require 150329 bytes of code in total, and are written in Java 2 MicroEdition (Connected Device Configuration/Personal Profile). SATIN and MiToolkit enabled us to build a number of middleware services, which are outlined below.

1.4.2 A Dynamic Advertising and Discovery Framework

One of the fundamental services of mobile computing middleware systems is the ability to reason about the environment. The environment is defined as the network of devices that can, at a specific point in time, communicate with each other. The devices can be both mobile and stationary — with the presence of mobile devices, however, the environment can be rapidly changing. In order to adapt, a mobile system needs to be able to detect changes to its environment. As the device itself is part of that environment, it needs to advertise its presence. A mobile device, however, may be able to connect to different types of networks, either concurrently or at different times, with different networking interfaces. There are also many different ways for advertising and discovery. Imposing a particular advertisement and discovery mechanism can hinder interoperability with other systems, making assumptions about the network, the nodes and environment, which may be violated at some later stage or simply not be optimal in a future setting — something which is likely to happen, given the dynamicity of the target area of this chapter.

From the point of view of SATIN, the ability to reason about the environment is translated into the ability to discover components currently in reach, and to advertise the components installed in the local system. This is achieved via the use of *Remote* and *Discovery* components, as well as *Advertiser, Advertisable, DiscoveryFacet,* and *ComponentListener* facets. This is described in detail bellow.

Components that wish to advertise their presence in the environment must implement the `Advertisable` facet. Examples of advertisable components include codec repositories, file transfer services, or sensing capabilities. The `Advertisable` facet exports a method that returns a message that is used for advertising; thus, the advertising message allows the Advertisable component to express information that it requires to be advertised.

An advertising technique is represented by an *Advertiser* component, which implements the `Advertiser` facet. An advertiser component is responsible for accepting the message of advertisable components, potentially transforming it into another format, and using it to advertise them. An advertiser allows components that wish to be advertised to register for advertising. The combination of component availability notification and advertiser registration allows an advertisable component to register to be notified when specific advertisers are added to the system. The advertisable component can then register to be advertised by them. Moreover, an advertisable component can express that it requires a particular advertiser in its dependencies. Thus, the semantics of the advertisable message are not defined and depend on the advertisable component and on the advertising technique (i.e., the advertiser component) used. Note that a component can implement both the `Advertiser` and the `Advertisable` facets. This allows for the advertising of advertising techniques; in this way, for example, the existence of a multicast advertising group can be advertised using a broadcast advertiser. Combined with the use of logical mobility primitives, this allows a host to dynamically acquire a different advertising and discovery mechanism, for a network that was just detected. Upon approaching a Jini network [22], for instance, a node can request and download the components that are needed to advertise to, and use functionality from, the network.

Similarly, discovery techniques are encapsulated by *Discovery* components, which implement the `DiscoveryFacet` facet. There can be any number of discovery components installed in a system. A discovery component acts as registry of advertisable components located remotely. The middleware system defines the `RemoteComponent`. It is used to represent components, which have been found remotely. A remote component is an immutable component that cannot directly export any functionality to local components. It only exports methods needed to access its properties, location, and advertising message. Hence, discovery components act as a collector of remote component references, which can be added and removed dynamically, as they are discovered. Discovery components emit events representing the availability of remote components. Local components can register a `ComponentListener` with a discovery component, to be notified when components satisfying a given set of attributes are located.

`ComponentListener` is represented as a SATIN *facet*. The advertising and discovery framework allows for remote components to be requested and deployed locally. Moreover, the logical mobility functionality encapsulated by SATIN allows the framework and instantiations of the framework to be dynamically deployed and used.

The framework itself requires 6668 bytes of code. We have implemented a centralised publish discovery and a multicast implementation of the framework. Those require 22797 and 22161 bytes, respectively.

1.4.3 A Dynamic Pervasive Security Framework

The project SEINIT [20], funded within the European programme IST, focused on the area of pervasive computing security. It involved the development of a trusted and dependable security framework, that could operate across multiple devices and heterogeneous networks. The main challenge was to provide an environment within which users could communicate securely while moving through multiple different security domains, without having to worry about continually keeping track of their changing environmental context. The aim of the project was to design an architecture that abstracts from low-level technology-specific security configurations. Users are able to define *security policies* per domain. The system is responsible for translating a policy to a concrete technology, and using it to transparently maintain the appropriate level of security per domain.

Fig. 1.6 provides a high-level illustration of SEINIT's overall architecture. The SEINIT middleware is composed of three main building blocks: Information, Decision, and Action. Any information that is gathered and stored within the Information module (such as trust status, user policies, dynamic environmental context, and device configuration), is then analysed and processed within the Decision module. This generates and establishes the necessary security policy and configuration ready for enforcement. The Action module is responsible to invoke the appropriate functionality required to enforce the relevant policies.

As users move through different security domains, they may find themselves connected to a security domain that is either untrusted, or in which some malicious activity is suddenly detected. At this stage, security renegotiation takes place automatically, to ensure that communication remains appropriately secured to the level defined in the security policies. However, this inevitably means that users may find themselves unable to communicate securely, unless they have a specific security technology's functionality available on their devices (e.g., IPSec configuration). To overcome this situation, SEINIT defines a class of code units called *TechnoWrappers* (Technology Wrapper). Each TechnoWrapper encapsulates the functionality and configuration required to invoke a specific security technology. A TechnoWrapper is defined as a SATIN component that implements the `Advertisable` facet.

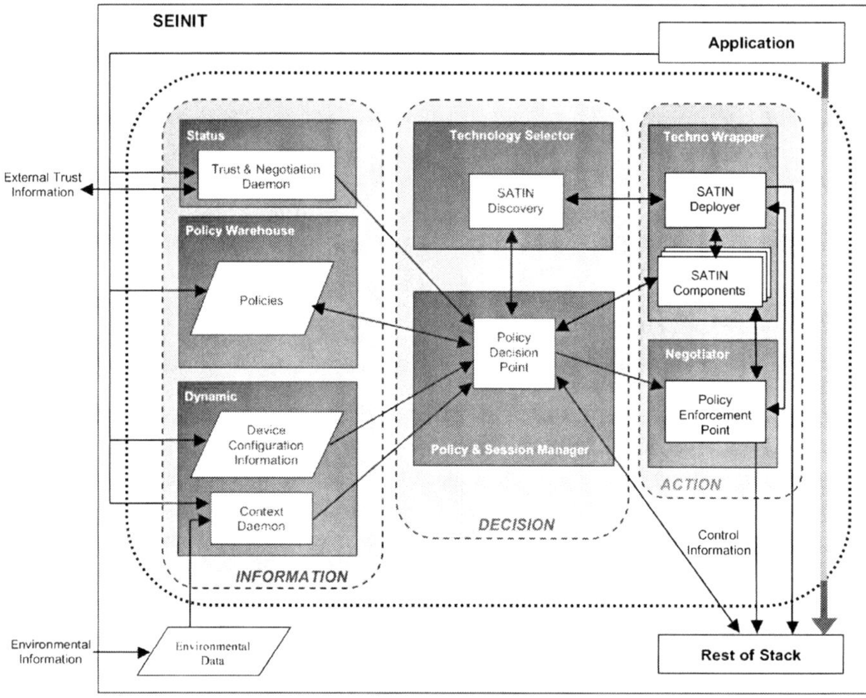

Fig. 1.6: The SEINIT architecture

Once security re-negotiation has determined the specific technology required to secure the user's communication through a given security domain, the SATIN Discovery process is used during decision-making, to discover and request the appropriate TechnoWrapper code unit from a SATIN advertising server. On retrieval, action is taken by the SATIN Deployer to fetch and deploy the appropriate TechnoWrapper code unit ready for invocation of its functionality.

Each TechnoWrapper code unit encapsulates into an LMU both its own code, and the data that it requires. This LMU is then transferred to the user's local device. On receipt of the LMU, the operations that permit inspection of an LMU's contents are used to extract the TechnoWrapper code, instantiate it, and initialise it with the appropriate data ready for use on the user's local device.

The SEINIT TechnoWrapper code unit hierarchy consists of an abstract class `TechnoWrapper`, written in 315 physical source lines of code, and containing functionality to encapsulate common data files into an LMU for transfer, undertake common policy transformations, and initiate basic activation of the security technology. The subclasses `TechnoWrapperIPSec` and `TechnoWrapperPANA` are written in 78 physical source lines of code.

They contain specialised functionality to encapsulate the TechnoWrapper code and any technology-specific data files into the LMU for transfer. The transfer has been noted to occur in the order of tenths of a second across a local-area network.

The use of SATIN in SEINIT shows the suitability of logical mobility techniques in the security domain. As mobile devices cannot have all the code needed to communicate securely in any possible security domain, logical mobility techniques, and SATIN in particular, are used to discover and download the code when needed.

1.4.4 Miscellaneous Services

We conclude this section by showing how two smaller-scale dynamically deployable services were built: a scripting service, and a service allowing the deployment of system (i.e., native) software packages.

The SATIN Scripting Framework

BeanShell [11], an open source Java source interpreter and scripting mechanism, was adapted to run as a SATIN component. This allows SATIN components to use scripts and to be scripted. A "Shell" application was created for SATIN using the BeanShell, which allows developers to manipulate the container and its contents by typing Java statements at runtime. The Shell component and BeanShell encapsulation require 10028 bytes of code. Logical Mobility Units can be used to dynamically deploy both the scripting framework and scripts.

System Package Management

In ongoing work to use SATIN in a programmable active networking project, we required the ability to interface with the underlying operating system, and in particular with the local package manager. In particular, we needed to create SATIN components that used system software to do intelligent packet dropping (using netfilter/IPTables) and media transcoding (using VideoLAN). The idea behind the project is that, if there is not enough bandwidth, active networking SATIN components will be deployed to routing nodes in the network and either drop packets, or dynamically transcode a media-stream to a lower bitrate. As such, we needed the ability to express dependencies on system packages and to dynamically ship and deploy them. Hence, we needed to interface with the underlying package manager.

Using the DataType LME specialisation, we encapsulated RPM packages into LMUs. We created `RPMPackageManager` as a component that tries to install any RPM packages included in an incoming LMU to the underlying system. It can also query the RPM database on the existence of particular packages. `RPMPackageManager` was written in 135 physical source lines of code, and exports functionality through an 8 line facet. Using it, we can send, receive, and deploy RPM packages through the code mobility service. Moreover, the `RPMPackageManager` itself can be sent and received dynamically.

1.5 Related Work

In this section, related approaches in conceptualising logical mobility are briefly discussed, and mobile middleware systems that employ its use are critically outlined. The paper [5] provides a conceptual framework for reasoning about code mobility, on which this chapter is based. It discusses code mobility and its applications, defines a mobile code system, as shown in Fig. 1.1, and examines the paradigms of code mobility as discussed in Section 1.2.2. It also provides a survey of mobile code toolkits, and outlines some application domains for the use of mobile code. There has also been research in evaluating the performance of logical mobility. In [6], a UML-based methodology for performance analysis of logically mobile software architectures is described. UML sequence and collaboration diagrams are annotated with mobility-related stereotypes, allowing the developer to model the code migration aspects of a system. The diagrams are then annotated with probabilities and cost information, and a performance model of the application is obtained, allowing the designer to evaluate the choices made. In [16], the notion of *location* in a Mobile Unity [18] program is used to model the various paradigms of transferring code between nodes. Although similar to what presented in this chapter, the main difference is the transfer of code, rather than the framework required to transfer it. μCode [15] is a light-weight Java library which provides a minimum set of primitives allowing code mobility. The framework presented in this chapter shares μCode's objective to offer a very light-weight set of primitives to support code mobility. Its non-obtrusiveness allows it to be easily integrated with various middleware systems, and its small size makes it suitable for mobile middleware. The framework described in this chapter differes in that it is modular, with each module modeled, and addresses concerns of heterogeneity through the use of metadata. There has also been a number of mobile middleware systems, that employ some logical mobility techniques. The difference with our system is that it is more general; other approaches offer the use of particular paradigms to solve problems which are limited in scope, while others only use logical mobility internally, hiding it from the application developer. Examples include Lime [9], where mobile agents are used to share data, PeerWare [4], where remote evaluation is used to perform operations on remote data sites, and Jini [1], where code on demand is used to offer services.

1.6 Conclusions

In this chapter it is argued for the utility of adaptable mobile middleware services, which can dynamically be deployed into a running system. Logical mobility was presented as a technique for adaptation, its advantages for physical mobility were outlined, and a conceptual framework that allows its flexible use by mobile systems was detailed. The instantiation of this framework in the SATIN middleware system was described, and its suitability was demonstrated, by illustrating a number of dynamically deployable services that were built using it.

We are currently experimenting on using aspects of this framework in the RUNES project [3], which would allow RUNES middleware instances to dynamically adapt. Furthermore, we are currently porting the framework to small, autonomous minirobots with sensors and wireless communication ("motes" or "smart dust"), using the Contiki operating system. Mitoolkit[1], SATIN, and a number of applications have been released as open source at [17].

Acknowledgments

The authors would like to thank Stephen Hailes, Peter Kirstein, Daniele Quercia, and James Skene for their help and suggestions in producing this work.

References

1. Arnold K, O'Sullivan B, Scheifler RW, Waldo J, Wollrath A (1999) The Jini Specification. Addison-Wesley
2. Carzaniga A, Picco GP, Vigna G (1997) Designing Distributed Applications with Mobile Code Paradigms. In: Proc 19th Int Conf on Software Engineering, Boston, MA, ACM Press, pp 22–32
3. Costa P, Coulson G, Mascolo C, Picco GP, Zachariadis S (2005) The RUNES Middleware: A Reconfigurable Component-based Approach to Networked Embedded Systems. In: Proc 16th IEEE Int Symp on Personal Indoor and Mobile Radio Communications, Berlin
4. Cugola G, Picco G (2002) Peer-to-Peer for Collaborative Applications. In: Proc IEEE Int Workshop on Mobile Teamwork Support, pp 359–364
5. Fuggetta A, Picco G, Vigna G (1998) Understanding Code Mobility. IEEE Trans Software Engineering 24 (5): 342–361
6. Grassi V, Mirandola R (2002) PRIMAmob-UML: a Methodology for Performance Analysis of Mobile Software Architectures. In: Proc 3rd Int Workshop on Software and Performance, ACM Press, New York, pp 262–274

[1] A mobile code toolkit developed by a master student at University College London

7. Ijaha M (2004) Mitoolkit. Master's thesis, University College London
8. Magee J, Kramer J (1999) Concurrency: Models and Programs — From Finite State Models to Java Programs. John Wiley
9. Murphy A, Picco G, Roman GC (2001) Lime: A Middleware for Physical and Logical Mobility. In: Proc 21st Int Conf on Distributed Computing Systems, IEEE Computer Society Press, Los Alamitos, CA, pp 524–536
10. Necula GC (1997) Proof-carrying Code. In: Proc 24th ACM SIGPLAN-SIGACT Symp on Principles of Programming Languages, ACM Press, pp 106–119
11. Niemeyer P (1997) BeanShell — Lightweight Scripting for Java
12. Object Management Group (2000) Meta Object Facility (MOF) Specification. Technical report
13. Object Management Group (2003) OMG Unified Modeling Language. Version 1.5. `http://www.omg.org/docs/formal/03-03-01.pdf`
14. Picco GP (1998) Understanding, Evaluating, Formalizing, and Exploiting Code Mobility. PhD thesis, Politecnico di Torino
15. Picco GP (1998) μCODE: A Lightweight and Flexible Mobile Code Toolkit. In: Rothermel K, Hohl F (eds) 2nd Int Workshop on Mobile Agents, LNCS 1477, Springer, Berlin, pp 160–171
16. Roman GC, McCann PJ (2001) Reasoning About Code Mobility With Mobile UNITY. ACM Trans Softw Eng Methodol 10 (3): 338–395
17. The SATIN Open Source Project (2005) The SATIN Component Model. `http://satin.sourceforge.net/`
18. Roman GC, McCann PJ, Plun JY (1997) Mobile UNITY: Reasoning and Specification in Mobile Computing. ACM Trans Softw Eng Methodol 6 (3): 250–282
19. Roman GC, Murphy AL, Picco GP (2000) Software Engineering for Mobility: A Roadmap. In: The Future of Software Engineering — 22nd Int Conf on Software Engineering, ACM Press, pp 243–258
20. The SEINIT Project (2003) Security Expert Initiative. `http://www.seinit.org`
21. The Distributed.net Project (1995) Distributed.NET. `http://www.distributed.net`
22. Waldo J (1999) The Jini Architecture for Network-centric Computing. Comm ACM 42 (7): 76–82
23. Zachariadis S, Mascolo C, Emmerich W (2002) Exploiting Logical Mobility in Mobile Computing Middleware. In: Proc IEEE Int Workshop on Mobile Teamwork Support, pp 385–386
24. Zachariadis S, Mascolo C, Emmerich W (2004) SATIN: A Component Model for Mobile Self-Organisation. In: Proc Int Symp on Distributed Objects and Applications, Agia Napa, Springer, 170–179

2

Towards Self-healing Composition of Services

Luciano Baresi, Carlo Ghezzi, and Sam Guinea

Dipartimento di Elettronica e Informazione
Politecnico di Milano, Piazza L. da Vinci 32, 20133 Milano, Italy
{baresi|ghezzi|guinea}@elet.polimi.it

Summary. Service-oriented architectures are emerging as a solution to integrate components in highly dynamic and evolving contexts. The discovery phase supports flexible binding regimes among components. Bindings can take place either at deployment time or at run-time. Due to of dynamism, however, bindings can fail. The main faults of service-oriented systems are identified and classified, and some solutions to make compositions self-healing are sketched. It is proposed to augment service compositions in such a way that they can be suitably monitored and re-planned according to changing contexts. Proposed solutions are exemplified in a simple case study.

Key words: Self-healing systems, web services, monitoring, reaction strategies, Business Process Execution Language

2.1 Introduction

Service-oriented architectures (SoAs) [5] define a new flexible co-ordination paradigm as a solution to integrate components (services) in highly dynamic and evolving contexts. In this architectural style, components export the services they provide, and clients can discover the services that meet their quality requirements. Once service requests match provisions, point-to-point interactions occur between clients and service providers. To support the matching process, service brokers are placed between clients and providers to collect and advertise available services and facilitate the interaction between them. Services are usually selected both based on their functional capabilities and on the quality of service (non-functional features) they offer.

The discovery phase can occur at different times. This impacts the degree of dynamism and flexibility embedded in these applications. If discovery is performed at design time, the designer selects the services, and composes them by hand, as in the case of Business Process Execution Language (BPEL) processes. If the discovery phase takes place at deployment time, a service

L. Baresi et al.: *Towards Self-healing Composition of Services*, Studies in Computational Intelligence (SCI) **42**, 27–46 (2007)
www.springerlink.com
© Springer-Verlag Berlin Heidelberg 2007

broker is used to "configure" the application. The set of bound services does not change dynamically and the binding between service requests and actual services is established once for all. Discovery and selection can also occur at run-time. This enables a kind of autonomic [8] and context-aware behaviour of service compositions that can yield beneficial effects: whenever a request for a service is issued from the workflow, different sets of services can be discovered and selected dynamically.

If the system fails, because the selection does not find suitable services, the interaction aborts, or the composition does not behave as expected (i.e., some functional and non-functional requirements are not met), simple shutdown is not a solution. The system should react by trying to keep the execution alive and on the right track; suitable recovery actions must be adopted to cope with identified faults. This means that compositions should display a *self-healing* behaviour [11].

To this end, in this chapter the main faults of service-oriented systems are identified and classified, and a solution towards self-healing composition of services is proposed. Since erroneous behaviours are unforeseeable at design-time, the solutions presented here heavily rely on the closed-loop approach typical for control systems. We propose *monitoring* (probing phase), to check whether the system behaves correctly, and whether functional and non-functional expectations are met, and *reaction strategies* (reaction phase) as means to recover from erroneous situations. Since monitoring might heavily impact performance and can be subjective, we propose aspect-orientation to blend monitoring and business processes. Designers specify *monitoring rules* as separate and external aspects of their compositions. The deployment-time weaving blends the business process with the selected set of rules; at run-time suitable parameters allow the user to switch them on and off dynamically and, thus, adjust the actual ratio between performance and ability to be self-healing.

Monitoring rules consist of monitoring expressions, which state the constraints that must be satisfied during execution, reaction strategies, which specify what the system must do to reach its goals regardless of faults, and parameters, which let the user fine-tune the amount of monitoring at run-time. Monitoring expressions define pre- and post-conditions, timeouts, and exception handling. Expressions can build predicates on both internal and external values, i.e., they can constrain both process variables and data gathered from the context. Reaction strategies can be as simple as recalling the same faulty service, or they can exploit dedicated exception handlers or external services. Parameters rank the different monitoring rules and allow the run-time infrastructure to decide whether to execute or skip them according to priorities and other quality indicators. The proposed initial solution is exemplified by the case study of a toy *pizza delivery* system.

The main body of this chapter is organised as follows. After introducing the case study, Section 2.2 identifies the main faults that characterise service-oriented applications and sketches some possible recovery actions.

Section 2.3 introduces the approach, Section 2.4 explains and exemplifies monitoring rules, and Section 2.5 presents the monitoring manager, which is the key component of our framework. Section 2.6 surveys the state of the art, and Section 2.7 concludes the chapter.

2.1.1 Running Example

The example presented in this section will be used throughout the chapter to introduce the problems that can arise with open-world SoA scenarios and the solutions we propose to cope with them. The *Pizza Company* example was first introduced in [7]; here we present a slightly modified version. Following the process definition of Fig. 2.1, we can informally state the requirements of the appplication.

Suppose a client wants to eat pizza. With a WAP[1] enabled mobile telephone, the client dials the *Pizza Company* and, after suitable identification (`Authenticate service`), his/her profile (`Profile Web Service`) determines which kinds of pizza the client likes. The `Pizza Catalogue Service` then offers the client four kinds of pizza; after selecting the favourite one (Double Cheese), the client provides his/her credit card number (included in the client's profile), which is validated by the `Credit Card Validation Web Service`. If everything is correct, the client's account is debited and the pizza company's account is credited. Meanwhile, the pizza baker is alerted about the new order, because after selection the pizza appears in his/her browser, which is integrated in the cooking gear.

At this point, the `Phone Company Web Service` is used to obtain the address of the client, by using his/her telephone number. The `GPS Web Service` is then called to obtain the co-ordinates of the delivery point. These co-ordinates are then passed on to a `Map Web Service`, which processes them and sends a map with the exact route to the pizza delivery agent on his/her PDA. The agent then only needs to deliver the pizza. In the meantime, the client is sent an SMS text message on his/her mobile telephone to alert about the delivery of the pizza within 20 minutes.

In the sequel, we assume that the example is implemented using an orchestrated approach, where the business process is specified using an extended version of BPEL, which supports dynamic binding through a dedicated proxy that is in charge of handling the discovery of services and the interactions with them.

2.2 Faulty Behaviours

Service-oriented applications can be characterised as distributed systems providing services that can be composed by means of *brokers*, which let

[1] Wireless Application Protocol, a collection of protocols enabling access to web content through mobile devices and communication technologies

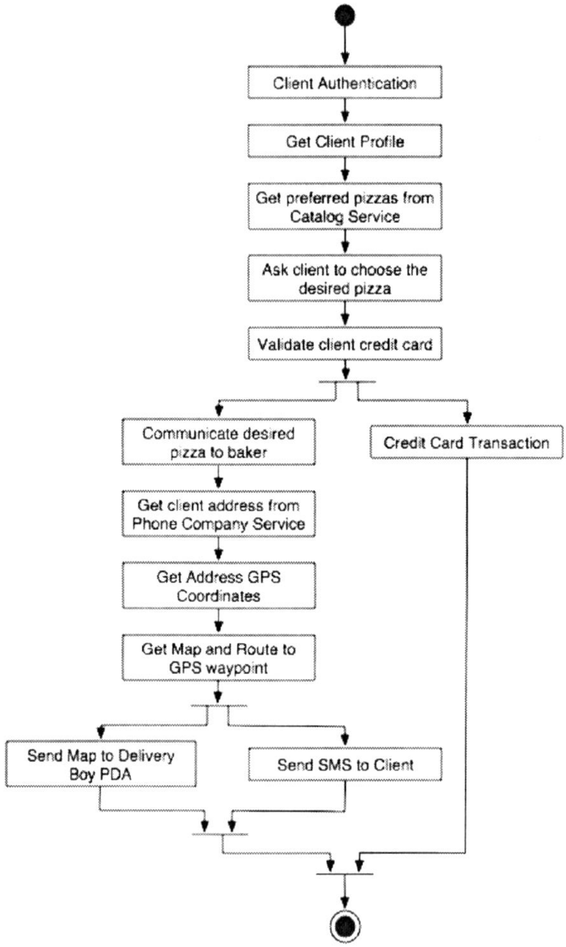

Fig. 2.1: The *Pizza Company* business process

clients discover and select available services, based on their quality requirements. In this framework, all the typical faulty situations that can occur in a distributed application may arise. Additional problems may also occur in the discovery/selection phase.

Service compositions are specified as BPEL-like workflows that fully model the execution flow between the different operations, and specify the contracts that must be satisfied to invoke the operations. The workflow, however, might not be bound to the actual service providers that supply such operations. If the binding is not set, it is left as a responsibility to the discovery/selection phase.

For simplicity, let us first assume that all bindings to external service providers are set at deployment-time. The following situations may then occur:

- The workflow, after successfully binding to a service, does not receive answers to service requests. This can be discovered by setting a timeout: if the answer does not arrive within the set time-frame, the service is considered to be unavailable. In our example, the service used to send the SMS message may not acknowledge the delivery of the message within 5 seconds. In this case, we could retry invoking the same service, if the failure is considered to be transient. If the problem persists, we could try to select another service that offers the same functionality, or we could even modify the workflow to achieve the same goal in a different manner.
- Services may answer by raising an exception, and the process might be specified in such a way that exceptions are caught. The recovery action may be the same as before: we can re-try the same service (to cope with transient failures), try a new one, or modify the workflow to obtain the same answer in a different way.
- Services may deviate from the contract imposed by the workflow, i.e., they do not behave as promised. Given the distributed nature of these applications and the openness of the environment, we cannot always be sure that what is promised — and agreed upon during look-up — corresponds to what is actually supplied. We should be ready to react to providers which fail to supply what they promised. In the worst case, providers might deliberately cheat on us. In a more co-operative setting, providers might expose a new version of a service, which supposedly improves the previous version, but for some reasons it fails to comply with the expected behaviour. For instance, the owner of the service that supplies city maps may improve its service by providing maps with higher resolution. This is an improvement in general, but may result in a more expensive service if the maps have to be transferred to a PDA using GPRS/UMTS technology. In these cases, we cannot re-try the same service. The only feasible solution is a new look-up for a different service, or a new composition that supplies the same functionality. We could also re-negotiate the QoS parameters with the provider, but negotiation is not addressed in this chapter.

If we move to workflows that discover their services at run-time, then we might also have:

- Selection steps that result in failures if no services with the required features are available. We assume that service selection is not only based on a purely syntactic match of required and supplied interfaces, but also envisage contracts that specify functional pre- and post-conditions and QoS. Given the example of Fig. 2.1, we could imagine that the system is unable to find a service that provides the maps of the city for free and/or with a resolution suitable for PDAs. Identifying the best route with respect to traffic conditions might fail as well. If the look-up fails, the only possible recovery activity is to modify the process to obtain acceptable results, i.e., through an on-the-fly composition that mimics the service that is not available.

2.3 A Novel Approach

The approach presented in this chapter is based on ideas coming from the concept of design-by-contract [17] and assertions in the programming language Eiffel [16] (and also in assertion languages like Anna [12] and JML [6]), which let the user set constraints on program execution and also identify possible reactions if they are not satisfied. Similarly, we propose monitoring rules to oversee the execution of BPEL workflows.

Monitoring rules are blended with the BPEL workflow at deployment-time. The explicit and external definition of monitoring rules allows us to keep a good separation between business and control logics, where the former is the BPEL process that implements the business process, and the latter is the set of monitoring rules defined to control — and modify, if needed — the execution. This separation of concerns lets designers produce BPEL specifications that only address the problem they have to solve, without intertwining the solution with awkward pieces of defensive programming. Different monitoring rules (and, thus, reaction strategies) can be associated with the same BPEL workflow to allow the designer to tailor the degree of control to the specific execution context without any need for re-working the business process. Moreover, a good separation of concerns allows for a better management of monitoring rules, and is an effective way to find the right balance between self-healingness and performance.

The approach was also conceived with the goal of adapting existing service-oriented technology to foster the adaptation of self-healing techniques. All these reasons led to the approach summarised in Fig. 2.2. It starts as soon as a BPEL workflow exists (or the designer starts working on it):

- Monitoring rules are always conceived either in parallel with the business process or just after designing it. These rules are associated with specific elements (e.g., invocations of external services) of the business process.
- When the designer selects the rules to use with a specific execution, $BPEL^2$ instruments the original BPEL specification to add the monitoring rules. This weaving introduces a proxy service, called *monitoring manager*, which is responsible to understand whether a monitoring rule must be evaluated, and, if it is the case, to react as stated by the reaction strategies.

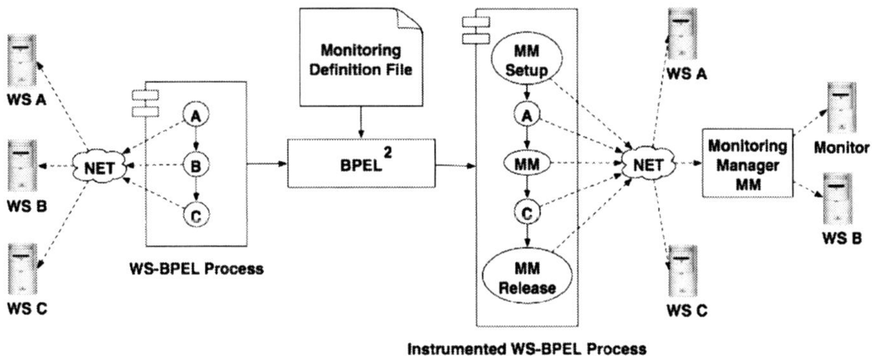

Fig. 2.2: The approach in a nutshell

- When the instrumented BPEL process starts its execution, it calls the monitoring manager whenever a monitoring rule has to be considered. The actual evaluation depends on the values of the parameters associated with the rule and on the context in which it is executed. For instance, one of these parameters is *priority*, and the context comprises the *global priority* set for the execution of the workflow at start-up time. A rule with priority lower than the global one would be skipped, and the monitoring manager would call the actual service directly.
- The user has a special-purpose interface to interact with the monitoring manager and change its status. This happens when the designer wants to change the impact of monitoring at run-time without re-deploying the whole process.
- If some constraints are not met, the monitoring manager activates the reaction strategies, which are part of the monitoring rule, to try to keep the execution alive. If no strategies are associated with the rule, the monitoring manager simply notifies the BPEL process by raising a pre-defined exception.

2.3.1 Weaving

Code weaving is performed by the $BPEL^2$ pre-processor. It parses the monitoring rules associated with a particular process and adds specific BPEL activities to the process to oversee its execution. For instance, if a rule embeds a post-condition to the invocation of an external web service, $BPEL^2$ substitutes the BPEL invocation activity with a call to the monitoring manager, preceded by BPEL assignment activities to prepare the data that have to be sent to the monitoring manager, and followed by a switching activity which checks the monitoring manager's response (i.e., the exception raised if the monitoring manager cannot maintain the execution). The monitoring manager is then responsible to invoke the web service being monitored, to check its post-condition with the help of an external data analyser, and to trigger defined reaction strategies, if needed.

Pre-conditions are treated the same way, except that the monitoring manager checks the condition — and, if needed, triggers the recovery strategies — before calling the actual service. Invariants are treated by adding an implicit post-condition to each BPEL activity defined in the scope with which the invariant is associated. With timeouts the monitoring manager wraps the actual invocation to detect if the execution takes too long; if so, it raises a known exception.

$BPEL^2$ always adds to the BPEL process an initial call to the monitoring manager to send the initial configuration such as the monitoring rules and information about the services it will have to collaborate with (see `MM Setup` in Fig. 2.2). $BPEL^2$ also adds a "release" call to the monitoring manger to communicate that it has finished executing the business logic (see `MM Release` in Fig. 2.2). This permits the monitoring manager to discard any configurations it will not need anymore. Every call to the monitoring manager (which is not a set-up or a release call) is also signed with a unique identifier. This is used to match the manager call with the specific rules and the data stored in the monitoring manager during set-up.

This solution does not require any particular tool to run and oversee BPEL processes. Once the weaving of rules has been performed, the resulting process continues to be a standard BPEL process which simply calls an external proxy service, which selectively applies the specified monitoring rules.

2.4 Monitoring Rules

Monitoring rules reflect the "personal" monitoring needs that individual users of BPEL processes may have. Any time a BPEL process is run, different monitoring activities should be enacted, depending on who has invoked the process and how they want to oversee its execution. This requires the ability to define and associate monitoring activities to each single BPEL workflow instance.

The definition of a set of monitoring rules follows the structure illustrated in Fig. 2.3. It is mainly a container for the definition of the monitoring rules that are to be executed at run-time, and the conditions under which they can be ignored. Obviously, this leads to the need for specific languages to identify the locations in the workflow where the rules have to be executed, and define the expressions and reaction strategies embedded in the rules.

The information supplied for any set of rules is organised in three main parts: *General Information*, *Initial Configuration*, and *Monitoring Rules*. The first part identifies the BPEL workflow to which the monitoring rules will be attached. The second part contains the initial values of the context variables associated with the workflow (e.g., global priority). The third part, the mon-

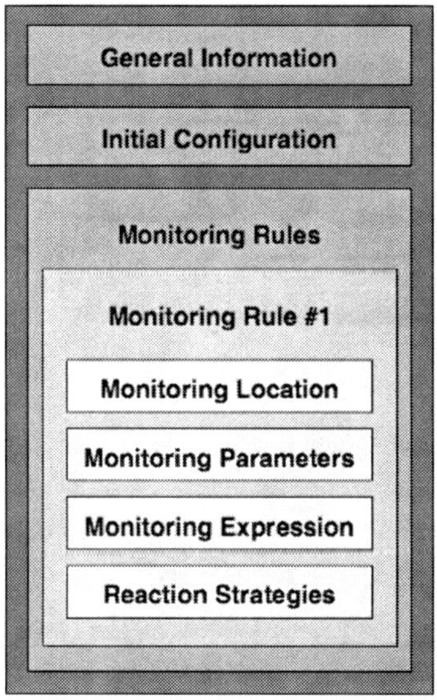

Fig. 2.3: Set of monitoring rules

itoring rules, represents the core of the self-healing approach. It is organised in *Location*, *Parameters*, *Expressions*, and *Reaction Strategies*.

2.4.1 Locations

While defining locations, we specify two things: the kind of condition we want to monitor, and in which lexical point of the workflow definition we want to monitor it. For the first part, we use a keyword indicating whether the monitoring rule specifies a pre- or post-condition, an invariant, a timeout, or an exception handler. The last two operations might be defined in the BPEL process directly, but we would loose the capability of working at a higher abstraction level and also the homogeneity with respect to the other expressions.

For the second part, we use an XPATH[2] selection condition whose evaluation yields the lexical point where the rule has to be checked in the process. Usually, the XPATH selection condition indicates the BPEL invocation activity to which we associate the rule, but in the case of an invariant it indicates the BPEL scope to which we associate it. Regarding pre- and post-conditions, we are only interested in attaching monitoring rules to BPEL activities that can modify the contents of the process' internal variables. We assume that pre- and post-conditions can be attached to BPEL invocation activities, post-conditions to receive activities, and pre-conditions to reply activities.

[2] XML Path language, specifically designed to address parts of documents written in the Extensible Mark-up Language XML

Recalling the example, for instance, if we want to define a post-condition on the invocation of the operation named `getMap` published by the `MapWS` web service and linked to the BPEL process through partnerlink `MapServicePartnerLink`, we would define the location as[3]:

```
type = "post-condition"
path = "//:invoke[@partnerLink="lns:MapServicePartnerLink" and
                 @operation="getMap"]"
```

2.4.2 Parameters

Parameters are meta-level information that define the properties of a monitoring rule itself. They influence the actual evaluation of the rule, and can even impede its run-time checking. We currently consider three parameters (but many others could easily be added[4]):

Priority is an integer between 1 and 5 indicating the level of importance that is associated with a rule (5 indicates the highest priority). As we said, the execution of a workflow is associated with a priority, and a monitoring rule is executed only if its priority is higher than that of the workflow. This makes it possible to execute the same business logic with different degrees of monitoring.

Validity is a time-frame associated with a monitoring rule: when a process execution occurs within this time-frame, the monitoring rule is checked, otherwise it is ignored. This can be useful when a service invocation must initially be monitored for a certain amount of time before deciding that it can be trusted.

Certified providers is a list of providers for which the monitoring activity does not have to be executed. This eliminates the overhead caused by monitoring in the case of trusted service providers.

2.4.3 Expressions

Monitoring expressions work with an abstraction of the Web Services Description Language's (WSDL) definitions of the services the BPEL process does business with. Depending on the degree of dynamism, these could be the actual services used by the application, or abstract descriptions of the services the process would like to bind to. To do this, we use a tool based

---

[3] This is what the system produces, but the user defines locations by pointing to the specific BPEL elements directly in the graphical editor, and by choosing the annotation type.

[4] Parameters may also provide contextual information, such as the physical location in which the process is executed, or the device on which the process executes.

on Apache AXIS WSDL2Java[5] [3], which permits us to reason on stereotyped class diagrams that represent the classes automatically extracted from a WSDL service description. In the tool, a web service, used by the BPEL workflow, becomes a **service** class that provides one public method for each service operation and no public attributes. Similarly, for each message type used in the WSDL definition, we introduce a **dataType** class, containing only public attributes and no methods. Fig. 2.4 shows a `MapWS` **service** class that provides a single method called `getImage`. The exposed method takes a `GetImageRequest` **dataType** as input and produces a `GetImageResponse` **dataType** as output. This way, we can state our expressions by referring to these classes. Pre- and post-conditions, timeouts and exception handlers are associated with single operations (methods), while invariants express conditions on variables visible within the scopes to which they are attached (and on data collected from the context, if needed).

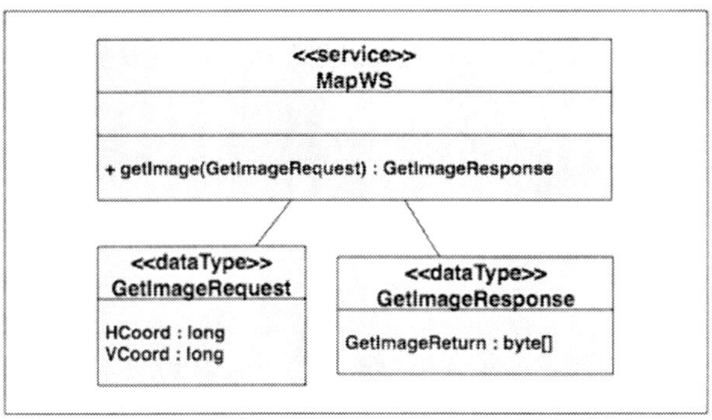

Fig. 2.4: Web service MapWS

Expressions are defined using WS-CoL, our assertion language for web services inspired by the Java Modeling Language (JML) [6]. WS-CoL complements standard assertion checking features with the capability to gather data from external sources. This is to support the case in which a monitoring rule defines a relationship that must hold between data existing within the process and data obtained by interacting with external data collectors.

WS-CoL adds a set of keywords that represent ways to obtain data from external data collectors. A different extension is introduced for each

[5] Apache Axis is an open source implementation of the Simple Object Access Protocol (SOAP), which is part of the web service standard; WSDL2Java is a tool that takes a description of a web service written in the web service definition language (WSDL) and generates Java code to access the web service form within a Java program.

of the standard XSD[6] types that can be returned by external data collectors: `\returnInt`, `\returnBoolean`, `\returnString`, etc. Therefore, while defining a monitoring expression, we can use these extensions. All follow the same design pattern. They take as input all information necessary to interact with the external data collector, such as the URL location of its WSDL description, the name of the operation to be called upon it, and the parameters to be passed to the data collector service. For instance, if we want to specify a post-condition for the `getImage` operation in Fig. 2.4 and state that the returned map must have a resolution less than 80×60 pixels, we would define the expression as:

```
\returnInt(wsdlLoc, getResolution, 'image',
       GetImageResponse.GetImageReturn, HResolution) <= 80 &&
\returnInt(wsdlLoc, getResolution, 'image',
       GetImageResponse.GetImageReturn, VResolution) <= 60;
```

In this example, a `getResolution` operation is invoked on a service that publishes its interface at the URL `wsdlLoc`. The array of bytes `GetImage-Return` (see Fig. 2.4) is passed as an input value and mapped onto the `image` message part defined at `wsdlLoc`. `HResolution` and `VResolution`, on the other hand, are the message parts defined in the output message at `wsdlLoc` that should be returned as integers. These returned values are compared with the desired resolution (80 pixels for the horizontal dimension and 60 pixels for the vertical dimension).

The use of external data collectors also allows to check whether look-up services work as expected or not. The invocation of the service in charge of finding a good candidate[7] can become a special-purpose data collector, which is managed with the dedicated XSD constructs described above.

2.4.4 Reaction Strategies

Reaction strategies represent the actions we want the system to take whenever expressions are not verified. This is a tentative partial list of possible strategies:

Re-try re-executes the same invocation with exactly the same parameters and contracts. This is a possible solution when a fault is transient.

Re-bind tries to select another service with the same characteristics as the faulty one. This is needed if the previously selected service does not answer anymore.

Re-organise creates an on-the-fly composition of available services to mimic the behaviour of a faulty interaction. This might be a solution when there are no services that supply the required functionality.

[6] XML Schema Definition, a language to define the structure and elements of XML documents

[7] For instance, the discovery framework developed within the SeCSE project [21]

Change monitoring rule alters the way the process is monitored by modifying one of the monitoring rules. This might help when a contract is too strict, and the designer wants to adopt a looser version, or when the designer wants to better understand what is going on, and needs a more restrictive expression or a different reaction strategy.

Change monitoring parameters allows parameters (e.g., priority or certified providers) to be modified dynamically, thus supporting different tradeoffs between self-healigness and performance. For instance, it can be used both to lower the amount of probing and, thus, the capability of reacting to unforeseen situations, and to increase it to adopt a more defensive approach.

Call handlers allows the designer to call both internal and external handlers to react to a bad situation. Internal handlers must be part of the process when it is deployed, while external handlers can be any external service that does something. The main difference between the two options is that the first type of handlers must be blended with the deployed business process, and can only be changed by re-deploying it, but they share the same internal variables as the business process they have to recover. This is achieved by hard-wiring the reactive actions directly into the business process prior to its deployment by using BPEL event-handling features. In other words, the designer decides upon a library of already existing reaction strategies from which specific actions may be chosen. In the latter case, we have much more flexibility, used services can change freely, but they do not share variables with the process they refer to. Interactions can only be carried out through exchanged messages.

Notify can be seen as a special case of the previous option, as it simply notifies the user that there is a problem. It does nothing to change the current situation, but it only warns the user that one of the expressions was violated. This option might be a solution when a given contract is violated and the designer wants to be informed, without changing the process.

Warn and stop is the last option. It simply stops the execution and notifies the user of the shut-down.

The above building blocks can be composed to define complex and hierarchical reaction strategies. The strategy **re-try→re-bind→re-organise**, for instance, means that the designer wants to recall exactly the same service; if the problem persists, then he would like to find an alternative service that offers the same functionality; and, as last option, he would consider substituting the single invocation with a dedicated subprocess. If the last option also aborts, then the monitoring manager would raise an exception for the main BPEL process.

Reaction strategies are delegated to the monitoring manager. Since the monitoring manager is a proxy capable of invoking any kind of external component that offers a WSDL interface, we can re-use this capability to take reactive actions. The syntax we use to specify the reaction strategy is:

```
Call <methodName> <parameters> <wsdlEndPoint>
```

where `methodName` indicates the name of the web method that should be invoked, `parameters` is a sequence of parameters separated by commas that should be sent to the web service, and `wsdlEndPoint` is the URL that indicates where to find the WSDL definition of the web service to be invoked. Since the monitoring manager invokes anything provided with a WSDL interface, the designer might also specify complex strategies by means of dedicated BPEL fragments.

The **re-organise** strategy we consider here provides a *local re-organisation*. In this case, we consider the BPEL process definition as a direct graph, and apply graph transformation rules to modify its topology. Rules are simple and work only locally. Each rule states a predicate on the faulty node and its neighbours. These rules can transform a single node into a sequence of two or more nodes, into a parallel composition of two threads (nodes), or into a branch. Rules can be applied only if their pre- and post-conditions satisfy certain constraints that ensure substitutability. More complex non-local reorganisation strategies may also be conceived. They will be investigated in future work.

For instance, we can split a single node n of an abstract process into a sequence of two nodes n_1 and n_2, if the pre-condition of n_1 implies that of n, the post-condition of n_2 implies that of n, and the post-condition of n_1 implies the pre-condition of n_2. The other rules are governed by more complex application conditions; the whole set is defined in [10].

The rule that splits a node into a sequence is applied in Fig. 2.5 to an excerpt of the example process of Fig. 2.1. Let us assume that the operation `Get Map and Route` cannot be supplied by any actual service. The service broker offers a service that returns maps and routes with higher resolution, and also a service that can filter maps to lower their resolution. The simple transformation of the process, along with the pre- and post-conditions of required and supplied operations is shown in Fig. 2.5, where `r` is the result produced by the operation; `hres` and `vres` return the horizontal and vertical resolution of a map. This re-organisation results in a new composed service that is called by the monitoring manager instead of the original one.

2.5 Monitoring Manager

The *Monitoring Manager* is the key component of our solution and Fig. 2.6 illustrates the architecture of a prototype implementation we developed. The manager is capable of interpreting monitoring rules, keeping trace of the configuration with which a user wants to run a workflow, interacting with external data collectors to obtain additional data for monitoring purposes, invoking external monitor services, and applying reaction strategies. This section describes the problem of analysing (and reacting to) a post-condition; the other cases are similar.

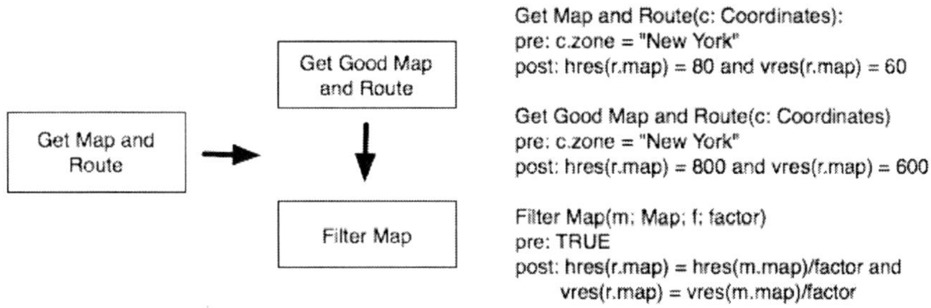

Fig. 2.5: Example of local process re-organisation

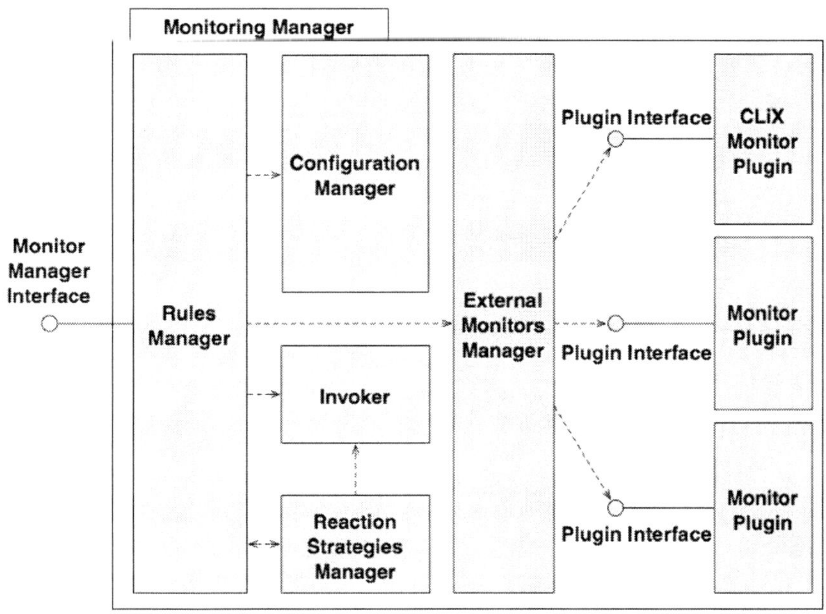

Fig. 2.6: Monitoring Manager

The manager is constructed to keep a configuration table for each process execution. These configurations are managed by the *Configuration Manager*. In particular, the manager needs to know: the initial overall process configuration, the monitoring rules, and all information necessary to interact with the external services (the service being overseen, the external data collectors, the external monitor services, and reaction elements). Most of these data can be sent to the manager at process start-up by invoking the set-up method published by the manager (see Fig. 2.2). In particular, everything except all data that originate within the process can be sent at process start-up, before executing the real business logic. This solution is preferable, with respect to sending all the data any time the process needs to interact with the manager, since an initial slow-down is certainly better than slowing down all intermediate steps. All information sent during the set-up phase is stored in the *Configuration Manager*, and is associated with a process execution through the unique identifier produced by the BPEL engine. When the workflow terminates, the manager is notified and the corresponding configuration table is de-allocated.

The manager also supplies a graphical interface to the user. It permits run-time consultation and modification of the values contained in the configuration table. For example, it is possible to modify the priority level at which a process is being run, or to add a new provider to the list of certified providers that are associated with a given monitoring rule.

The interaction between the components that co-operate to execute the service of Fig. 2.4 and to check its post-condition (Section 2.4.3) proceeds as follows. Initially, the BPEL process sends the data that will be necessary to the manager. Since no pre-condition needs to be checked, the *Rules Manager* asks the *Invoker* to invoke the external web service (in our case service `MapWS`). When the *Rules Manager* receives the results of the service invocation, it interacts with the *Configuration Manager* to retrieve the expression of the monitoring rule (i.e., the post-condition) that has to be checked. By examining the monitoring parameters attached to the rule, the *Rules Manager* dynamically decides if the rule is to be checked or not. Suppose that the associated priority parameter is 4, and that the process is run with a priority value of 3. The rule is checked, since its priority parameter is higher than the process's priority.

Then, the *Rules Manager* decides whether additional data are required from external data collectors. If this is the case, it calls the *Invoker* to obtain them. This component is built around the Apache Web Services Invocation Framework (WSIF) [2], and is capable of invoking a web service without previously creating client-side stubs, but by dynamically interacting with the service through its WSDL description. The *Invoker* can be used to invoke any service provided it knows: the URL of the service's WSDL description to be invoked, the name of the operation that is to be invoked on that service, a list of keys that help map the operation's input values onto the operation's message parts as defined in the WSDL description, a list of input values for the operation to be invoked, and a list of keys to indicate the parameters (as indicated in the output message parts contained in the WSDL description) we want to receive as output.

Once all data necessary are available, the *Rules Manager* begins its interaction with the *External Monitors Manager*. This component is responsible to manage the different kinds of external monitors that the manager can work with. In particular, it manages the set of plug-ins that contain the logic necessary to convert the WS-CoL syntax used to define the monitoring expressions into the proprietary syntax used by each external monitor. A monitor plug-in also prepares the data that must be sent to the monitor by formatting them in a way that can be interpreted by the monitor. The prototype we illustrate here uses an interpreter called `XlinkIt` [18]. Therefore, the WS-CoL expressions to monitor must be re-written as rules in the format requested by XlinkIt, called CLiX, and the data expressed as XML fragments. CLiX is constraint expression language used to constrain the content of XML documents. CLiX uses a mixture of first-order logic and XPath expressions. When the *External Monitors Manager* has finished to adapt the monitoring data and the monitoring rules, the *Invoker* is called to invoke the external monitor. If the monitor's response is that the condition is satisfied, the manager returns the service response to the BPEL process. If the monitor responds with an error (i.e., the condition is not satisfied), the *Reaction Strategies Manager* retrieves the associated strategy from the *Configuration Manager* and applies it. If we consider the strategy presented in Section 2.4.4, for instance, the manager uses the *Invoker* to try to recall the same service. If the answer is negative, it tries to find another service, by invoking the discovery service, and calls it. If also this option does not work, the *Reaction Strategies Manager* calls the external service that creates an on-the-fly composition, and the *Invoker* calls the newly created service. If no strategy works, the manager returns a standard fault message, as publishes it in its WSDL description.

2.6 Related Work

The research initiatives in the field of self-healing web services can be classified in two main categories: those that mainly concentrate on monitoring, and those that mainly address automatic and dynamic compositions.

In the former group, all proposals share the common goal of discovering erroneous situations during the execution of services, but they differ in a number of ways: degree of invasiveness, abstraction level at which they work, reactiveness or pro-activeness, and others.

Spanoudakis and Mahbub [14] propose a framework to monitor requirements of BPEL-based service compositions. Their approach uses event-calculus to specify the requirements that must be monitored. Requirements can be behavioural properties of the co-ordination process, or assumptions about the atomic or joint behaviour of the deployed services. Events, produced by the normal execution of the process, are stored in a database and run-time checking is performed by an algorithm based on integrity constraint checking in temporal deductive databases. This approach is less intrusive than ours: it proceeds in parallel with the execution of the business process and, thus, has a lower impact on performance, but it is also less responsive in discovering erroneous run-time situations.

Lazovik et al. [9] present an approach based on operational and actor assertions. The former can be used to express properties that must be true in one particular state of execution, invariants that must hold throughout all execution states, and properties on the evolution of process variables. The latter can be used to express a client request regarding the entire business process, regarding all providers playing a certain rôle in process execution, or regarding a specific provider. This approach and ours are both assertion-based. Once assertions are defined, it is completely automatic in its set-up and monitoring, but it lacks the possibility to dynamically modify the degree of monitoring. It also lacks adoptability.

The Cremona framework [13] proposes a framework to create and monitor WS-Agreement [1] contracts. WS-Agreement is a standard that defines a protocol to establish agreements between clients and providers. It also defines an interface to extract the state of agreements at run-time, thus supporting their run-time monitoring. Cremona implements all WS-Agreement interfaces, provides management functionality, and defines abstractions for service providing systems that can then be implemented in domain-specific environments. The Cremona framework does not address the problem of reaction strategies.

In the latter group of research initiatives, the proposals are less mature. For instance, Web Service Composition Graph (WSCG) [20] uses graph theories to supply a transformation engine for those nodes of a graph that represent single services.

Some approaches are based on the use of Petri nets. For example, Pernici et al. [15] propose a framework, based on proprietary service descriptions, to orchestrate dynamically composed services by means of Petri nets.

Pautasso et al. [19] provide an autonomic re-configuration component for JOpera, their distributed service composition and deployment platform. It allows the system to automatically re-configure its deployment strategy in the wake of QoS problems that can arise due to excessive workload. Recovery strategies are chosen by the autonomic controller according to a number of possible goals (e.g., minimisation of resource allocation or response time). In practice, they consist in a modification of the number of nodes among which the deployment infrastructure is distributed.

Berardi et al. [4] propose an approach to the automatic composition of web services within their Colombo framework. In this framework, services are characterised in terms of message exchanges, data flow, and effects on the real world (represented by a relational database). In particular, services' behaviour is specified by a finite state transition system. Therefore, the goal of the automatic composition is given in form of the definition of a transition-based service and of its behavioor. The composite service is then automatically synthesised by Propositional Dynamic Logic. The result is a mediator that, through message exchanges, interacts with existing services to mimic the goal of the automatic composition.

2.7 Conclusions and Future Work

In this chapter, we presented an approach to deal with faulty behaviours of service-oriented systems. The dynamism and openness that characterise these systems can lead to problems both in the process of discovering and selecting services, and in the "normal" execution of the system. A service might not answer to its clients, or it could violate the contract — functional requirements and QoS — agreed with the client. The proposed solution is based on monitoring the execution of a service composition and triggering suitable reaction strategies to make the system continue its execution. This is a step towards *self-healing* compositions of services. A prototype tool has been illustrated here.

Our future work includes a better assessment and fine-tuning of the approach. More precisely, we still need to thoroughly investigate reaction strategies and the rule-based approach to re-organise compositions. In the meantime, we are applying these ideas to new and more complex case studies, and are enhancing our prototype execution environment.

References

1. Andrieux A, Czajkowski K, Dan A, Keahey K, Ludwig H, Nakata T, Pruyne J, Rofrano J, Tuecke S, Xu M (2005) Web Services Agreement Specification (ws-agreement). `http://www.ggf.org/Public_Comment_Docs/Documents/Oct-2005/WS-AgreementSpecificationDraft050920.pdf`
2. Apache Software Foundation (2003) WSIF: Web Services Invocation Framework. `http://ws.apache.org/wsif/`
3. Apache Software Foundation (2005) Web Services — Axis. `http://ws.apache.org/axis/`
4. Berardi D, Calvanese D, De Giacomo G, Hull R, Mecella M (2005) Automatic Composition of Transition-based Semantic Web Services with Messaging. In: Proc 31st Int Conf on Very Large Data Bases, ACM, pp 613–624
5. Booth D, Haas H, McCabe F, Newcomer E, Champion M, Ferris C, Orchard D (2004) Web Services Architecture. World Wide Web Consortium. `http://www.w3.org/TR/2004/NOTE-ws-arch-20040211/`
6. Burdy L, Cheon Y, Cok DR, Ernst MD, Kiniry JR, Leavens GT, Leino KRM, Poll E (2005) An Overview of JML Tools and Applications. Int J on Software Tools for Technology Transfer 7 (3): 212–232
7. Duivestein S (2001) Web Services and Workflow — Organizing Web Services. `http://www.webservicesarchitect.com/content/articles/sander01.asp`
8. Kephart JO, Chess DM (2003) The Vision of Autonomic Computing. IEEE Computer 36 (1): 41–50
9. Lazovik A, Aiello M, Papazoglou MP (2004) Associating Assertions with Business Processes and Monitoring their Execution. In: Proc 2nd Int Conf on Service Oriented Computing, ACM, pp 94–104
10. Lehner T (2004) Dynamic Reconfiguration of BPEL Processes. Master's thesis, Politecnico di Milano and Universität Passau

11. de Lemos R, Gacek C, Romanovsky AB (2003) ICSE Workshop on Software Architectures for Dependable Systems. In: Proc 25th Int Conf on Software Engineering, IEEE Computer Society Press, pp 781–782
12. Luckham DC, von Henke FW (1985) An Overview of Anna, a Specification Language for Ada. IEEE Software 2 (2): 9–22
13. Ludwig H, Dan A, Kearney R (2004) Cremona: an Architecture and Library for Creation and Monitoring of WS-agreements. In: Proc 2nd Int Conf on Service Oriented Computing, ACM, pp 65–74
14. Mahbub K, Spanoudakis G (2004) A framework for Requirements Monitoring of Service based Systems. In: Proc 2nd Int Conf on Service Oriented Computing, ACM, pp 84–93
15. Mecella M, Parisi-Presicce F, Pernici B (2002) Modeling E-service Orchestration through Petri Nets. In: Proc 3rd Int Workshop on Technologies for E-Services, LNCS 2444, Springer, pp 38–47
16. Meyer B (1992) Eiffel: The Language. Prentice-Hall
17. Meyer B (1997) Object-oriented Software Construction. 2nd edn. Prentice Hall
18. Nentwich C, Capra L, Emmerich W, Finkelstein A (2002) xlinkit: a Consistency Checking and Smart Link Generation Service. ACM Trans Internet Technology 2 (2): 151–185
19. Pautasso C, Heinis T, Alonso G (2005) Autonomic Execution of Web Service Compositions. In: Proc IEEE Int Conf on Web Services, IEEE Computer Society Press, pp 435–442
20. Ren ZH, Jin BH, Li J (2004) A New Web Application Development Methodology: Web Service Composition. In: Proc 2nd Int Workshop on Web Services, E-Business, and the Semantic Web, LNCS 3095, Springer, pp 134–145
21. SeCSE Crew (2004) Service Centric System Engineering — EU/IST Integrated Project. http://secse.eng.it/

3

Wireless Sensor Networks

Yusuf Öztürk

Electrical and Computer Engineering
San Diego State University, San Diego, CA 92182-1309, USA
yozturk@mail.sdsu.edu

Summary. This study introduces a new self-configuring, multirooted tree topology ad hoc network designed to offer robust and scalable sensor communication services. A generic sensor node is equipped with a short-range wireless transceiver, while border nodes acting as gateways between sensor clusters and cyberspace are equipped with an additional network interface. All traffic is assumed to originate from or terminate at the sensors. A new cost metric is proposed to partition a network of sensors into multiple trees each rooted from a gateway node. The novel cost function proposed here takes into account several metrics including link capacity, link reliability, number of hops to a gateway, and power constraints of a node. The contribution of each metric to the cost function is decided by weights assigned to each metric. A middleware offering services similar to Universal Plug and Play (UPnP) is developed. This middleware offers sensor discovery, sensor event notification, and sensor object access services for simple deployment and operation of sensor nodes. A network is modeled and simulated with Opnet Technologies' Opnet Modeler prior to implementation on a low power baseband processor and a low-bit-rate radio frequency (RF) transceiver. Simulations have revealed that the proposed architecture adapts well to changes in the state of links/nodes, and always converged to a valid solution.

Key words: Energy efficient sensor, wireless sensor networks, adaptive routing, transducer electronic data sheet, ad hoc networks, topology control, multi-hop communication, sensor access protocol.

3.1 Introduction

Remote sensing has been a centre of attraction since the invention of microcontrollers. Traditional sensors and control logic have been networked for more than two decades using a variety of communication technologies. Advances in processor, radio and memory technologies have made wireless access to sensor data possible and enabled wireless sensor networks. Academic research and industrial developments marked several milestones in wireless sensor networks during the last five years. The IEEE standard 1451 [14] was

Y. Öztürk: *Wireless Sensor Networks*, Studies in Computational Intelligence (SCI) **42**, 47–69 (2007)
www.springerlink.com © Springer-Verlag Berlin Heidelberg 2007

developed to describe a set of common network-independent communication interfaces to connect transducers to microprocessors, instrumentation systems, and control/field networks. The working group IEEE 802.15.4 defined the physical layer and medium access control layers of low-power networked sensors [13]. The Zigbee forum developed the network and transport layers for the IEEE 802.15.4 protocol stack, and is currently deploying systems in the field [32]. In addition to the Zigbee forum, several research groups are leading the way into building smaller-sized, low-cost, low-power, high-data-rate wireless sensor nodes which can be deployed in medical services, industrial control, home automation, and other areas [1] [22].

Several wireless sensor networks (WSN) have been deployed in a variety of applications. ZebraNet is deployed to observe the behaviour of wild animals within a spacious habitat [31] at the Mpala Research Centre in Kenya. Telenor's Electronic Shepherd project [27] and the Southern California Puma Project [29] of the University of California at Davis are using GPS-based sensors for animal tracking. In order to better understand the Earth's climate, subglacier environment is monitored at the Norwegian Briksdalsbreen [20] using a wireless sensor network. The Argo project [2] is using sensor nodes to obtain temperature, salinity, and current profile of the upper ocean. Argo is a global array of 3,000 free-drifting profiling floats that measure temperature and salinity of the upper 2000 m of the ocean. Wireless sensor networks are being designed for a variety of military applications ranging from active tagging to sniper localisation [26]. The adaptation of sensor networks to new application areas is taking place at a very fast rate.

A wireless sensor network may be composed of hundreds of thousands of sensor nodes deployed in a geographical region. Alternatively, it may be composed of a small number of radio-frequency (RF) devices worn as "wear and forget" apparel to collect biometric readings such as heart rate, chest actigraphy, chest temperature, wrist temperature, wrist actigraphy/sleep scoring [19], metabolic expended energy, core body temperature, and geolocation of an athlete, patient or soldier. The sensor nodes may be deployed randomly or installed at deliberately chosen locations. Deployment may be a one-time activity or a continuous process where faulty nodes are replaced or more nodes added to the network. Sensor nodes may be fixed or may change their location after initial deployment. Mobility may be a side-effect or a desired property of a system. Sensor nodes may be homogenous or may differ in the type or number of sensors attached sensors. Some nodes may computationally be more powerful while others may be equipped with special hardware such as GPS receivers to act as location identifiers for a sensor cluster. Some nodes may act as gateways to long-range data networks such as GSM or Internet. Some sensor nodes may use radio waves while others may use infrared light, laser or ultrasound for communications. A more thorough review of design spaces for sensor networks is given in [24].

Most sensor applications are domain-specific, and impose different constraints on the design of sensor nodes and, thus, the network. Among these constraints are limited bandwidth, constrained battery power, topological limitations, and limited transmission power. These constraints force designers to use domain knowledge to optimise system designs. Although there has been a notable progress in this area, several problems including scalability, fault tolerance, quality of service, security, interoperation with the Internet, power control, node co-operation, support for different routing protocols, interoperation with other wireless networks, and data aggregation pose significant challenges.

Research and industry trends revealed that the desired — if not required — characteristics of sensor nodes are:

Energy efficient Since it may not be practical to re-charge sensor nodes in the field due to cost or other constraints, energy is a major resource placing constraints on development and deployment of sensor networks.

Cheap If mass numbers of sensor nodes are deployed and if maintenance is not always cost efficient, sensor nodes must be cheap enough to be disposable.

Routing capability Not all sensors may be in the reach of a base station. Thus, sensor data may need to hop via other sensor nodes acting as relay devices between sensor nodes and base station or other nodes.

Easy to deploy In most cases a structured deployment of sensor nodes may not be possible and self-organisation of sensor nodes is desirable.

Other desirable features of sensor nodes are aggregation of data and delayed transmissions for energy conservation, local processing and validation of sensor data, and selective reporting.

Design and implementation of sensor nodes have been addressed in a number of studies. In the project "Smart Dust" the hardware development of a sensor system was investigated that can be integrated into a cubic millimeter package [30]. In the project TinyOS, carried out at the University of California at Berkeley, a small open-source operating system for wireless sensor nodes was developed. The project Terminodes addressed the self-organisation of large-scale ad hoc networks [3] for networking sensors. The project EYES [10] addressed self-organising and collaborative energy-efficient sensor networks. In the project Swarm-bots [28, 7, 8] funded by the European Union design and implementation of self-organising and self-assembling artifacts were studied. Several other studies emphasised the ability to dynamically modify network topologies, the ability to better utilise bandwidth available, the ability to operate with constrained battery powery, to provide robust, scalable service and, finally, to provide quality of service enabling high-priority events or commands to propagate in a network at the cost of blocking low-priority traffic.

Although the design space of sensor networks offers a wide variety of dimensions affecting design and implementation of sensor nodes, one design requirement is common in most sensor applications. Sensor nodes are planted in the field to harvest data. The data reported by a sensor node are collected at a data collection/control centre. Similarly, the commands addressed to sensor nodes originate from a control centre. Sensor-to-sensor communication is rarely required in most applications, unless a network validates data using multiple sensor readings, or fuses data on the network before transmitting them to the command centre. Data may be aggregated on the network before being transmitted to the command centre, in which case sensor-to-sensor communication is goal-driven rather than ad hoc. The main body of this chapter will discuss a multirooted sensor network designed with directed data movement in mind, i.e., data are moved between sensor nodes and control/command centre or vice versa. This sensor network will be referred to as SomsNet (Self-organising Multirooted Sensor Network) [18].

3.2 Self-organising Multirooted Sensor Network

Several studies proposed hierarchical topologies layering nodes into several levels to address the scalability problem. In some cases, the radio coverage of a sensor node has been the determining factor, in others power constraints, or simply communication policies. Researchers from the University of California at Berkeley and Intel Berkeley Research Laboratory demonstrated a network with 800 nodes and proposed a self-organising topology. Researchers of the Massachusetts Institute of Technology and Millennical Net Inc. presented an ad hoc self-organising network protocol which allows ultra-low power sensor devices called "i-Beans" to form an ad hoc self-organising network with optimised power efficiency [22]. In these studies, networks were organised into trees of clusters, scaling well up to a certain degree. The SomsNet development [18] took these studies one step further and offered a multirooted self-organising ad hoc network solution. SomsNet is a multirooted, clustered, tree-topology network featuring a scalable and robust architecture. It achieves scalability and robustness by dynamically partitioning a network into several overlapping spanning trees. Tree partitioning as well as clustering is achieved by a self-organising algorithm optimising the novel cost function given in Eq. (1). The cost function employed by SomsNet is a weighted average of the received signal's strength indicated by RSSI, the cumulative number of hops to the access point or the gateway nodes, the fitness value of the node, and the trust level of the path.

$$F_i = W_1 \times F_j + W_2 \times R_{j,i} + W_3 \times H_i + W_4 \times T_j + W_5 \times X_j \tag{3.1}$$

where

W_k Weight for cost factor k
F_j Fitness value of node j as a potential parent
$R_{j,i}$ Received signal strength indicator between $node_j$ and $node_i$
H_i Number of hops to root node

T_j Trust level of $node_j$
X_j User-defined cost factor for node j

SomsNet is a dynamic network, and will adapt to changes as new nodes are added to the system or removed from the system. It will also adapt to changes in link state or trust level of the links between sensor nodes. Moreover, SomsNet allows branches formed by more than one sensor to move between subtrees given in Fig. 3.1. The subtree migration between branches reduces convergence time.

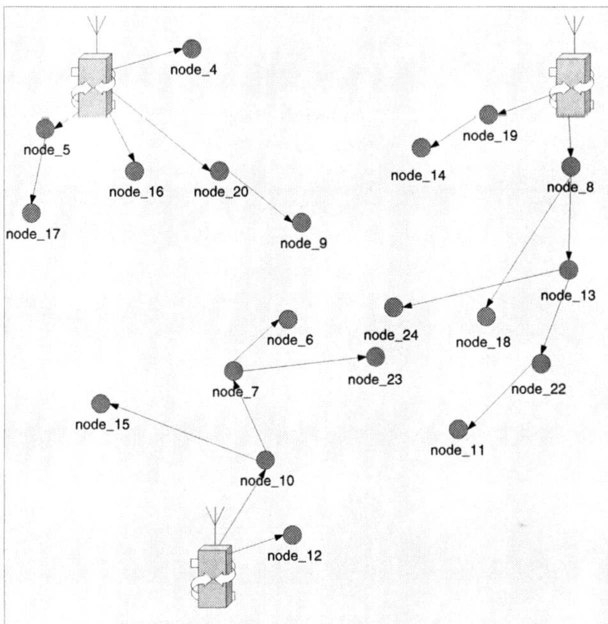

Fig. 3.1: Captured state of SomsNet

In Fig. 3.1, $node_{21}$ has not yet joined the network, and has more than one potential entry point to the network. Each one of the eligible nodes ($node_{18}$, $node_{24}$, $node_{11}$) will, in turn, send an invitation message containing its fitness metrics to $node_{21}$. Received signal strength will be measured by $node_{21}$ for each link, and a fitness value for each potential parent node will be computed using Eq. (1). The node with the highest fitness value will be chosen as the entry point of $node_{21}$ to the network. The fitness values are transmitted by all nodes as part of control messaging. A node receiving one such message can combine the fitness metrics propagated with the signal strength measured to change the entry point to the network. The weights given to each metric are application-dependent.

In addition to offering a scalable, power-aware medium access control (MAC) and network layer, the SomsNet architecture offers a standards-based solution to network configuration and management. The design of a single SomsNet node is depicted in Fig. 3.2. Management and configuration in SomsNet is performed by an integrated overlay network protocol family. In [11] a standard XML-based modeling language for wireless sensor networks is defined that enables seamless remote access to sensors. The Universal Sensor Access Protocol (USAP) is defined to bring easy-to-use, flexible, standards-based network connectivity to wireless sensors. It is a distributed, open networking architecture that leverages Internet protocols. USAP employs web technologies to enable automated control and data transfer among sensors. It is composed of Sensor Discovery Protocol (SDP), Sensor Access Protocol (SAP), and Sensor Event Notification Protocol (SENP). The three architectural units of USAP are sensor node, control node, and visual interface. A control node acts as a registry for sensor nodes, and provides gateway services between sensor nodes and visual interface units. Using SDP, a sensor advertises its services to control points on the network. SDP also allows a control point to search for sensors of interest on the network. In both cases, the discovery message contains specifics about the sensor and/or its services. In order to learn more about a discovered sensor and its capabilities, or to interact with the sensor, the control point retrieves the sensor's transducer electronic data sheet (TEDS) from the sensor. As an effort to enable access to web-resident sensors, USAP offers a novel framework to maximise the interoperability of networked sensors through standard web-based services.

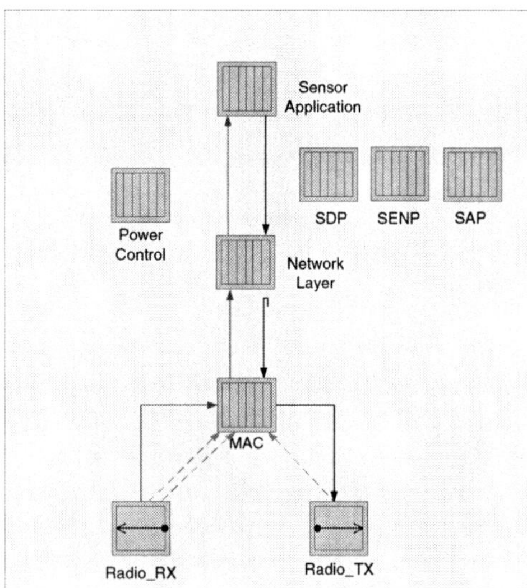

Fig. 3.2: The architecture of a sensor node

The Universal Sensor Access Protocol requires services of the TCP/IP protocol stack, and is defined over a reduced version of the TCP/IP protocol suite. The success of TCP/IP in wired networks led to utilise this well-understood protocol stack in wireless sensor network applications. Employing the TCP/IP protocol stack in embedded sensor platforms enables interfacing with a well-defined application programming interface (API). However, several problems inherent to wireless sensor networks do not exhibit themselves in wired environments and are, thus, not addressed by the standard TCP/IP protocol suite. Before it can be utilised in this new context, TCP/IP should be refined and simplified to be able to run in a resource-constraint network. Several lightweight TCP/IP implementations ranging from full protocol functionality to limited numbers of connections etc. have been proposed. Lightweight IP (lwIP) [9] is a full-scale but simplified TCP/IP implementation supporting the Internet Protocol (IP), the Internet Control Message Protocol (ICMP), the User Datagram Protocol (UDP), and the Transmission Control Protocol (TCP). Dunkels' open-source protocol suite Micro IP (uIP) [9] is designed to have only the absolute minimum of the features needed for the full TCP/IP stack. Yet another reduced IP implementation, nanoIP [25] implements a minimum subset of the TCP/IP stack, and can be implemented in 1500 bytes of ROM and 300 bytes of RAM. A functional, yet low-complexity and low-cost TCP/IP implementation is essential to implement USAP on sensor nodes.

3.2.1 SomsNet Sensor Node

SomsNet achieves a high degree of scalability and robustness by partitioning networks into several trees of clusters. In SomsNet, a node can only appear as an active participant on one subtree although there are multiple trees it can join. The cost metric used to build a network topology and routing can also be tuned to optimise battery lifetime for some nodes and better utilise memory for others. The cost metric given in Eq. (1) is local to a node and not global for a whole network.

Gateway nodes are doorways to the Internet, and can be placed at random as well as at predefined locations. Gateway nodes interface between sensor trees and external networks, such as the Internet. Sensor nodes may be distributed randomly in an area without awareness of each other's existence before a network topology is built. Alternatively, sensor nodes may be placed orderly in an optimised manner.

Each sensor node has a limited radio range. Therefore, communication between a certain sensor node and a gateway may not be possible without some nodes interacting as intermediate hops. Radio frames originating at a sensor node are forwarded by all intermediate nodes in a selected route and, finally, reach their destination. All wireless nodes use the same radio channel to communicate and, therefore, have to compete for the channel when sending a frame.

Design requirements for SomsNet nodes can be summarised as follows.

Automation: A network should be initialised and configured with minimum human interference.

Routing: Due to limited radio coverage, some nodes should be able to assume the functionality of a router to relay sensor-initiated or gateway-initiated messages over the network.

Convergence: A network configuration should converge to a valid solution in a limited time. It should not create isolated islands. Convergence time should be bounded.

MAC Access Time: In case a single-channel solution is implemented, all wireless nodes occupy the same frequency channel. The MAC is required to provide a fair access time among all ordinary nodes. Nodes with routing functionality may be given priority over ordinary ones.

End-to-end Delay: End-to-end delay is the difference between the time an event is triggered at a sensor node and the time the event is recorded at the control centre. The packet processing delay, transmission delay, queuing delay, and propagation delay contribute to delays observed by packets. Nodes should demonstrate the best effort to minimise end-to-end delay.

Fault Tolerance: A sensor network should be tolerant to partial network failures, including failure of sensor nodes and the gateway node. The network should have self-healing abilities. Recovery from node and link failures should be possible. There may not be a notification for the absence of any node, thus the network should be able to detect the failure and always recover a valid solution.

Power Awareness: Since power is limited in most sensor network applications, power should be conserved by any means.

3.3 The Need for Power-Aware Protocols

The key issue here is wireless. When there is no wire to carry data or to supply power, it is extremely important to preserve the available power and/or harvest energy. To exemplify the power requirements of a sensor node, we shall discuss power consumption of a commercial low-power transceiver. The Texas Instruments TRF6903 is a single chip transceiver with several properties comparable to those of IEEE 802.15.4 (Zigbee) transceivers. The transmit chain, when enabled completely, draws 35 mA at 0 dB attenuation. The receiver chain, when enabled, drains 18 mA. When in standby mode, the power consumption is 0.6 μA for both transmit and receiver chains. The number of hours the battery can sustain a drain of 1 mA is a measure of how much power the battery has, and its unit is mAh. Then, a 1300 mAh battery can supply 1 mA for a period of 1300 h. If this battery is used to drive the transceiver above and the transmit chain is always active, then it is evident that the battery will run out in less than 38 h. Now assume the target lifetime of a sensor before a battery replacement is one year (365 · 24 h = 8760 h). If the sensor is fed from a battery pack with a capacity of 1300 mAh, the average power allowed over a one year period will be $\frac{1300\ mAh}{8760\ h} = 148\ \mu A$. Clearly, the average power consumption of 148 μA can only be achieved by designing intelligent medium access and routing protocols, which will let the device sleep for a major portion of its operational life and only wake up for short periods when it needs to be active. Now let us consider a 38.4 kBd channel employing TRF6903 transceivers. Assume a preamble length of 40 bits and an average packet length is 32 bytes including header and payload. A sensor generates a reading every second. The intelligent MAC algorithm keeps the transceiver on the same sensor node in standby mode except when a scheduled receive operation is being performed or when it is reporting a sensor reading. The MAC algorithm enables the transmit and receive chains once every second for reporting the sensor readings and receiving acknowledgments. The average time the transceiver chain will be active is 7.7 ms over a cycle of 1 s. Similarly, the total time the receiver chain will be active is 7.7 ms over an interval of 1 s.

$$ActiveTime = \frac{40\ bits\ preamble + 256\ bits\ data}{38400\ Bd} = 7.7\ ms \tag{3.2}$$

The transceiver will be in standby mode for 982.3 ms over a one-second period. When in standby mode, the average current drawn from a 2.2 V power supply is 0.6 μA. Therefore, the current drawn under the stated operating conditions can be computed as 85 μA. A battery with a charge of 1300 mAh can drive the sensor node for as long as 1.7 a. Clearly, a power saving MAC algorithm can extend the lifetime of a sensor.

The transceiver can deliver power to the antenna with attenuations of 0 dB, 10 dB, or 20 dB. The corresponding current drains from a 2.2 V battery are 25 mA, 28 mA and 27 mA, respectively. Clearly, the choice of an attenuation factor can be a power-saving factor, but it limits the propagation distance. To explore the effect of transmission power on power budget and routing design decisions that may extend battery life, let us analyse the link budget and the free space path loss equations given below.

$$S_i = P_o - C_t + G_t + G_r + C_r \tag{3.3}$$

$$P_l = 32.4 + 20 \cdot logF\ [MHz]\ + 20 \times logR\ [km] \tag{3.4}$$

where

S_i Received power [dBm]
P_o Transmitted power [dBm]
C_t Transmitter cable attenuation [dB]
G_t Transmitter antenna gain [dBi]

P_l Path loss
G_r Receiver antenna gain [dBi]
C_r Receiver cable attenuation [dB]

Ignoring the cable attenuation and antenna gains both at the transmitter and receiver, the receiver sensitivity equation reduces to the difference of output power and path loss. Since path loss is a function of frequency and distance, the above equation states that the more the output power is, the longer the transmission distance is. To output more power, however, we need to drain more current from the battery, which reduces its lifetime. A power-aware sensor network protocol should choose a power level accordingly to enable communication and minimise power consumption.

Lastly, communication between two nodes can be accomplished by bursting power into a single-hop transmission, or using a multi-hop transmission employing a smaller amount of transmission power for each hop. Reducing the power level also reduces the broadcast domain and, thus, allows better spatial re-use of the frequency band.

3.4 SomsNet Medium Access Layer

Wireless sensor networks utilise some form of Time Division Multiple Access (TDMA) or Carrier Sense Multiple Access Collision Avoidance (CSMA-CA) protocols to control channel access. TDMA has strict timing requirements and adds control traffic overhead as it requires synchronisation between nodes; nevertheless, it is very appropriate for wireless sensor networks where a control node schedules all communication on a channel. TDMA offers deterministic channel access and quality of service. On the other end of the spectrum, CSMA-CA has no deterministic transmission characteristics and enables best effort services. The protocol overhead is lower than that of TDMA protocols. To reduce the probability of collisions, several versions of CSMA have been proposed in the literature including 1-persistant, non-persistent, and p-persistent CSMA. In a "p-persistent" version, a station decides to transmit with a probability p and defer transmission until the next slot with a probability $(1-p)$ occurs. The optimum value of p can analytically be shown to be equal to $\frac{1}{N}$, where N is the number of stations. A dynamic value of p is used in most implementations. The value of p is lowered as the number of unsuccessful transmissions increases and vice versa.

The SomsNet MAC employs a modified p-persistent CSMA-CA algorithm with dynamic choice of the value p. Clear Channel Assessment (CCA) is used to determine the state of the medium before a transmit attempt is made. Clear channel assessment is based on the measurement of the received signal strength. A threshold value determined experimentally is used to determine whether a channel is busy or idle. As signal strength varies with time, the time-domain variation of the received signal's strength is used for threshold adaptation.

If the Received Signal Strength (RSS) is measured to exceed the threshold, "busy" status is reported to the MAC layer. Then, packet transmission will be delayed until the RSS value reduces below the threshold indicating a clear channel. Here, the carrier could be the data themselves for base-band transmission, the modulating carrier for narrow-band transmission, or the PN code for spread-spectrum transmissionm, a form of wireless communications in which the frequency of the transmitted signal is deliberately changed over time, which results in much greater bandwidth.

The CSMA protocol cannot completely eliminate collisions, even when the channel status is detected precisely. This is due to sensors being independent (no co-ordination function), and the propagation delay being determined by medium and transmission distance. The SomsNet MAC protocol employs link layer acknowledgments to guarantee message delivery at the link layer. A packet will be re-transmitted if its sender does not receive an acknowledgment from the receiver within a specified period.

Wireless networks using the CSMA protocol also suffer from the "hidden" station problem where two nodes outside the radio distance of each other both send a message to a third node within the radio distance of both. In this case, a collision takes place at the destination. The hidden station problem cannot be solved by the simple curvilinear component analysis (CCA) algorithm [6]. A typical solution to the problem is to send shorter request and acknowledge packets to reserve the channel before any data packet transmission. A common protocol used in wireless networks sends a shorter Request-to-Send (RTS) message and expects a Clear-to-Send (CTS) message to reserve the channel for the duration of the data transmission. This mechanism is, however, very inefficient when the data packets are short. Since sensor messages are short messages, RTS-CTS is not implemented in SomsNet.

Saving power is essential in sensor networks, and MAC protocols have a great impact on the efficiency of power usage. Hence, sensors have to sleep for most of the time and only wake up to perform data acquisition and transmission. In addition, sensor nodes may have routing functionality and, therefore, may be intermediate targets of messages between sensor nodes and gateways. A sensor node may also be addressed by its gateway for calibration or immediate query. It is, therefore, essential that the sensor node's sleep and wake-up pattern is known to its neighbours. In SomsNet, any node notifies its parent node in the tree and its child nodes about its sleep pattern. All SomsNet nodes update their child nodes and their immediate parent nodes in every sleep cycle defined in the protocol. All nodes keep track of the sleep patterns of their immediate parents and their children to schedule their own sleep patterns accordingly. The sleep cycle duration can be a major factor in battery life; an extended sleep cycle, however, will result in unwarranted delays. The sleep cycle decisions should be based on the real-time requirements of sensor data transmission.

3.5 SomsNet Network Layer

The traffic in sensor networks is directed from sensors nodes toward the control centre via intermediate routers and gateways. Similarly, the traffic from the control centre is directed to the sensor nodes. When a network is designed with this fact in mind, and its topology is built as a multirooted tree, routing is considerably simplified. The need for complex routing protocols disappear, and the routing table computations become very efficient.

Routing in SomsNet is implemented as a "limited gossip protocol": *Only spread the word if you heard it from your children, or if it is addressed to one of them.* This rule requires that, during tree construction, every node registers with a gateway node or parent node. If other sensor nodes are registered to the network through $Node_{registered}$, then $Node_{registered}$ should keep a list of its children and report this list to its immediate parent. Then, each node has to keep a list of all its children, whether they are immediate children or second or third layer children. The main disadvantage of this model is increased memory usage closer to the tree's root. When a message is received at a sensor node, and if it is not addressed for this node, the sensor node will compare the destination address and the message's source with its list of children. If the message's source or destination address is included in the children list, the message will be re-transmitted. As the topology of the network changes due to link and node failures or mobility, the node's routing table (children list) is updated.

Since network topology is optimised using Eq. (1), routing in SomsNet is selecting a fixed route which traverses the tree upstream to carry sensor data to the control centre, or downstream to carry control and calibration information to the sensors.

3.6 Deploying and Managing Sensor Networks

Middleware is a common name for software components that lie between operating system and applications. The main purpose of middleware in wireless sensor networks is to support development, maintenance and deployment of sensor nodes. Several research projects challenged the development of middleware to manage wireless sensor networks. Data Service middleware (DSWare) [17], for instance, is a specialised layer that integrates various real-time data services for sensor networks and provides a database-like abstraction to applications. The real-time event detection service is one of most the important components of DSWare, which supports confidence functions based on data semantics, including relative importance of subevents and historical patterns. The Cougar [4] sensor database system applies a database approach to assign sensor networks tasks through declarative queries. In Cougar, a query layer was proposed consisting of a query proxy on any sensor node. The query proxy layer sits between network layer and application layer, and provides higher-level data communication services through queries. Members of the Open GIS Consortium, an association of geographic information system providers, IT services and database companies, universities and U.S. government bodies like the National Aeronautics and Space Administration (NASA), the National Imaging and Mapping Agency, and the Environmental Protection Agency, are developing a standard XML encoding scheme for metadata describing sensors, sensor platforms, sensor tasking interfaces, and sensor-derived data.

The sensor management middleware of SomsNet is the Universal Sensor Access Protocol (USAP). The latter is composed of a sensor definition language, a sensor query language, and a new protocol family to manage wireless sensor networks. The main purpose of USAP is to provide tools to manage sensor nodes. An SQL-like query language is used to issue tasks to wireless sensor nodes and collect responses from them. USAP provides a seamless way to access wireless sensor nodes and to integrate data collected by sensor nodes through the dedicated XML-based Sensor Access Protocol and web-service-enabled middleware. The sensor definition language is based on the IEEE 1451 Transducer Electronic Data Sheet (TEDS) which was proposed to meet the demand for a universal standard for smart transducers. The objectives of IEEE P1451 are to define a set of common communication interfaces for connecting transducers to microprocessor base-systems and field networks. IEEE P1451.4 was developed to bring the fundamental and most important element of intelligence and self-identification to the prevalent analogue sensor. Compared with more traditional smart sensor technologies, the mechanisms of IEEE P1451.4 to enable analogue plug-and-play sensors are inexpensive, simple, and compatible with existing measurement systems. The kernel of the standard, and all of the IEEE 1451 family of standards for smart transducer interfaces, is the transducer electronic data sheet containing the information required by a plug-and-play connectivity system. Here, the TEDS is adapted from IEEE P1451.4 to define sensor objects.

The first portion of the transducer electronic data sheet, called the basic TEDS, contains identification information, including manufacturer ID, model number, and serial number of a transducer. The basic TEDS is followed by one or more IEEE standard TEDSs, which contain the technical information on the transducer. This section typically provides the information needed to properly configure the electrical interface and convert the measurement data into engineering units. Typical TEDS parameters include measurement range, electrical output range, sensitivity, power requirements, and calibration information.

Sensor networks are deployed to make measurements and provide data to control modules. It is, therefore, necessary for a sensor node to store its data in formats that are simple, easy to use, flexible to transfer, and allowing future enhancements. Inevitably, XML appears to be the best choice at this time as a universal language, since it does not depend on any operating system platform and programming language. Thus, any sensor node's TEDS and collected data are stored in corresponding XML files. The USAP family is defined in XML schemata. Therefore, using simple XML over HTTP provides more power and flexibility for communication between middleware and sensor nodes.

A three-tier model is used to control and manage wireless sensor networks. A relational database system is employed to store sensor information as well as measurements performed over time. Web servers are employed as the control nodes in sensor networks. All sensor nodes are registered with a respectiv control node, and will be accessible through it. Control nodes store sensor information and data in the relational database. The control interface to a sensor network is designed as a web function and is accessible using a standard browser. The web server embedded with the middleware provides a list of services to perform central control and management. First, it enables clients to initialise networks, configure network settings, and send network commands; access data are gathered by sensor nodes either from the database or from sensor nodes. Second, it integrates with an application server to easily access web services. Finally, it parses the data retrieved from the sensor node in XML format and publishes them on the web.

3.6.1 Sensor Discovery Protocol

The SDP enables sensor nodes to search for control nodes on the network and advertise sensor properties and services to the network. The fundamental exchange is a discovery message which is used to pull or push sensor identification information. For example, when a new sensor node is added to a network, SomsNet's self-organisation algorithm [18] will establish a link between the new sensor node and an existing network node. Once the physical connection is established, the USAP protocols come into play to register this sensor with a control node on the network. The TEDS of each sensor provides a list of attributes and methods publically announced to the control node. By querying the control node, a user can locate the sensors and be informed about the public methods and attributes of this sensor. The TEDS on the control node contains the command set, state parameters, and other information for control, event reporting, and presentation.

The discovery messages are be sent out in several other situations, including when a sensor node is updated or when a sensor is to be removed from a network. All messages are parsed using XML and delivered in HTTP format. The control node is implemented on the web server in Fig. 3.3. In Fig. 3.3a, the transaction sequence labeled 1 demonstrates sensor discovery and registration to a control node. The transaction sequence labeled 3 illustrates a query from the command interface searching certain sensor(s).

3.6.2 Sensor Event Notification Protocol

The goal of deploying sensor networks is to interact with the environment and be informed about the changes there. Monitoring the environment may

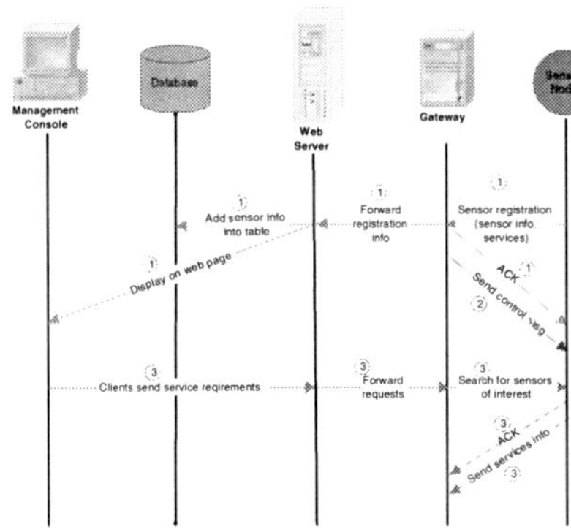

(a) Sensor Discovery Protocol

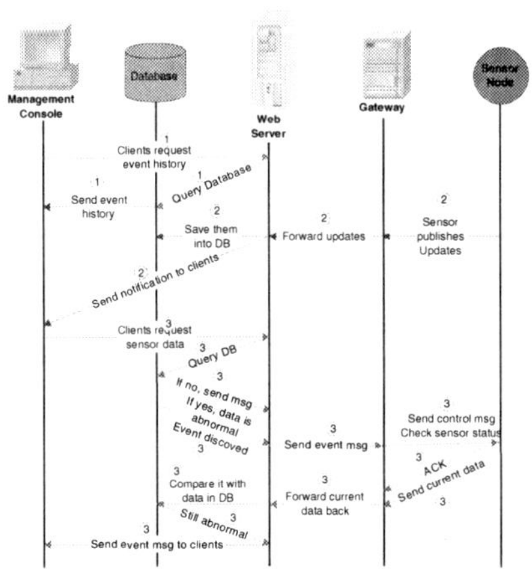

(b) Sensor Event Notification Protocol

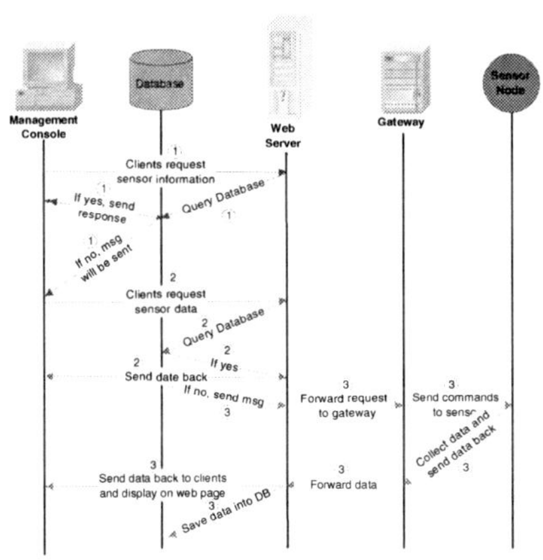

(c) Sensor Access Protocol

Fig. 3.3: Universal Sensor Access Protocol

or may not require real-time event notification. In home control systems, most events may be considered as real-time ones, while in ZebraNet events are not real-time, therefore, a store and report method is more appropriate for event notification. A real-time environment is dynamic and volatile, and requires protocols to respond within preset time limits. The Sensor Event Notification Protocol is designed to respond to real-time events and non-real-time queries; it is shown in Fig. 3.3

Events generated by sensors can be grouped into two different categories: atomic and compound ones. An atomic event refers to an event that can be detected and determined based on the data sensed by one sensor node. For instance, high temperature events, light events, or acoustic events occur when a sensor's measurement exceeds a certain threshold. A compound event is one that must be reported based on the occurrence of abnormalities detected by a group of sensor nodes, like an explosion event. It is defined as the simultaneous occurrence of the above three atomic events. SomsNet offers support for this by registering all variables that may change by either the execution of client commands or by sensor node events. A special initial event message is sent when a control node first subscribes, when a sensor experiences an atomic event or a compound event, or when the sensor nodes themselves meet any unexpected failure or removal from the network. The event messages are expressed in XML and formatted using SENP. In Fig. 3.3b, the transaction sequence labeled 1 illustrates the query of the database for historical sensor data. The transaction sequence labeled 2 illustrates notification of a real-time event by a sensor node, and transaction sequence labeled 3 illustrates a real-time sensor query initiated by the control node.

3.6.3 Sensor Access Protocol

Once a sensor is deployed in the field, it will operate as specified and report the data it collects. In time, however, it may be necessary to re-calibrate this sensor or to re-program it over the network. It should also be possible to access the sensor and change its operating parameters such as sampling frequency or reporting frequency. In addition, the user should be able to obtain sensor data on demand by querying the sensor. The Sensor Access Protocol presents the sensor as a network object, and allows access to it via the public methods it exposes using SDP. The SAP server on the sensor accepts SAP messages and invokes the requested functions on behalf of the user. USAP middleware manages all sensor nodes and presents the accessed data from sensor nodes or stored data from the database onto the client machines. SAP is illustrated in Fig. 3.3c. The transaction sequence labeled 1 demonstrates a sensor database query for sensor information, transaction sequence 2 illustrates a query of the database for sensor data, and transaction sequence 3 illustrates sensor command and control using SAP. The object diagram for the sensor is depicted in Fig. 3.4.

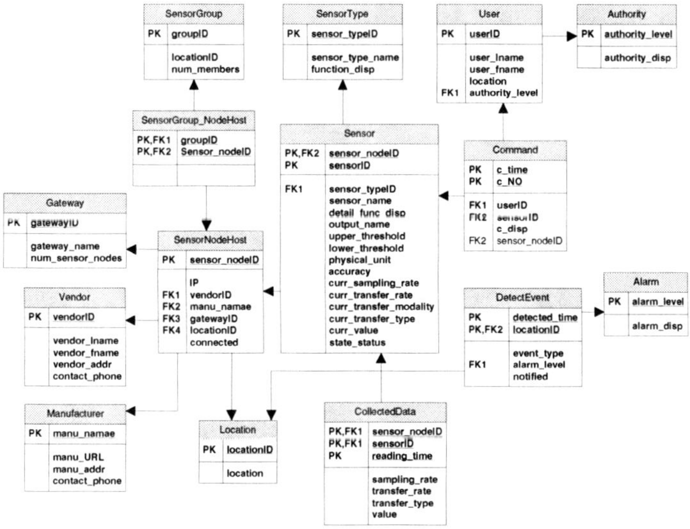

Fig. 3.4: Sensor database

3.7 SomsNet Model Simulations

The SomsNet protocol stack is implemented using the Opnet modeler supplied by Opnet Technologies[1]. Experiments were designed to verify the recovery of networks in case of node failures, estimate MAC channel access delays, estimate end-to-end delays, and verify fault tolerance. In simulations, a number of sensor nodes is distributed randomly in an area of 1000×1000 m. Gateway nodes are placed in the corners of the area where sensors are deployed, to serve as interface nodes between sensor cluster and Internet. The parameters of the wireless link are given in Table 3.1.

Table 3.1: Properties of wireless link

Minimum Frequency	Bandwidth	Transmitter power	Data rate	Coding	Terrain
2.4 GHz	22 kHz	1 mW	1 Mbps	BPSK	Free Space

In Fig. 3.5a three gateway nodes are placed at the corners of the deployment area and sensors are distributed at random. The radio perimeter is set to 150 m. Eq. (1) is used to construct the tree, and the network is observed to recover a solution where multi-hop data paths were needed to access sensors beyond radio perimeter of the gateway. In Fig. 3.5b, the radio perimeter

[1] `http://www.opnet.com/`, last visited 20 March 2006

is increased to 300 m, i.e., any radio packet transmission beyond 300 m will be treated as in-band noise. In Fig. 3.5b, we expect to see fewer hops in the tree topology at the cost of both increased transmission power and power consumption.

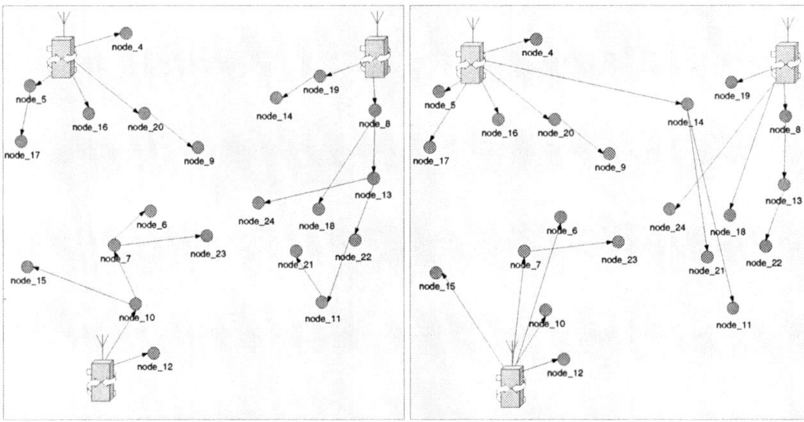

(a) 150 m radio threshold (b) 300 m radio threshold

Fig. 3.5: Network topology

The fault tolerance feature of SomsNet is presented in Fig. 3.6. To study the self-healing characteristics of the network, a gateway node is configured to "fail" during simulation, and to "recover" after a certain time. When the gateway is out of order, all sensor nodes rooted from this gateway node are expected to reset and join other trees in the network. After the gateway resumes its normal operational mode, the network should optimise Eq. (1), and a dynamic re-partitioning of the network should take place.

In Fig. 3.6a, the network is displayed at a stable state. Fig. 3.6b shows the gateway node $gateway2$ to fail and reset all nodes rooted from this tree, creating isolated nodes or islands. Nodes previously rooted from $gateway2$ will reset and join other subtrees as shown in Fig. 3.6b using the cost metric given in Eq. (1). In Fig. 3.6c, all abandoned child nodes have joined the tree under $gateway3$, which is also a valid, convergent-topology solution. Fig. 3.6d shows a failed gateway recovering and announcing its services. Using Eq. (1), each node in the radio distance of the recovered gateway re-evaluates its entry point to the network, and changes its root if it sees the latter fit. Note that the topology in Fig. 3.6d is different than the one in Fig. 3.6a, since the weights in Eq. (1) favoured RSSI over the number of hops in Fig. 3.6d. In Fig. 3.7, we illustrate the MAC channel access time and end-to-end delay. End-to-end delay is defined as the duration from a packet being generated

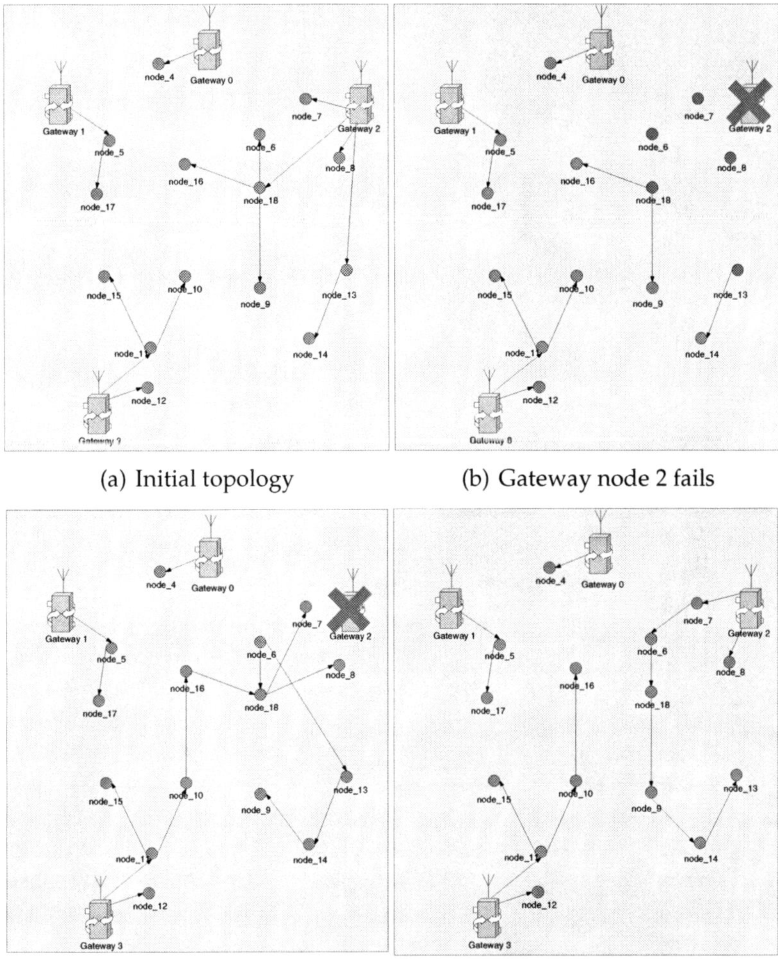

(a) Initial topology

(b) Gateway node 2 fails

(c) Topology change

(d) Final topology

Fig. 3.6: Fault recovery

until it reaches the destination edge node. It equals the sum of media access delay, transmission delay, propagation delay, and processing delay. Fig. 3.7 compares the average end-to-end delay for two nodes at different levels of the network. Sensor nodes staying at a higher level (less hops to gateway node) experience less delay as compared to nodes at lower levels. The MAC Access Delay is also given in Fig. 3.7a defined as the duration of packet reception from the network layer and beginning of actual transmission. This is observed to be in the order of tens of milliseconds.

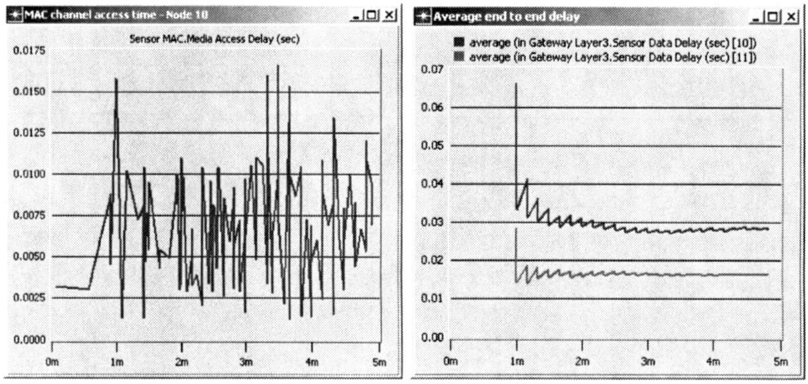

(a) MAC Access Delay (b) Sensor Data Delay

Fig. 3.7: Delay statistics

3.8 Applications

SomsNet offers a self-configuring and robust sensor and control architecture in several application domains. Here we discuss two such domains. The first one is in-home sensor and control systems. In the U.S.A., 70% of the returns in technology products are due to complexity of installation and operation. The customer does not want the burden of studying detailed user manuals nor has the time. Therefore, it is essential for a successful product to be simple to install and operate. The Universal Sensor Access Protocol (USAP) offers a simple, yet very user-friendly solution. We have implemented sensor nodes using a low-power processor (MSP430), and a low-bit-rate RF transceiver (TRF6903), both from Texas Instruments. Our home gateway interfaces to sensor networks via a standard serial connection, and serves as root node for sensor trees. Once powered up, sensor nodes join sensor trees and advertise their presence using the Sensor Discovery Protocol. The user is expected to provide additional data such as sensor location, upper and lower thresholds, sampling rate, and reporting interval. The meta-data entered by the user and attributes provided by sensors are associated in the database maintained on the home gateway. The home sensor network is implemented as part of a complete smart home solution, where the home gateway interfaces to home automation devices supporting the X10 communication system, and allows remote control of home appliances over wireless networks using cellular phones, and over wired networks using a standard browser.

Currently, companies frequently outsource maintenance and service. Heating, Ventilating, and Air-Conditioning (HVAC) systems are no exceptions. Most corporate offices or restaurant chains have their HVAC systems run and maintained by specialised companies. A currently deployed solution for remote management and diagnosis of HVAC clusters uses cellular or pager networks. However, the cost of deploying a radio system with cellular interface for each HVAC unit in a cluster is high, both from the point of view of initial installation and operation. SomsNet offers a unique solution where all HVAC units in a cluster can share a single cellular or wireline WAN connection to the operation centre. Each HVAC unit houses a SomsNet node to control HVAC functions as well as to collect sensory information. One or more SomsNet nodes also have a secondary connection to a wired or wireless network. These nodes acts as gateways to the control centre, and the sensor/actuator tree is grown off these nodes. The deployment of SomsNet results in reduced installation as well of operational costs.

3.9 Security in Sensor Networks

Wireless sensor networks, security is becoming more and more important [21, 16]. It is inevitable to use appropriate means to protect sensor devices from intrusion, and to make the communication between middleware and sensor nodes secure. The access to the SomsNet is limited by a simple authentication requesting a user's name and password in the user mode. The channel employed is clear and requires a simple yet efficient encryption to be employed.

Acknowledgments

The SomsNet model was implemented in part by Xin Liu, and the USAP model was implemented in part by Ying Fu. The Opnet modeler was provided by the Opnet Technologies Inc. university programme.

References

1. Akyildiz IF, Su W, Sankarasubramaniam Y, Cayirci E (2002) A Survey on Sensor Networks. IEEE Communications Magazine 40 (8): 102–114
2. ARGO (2006) Global Ocean Sensor Network. http://www.argo.ucsd.edu
3. Blazevic L, Buttyan L, Capkun S, Giordano S, Hubaux JP, Le Boudec JY (2001) Self Organization in Mobile Ad hoc Networks: the Approach of Terminodes. IEEE Communications Magazine 39 (6): 166–174
4. Bonnet P, Gehrke, JE, Seshadri P (2001) Towards Sensor Database Systems. In: Proc 2nd Int Conf on Mobile Data Management, LNCS, Springer, pp 3–14
5. Culler D, Estrin D, Srivastava M (2004) Overview of Sensor Networks. IEEE Computer Magazine, August, 41–49
6. Demartines P, Hérault J (1997) Curvilinear Component Analysis: A Self-organizing Neural Network for Nonlinear Mapping of Data Sets. IEEE Trans Neural Networks 8 (1): 148–154

7. Dorigo M (2005) Swarm-bot: A Novel Type of Self-Assembling Robot. In: Proc 3rd Int Symp on Autonomous Minirobots for Research and Edutainment, Springer, pp 3–4
8. Dorigo, M (2005) SWARM-BOT: An Experiment in Swarm Robotics. In: Proc IEEE Swarm Intelligence Symp, IEEE, Piscataway, NJ, pp 192–200
9. Dunkels, A (2003) Full TCP/IP for 8-Bit Architectures. In: Proc 1st Int Conference on Mobile Applications, Systems and Services, San Francisco, Usenix, pp 85–98
10. EYES Project (2006) `http://www.eyes.eu.org`
11. Fu Y, Öztürk Y (2004) Universal Sensor Access Protocols. Technical Report, San Diego State University
12. Hill J, Szewczyk R, Woo A, Hollar S, Culler D, Pister K (2000) System Architecture Directions for Networked Sensors. In: Proc ASPLOS-IX, ACM, Cambridge, MA, pp 93–104
13. IEEE (2003) Standard 802.15.4. `http://www.ieee802.org/15/pub/TG4.html`
14. IEEE (2002) Standard 1451. `http://ieee1451.nist.gov/`
15. Juang P, Oki H, Wang Y, Martonosi M, Peh LS, Rubenstein D (2002) Energy Efficient Computing for Wildlife Tracking: Design Tradeoffs and Early Experiences with ZebraNet. In: Proc 10th Int Conf on Architectural Support for Programming Languages and Operating Systems, ACM
16. Karlof C, Sastry N, Wagner D (2004) TinySec: A Link Layer Security Architecture for Wireless Sensor Networks. In: Proc ACM SenSys, Baltimore, MD, ACM, pp 162–175
17. Li S, Son SH, Stankovic JA (2003) Event Detection Services Using Data Service Middleware in Distributed Sensor Networks. In: Proc IPSN, Palo Alto, CA, LNCS 263, Springer, pp 502–517
18. Liu X, Öztürk Y (2004) A Self Configuring Multi-Rooted Tree Topology Sensor Network. In: Proc Opnetworks, Washington, DC. `http://www.opnet.com/opnetwork2004`
19. Lötjönen J, Korhonen I, Hirvonen K, Eskelinen S, Myllymäki M, Partinen M (2003) Automatic Sleep-Wake and Nap Analysis with a New Wrist Worn Online Activity Monitoring Device Vivago WristCare. SLEEP 26 (1): 86–90
20. Martinez K, Hart JK, Ong R (2004) Environmental Sensor Networks. Computer 37 (8): 50–56
21. Perrig A, Stankovic JA, Wagner D (2004) Security in Wireless Sensor Networks. Comm ACM 47 (6): 53–67
22. Rhee S, Liu S (2002) An Ultra-low Power, Self-organizing Wireless Network and Its Applications to Non-invasive Biomedical Instrumentation. In: Proc IEEE/Sarnoff Symp on Advances in Wired and Wireless Communications
23. Rhee S, Seetharam D, Liu S, Wang N (2003) i-Beans : An Ultra-low Power Wireless Sensor Network. In: Proc 5th Int Conf on Ubiquitous Computing Seattle, WA. `http://www.sokwoo.com`
24. Römer K, Mattern F (2004) The Design Space for Wireless Sensor Networks. IEEE Wireless Communications 11 (6): 54–61
25. Shelby Z (2003) NanoIP: The Zen of Embedded Networking. In: Proc ICC, IEEE, pp 1218–1222
26. Simon G, Ledezczi A, Maroti M (2004) Sensor Network-Based Countersniper System. In: SenSys'04, Baltimore, MD, ACM Press, pp 1–12

27. Thorstensen B, Syversen T, Walseth T, Bjornvold T (2004) Electronic Shepherd: A Low-Cost, Low-Bandwidth, Wireless Network System. In: Proc 2nd Int Conf on Mobile Systems, Applications and Services, ACM Press, pp 245–255
28. Trianni V, Dorigo M (2005) Emergent Collective Decisions in a Swarm of Robots. In: Proc IEEE Swarm Intelligence Symp, pp 241–248
29. University of California at Davis Wildlife Health Center (2004) Southern California Puma Project web site, `http://www.vetmed.ucdavis.edu/whc/scp/`
30. Warneke B, Last M, Liebowitz B, Pister KSJ (2001) Smart Dust: Communicating with a Cubic-Millimeter Computer, IEEE Computer 34 (1): 44–51
31. Zhang P, Sadler CM, Lyon SA, Martonosi M (2004) Hardware Design Experiences in ZebraNet. In: Proc SenSys, Baltimore, MD, ACM, pp 227–238
32. Zigbee Alliance (2006) `http://www.zigbee.org`

4

Common-interest Based Self-organising E-Learner Communities

Fan Yang, Bernd J. Krämer*, and Peng Han

Faculty of Electrical and Computer Engineering
FernUniversiät in Hagen, 58084 Hagen, Germany
{fan.yang|bernd.kraemer|han.peng}@fernuni-hagen.de

Summary. E-Learning has acquired great momentum in academia and industry as it provides unprecedented flexibility, timing, and other convenience to all participants involved in learning processes. However, in its current state it also exhibits several drawbacks including the dominance of technology-driven rather than pedagogically well-founded solutions, the introduction of impersonality, and the danger of social isolation, which may eliminate peer-to-peer learning and lead to a loss of student motivation.

Here, an innovative social software solution to form communities is proposed that is adaptive and relies on the concept of agents including agents for mobile devices to provide ubiquitous access to the system. This system purposefully provides access to appropriate learning resources based on learner interest, and facilitates regrouping of students guided by similar learning interest and preferences. The architecture and functionality of the proposed community building system's core components are presented. Also, a group formation algorithm and a novel self-organisation process, which uses custom-designed award adjustment and member exchange rules, are discussed in detail. Through simulated experiments the efficiency and scalability of the proposed mechanisms are demonstrated.

Key words: Information filtering, distributed application, self-organising community, computer supported co-operative work, multi-agent system.

4.1 Introduction

Technology-enhanced or electronic learning (short: e-learning) has acquired great momentum in academia and industry as it provides unprecedented flexibility, timing, and other convenience to people involved in learning processes. E-learning refers to different forms of Internet- and software-

* The second author's work was financially supported by the EU under contract numbers 2003-IRL/03/B/F/PP-153111 and EC-US U8041.

F. Yang et al.: *Common-interest Based Self-organising E-Learner Communities*, Studies in Computational Intelligence (SCI) **42**, 71–94 (2007)
www.springerlink.com © Springer-Verlag Berlin Heidelberg 2007

based learning processes comprising individual and collaborative learning activities, information exchange and interaction between people, learning and authoring techniques and tools, various forms of digital media, computers, and mobile devices. Like traditional distance learning, e-learning introduces time- and location-independence into educational processes and, thus, enables self-paced learning [7]. In addition, it changes the presentation and use of content, comes with programmed learning objects, supports new pedagogic scenarios, and bears the potential for personalised instruction. Current drawbacks of e-learning include: the dominance of technology-driven rather than pedagogically well-founded solutions; the introduction of impersonality; and the danger of social isolation, which may eliminate peer-to-peer learning and lead to a loss of student motivation.

This chapter addresses the issue of social isolation in e-learning settings, different facets of which were investigated in the last decade. A few researchers including Lave and Wenger [8], Putnam [10], and Suthers [14] have explored ways to adapt the concept of collaborative learning to e-learning scenarios with a geographically dispersed student population. Collaborative learning aims to make students work together in small groups toward a common goal. It includes three core elements: learner sharing, learner interdependence, and the active involvement of learners in learning activities [15]. So-called social software including Wikis, Web logs, forum software for the web, and collaborative learning environments aims to support human communication, interaction and distributed group work by means of computers and the Internet. Grouping and re-grouping are the core strategies in organising collaborative learning [3, 4] and in the formation of virtual learning communities (VLC). Both bring together learners with shared interests to engage them in computer- and Internet-mediated discourses that transcend issues of time and space [9]. Rothmeier et al. define a learning community as a community in which people are brought together by mutual interests to intensively examine a particular theme, and in doing so students are able to learn together, exchange existing knowledge, and construct new knowledge and meanings [12]. Thus, a community enhances the acquisition of knowledge and understanding, and it satisfies the learning needs of its members. In addition, a community can counteract the isolation of individual learners and the associated drop-out rate [11, 13].

Community-building has been explored extensively in various educational systems [6, 11]. Related research has generated various strategies for building communities in an on-line setting including effective communication strategies, strengthening of social ties, small-team collaboration, social networking, and collaborative knowledge construction [18]. However, most of these systems as well as all variants of social software only stress the implementation of collaborative learning activities inside a given group (e.g., collaborative learning or competitive learning). Little attention has been paid to the formation of groups [16]. Wang and Kang designed a "cybergogy" model [5] for fostering learning experiences that are cognitively, socially, and emotionally engaging for distance learners with diverse backgrounds [19]. At the core of cybergogy is engaged learning, in which learners establish their own goals, explore appropriate resources, work with others in groups, and construct knowledge in meaningful ways. But most of the group-forming and community-building strategies derived from this model are created for on-line courses of reasonable size (20 to 50 learners). Thus, these strategies do not always apply to the large-size on-line classes in the Chinese educational context from which this research grew out[2] [20]. In the distance courses offered through the Network Education College at Shanghai Jiaotong University, for example, instructors simultaneously deliver their presentations to about 100 learners face-to-face and, in addition, 300–500 learners on-line. This allows for very few opportunities of interaction such as spontaneous questioning and answering or interactive discussion.

In this chapter an innovative social software solution for forming com munities being adaptive and relying on the agent concept is proposed. This system purposefully provides access to appropriate learning resources based on learner interest, it facilitates their exchange, and enables collaborative learning. Section 4.2 introduces the architecture and functionality of the proposed community building system's core components, namely learner agents, which act on behalf of real persons, and group agents, which manage sub-communities. Section 4.3 sketches the underlying conceptual framework. In Section 4.4 we discuss in detail the group formation algorithm, the self-organisation process, which uses custom-designed award adjustment and member exchange rules, and a novel method to evaluate learning interest. Section 4.5 describes simulated experiments we conducted to demonstrate the formation of learner communities and to evaluate the efficiency and scalability of this mechanism. Section 4.6 summarises the chapter's contributions and draws some conclusions for future extensions.

4.2 Structure of a Self-organising Community Maintained by Group Agents

A key issue in the design of recommendation systems that are based on reciprocal community relationships is how to organise users with similar interests during the recommendation process, so that the users can also be assured to receive appropriate recommendations quickly and efficiently. Community organisation here includes three core elements: how to define similar interest, how to identify users with similar interest, and how to provide adequate recommendations.

Resources are the primary media of recommendation systems. These resources can be classified into different categories according to their context of

[2] This research extends the first author's doctoral thesis accepted jointly by Shanghai Jiaotong University and FernUniversität in Hagen under a bi-national agreement.

use, such as business, education, entertainment, health, law, literature, politics, science and technology, sports, or travel. In this chapter we concentrate on the education category, but our results can be applied to other categories as well. To identify users with similar interest, we introduce the concept of *recommendation*, which is expressed by a rating value a user assigns to a resource. We assume that each user of the recommendation system is willing to recommend a resource once she finds it interesting or is asked for a recommendation by others. Other users who receive a recommendation will check whether they were also interested in this resource and whether they assigned a similar rating value to it. If their rating is close to that of another user, we interpret this as a shared interest with respect to this resource. The interest sharing increases with every resource two or more users rate similarly. Users with a high interest overlap will be grouped into the same sub-community. Based on this model, we define a *reciprocal community* as a group of users who share common interests and are mutually satisfied with resource recommendations received by community members.

To track the behaviour of human users, we propose to assign a *user agent* to each person and let it handle the user's requests, and seek desired recommendations from the community. To exploit the potential of sub-communities composed of users with similar interest and to increase the efficiency of interest matches, we propose another kind of agent, called *group agent*, that serves as a broker for user agents. Their specific purpose is to provide a mechanism for appropriately grouping users with matching interest into small sub-communities. In particular, a group agent is responsible for locating users interested in recommended resources and managing the association of users to communities, which can interact with both the local user agents in their management domain and any group agent. The group agent also maintains a membership award for each user agent in its group. The membership award is a measure for the degree to which a user agent matches the interests of other users in the sub-community maintained by its group agent. The algorithms for similar-user matching and community self-organisation are supported by membership-award adjustment and member-exchange rules proposed in Subsection 4.4.2. The structure of a self-organising community linked through group agents is depicted in Fig. 4.1.

We provide further details about the architecture and functionality of user and group agents in the following Subsections 4.2.1 and 4.2.2, respectively.

4.2.1 Design of User Agents

A user agent operates on behalf of a real person and fulfils three primary functions:

1. **Behaviour tracking**
   A user agent track a user's behaviour when, e.g., reading or rating a resource, recommending an interesting resource, or seeking the suggestion of an unread resource.

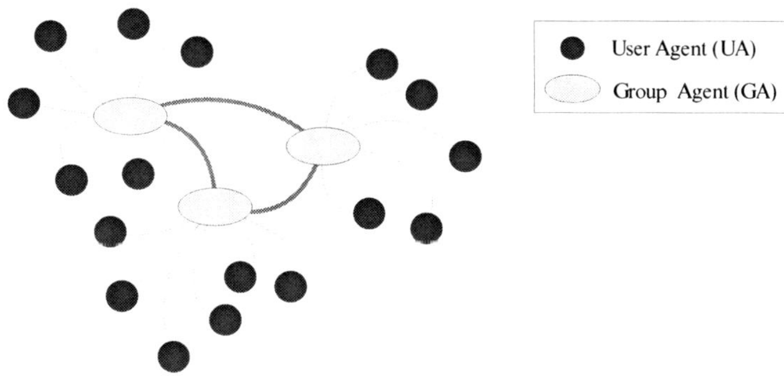

Fig. 4.1: Structure of a self-organising community linked through group agents

2. **Communication with the assigned group agent**
   When seeking users with similar interests or useful recommendations, a user agent must interact with its group agent, for example, to register personal information, submit recommendation requests to other users, or receive such requests form others.
3. **Profile maintenance**
   A user agent is also in charge of user profile generation, which maintains and updates a user's personal information, resource reading, and rating cases.
   In a recommendation system, users typically collect, evaluate and recommend resources according to their interests. In the prototype system we have developed, *learner interest* is modeled in terms of the learner's personal *preferences* and the *learning resources* he or she uses in the learning process. To express the learner's opinion about a resource she has access to, in particular, to what extent this resource supports her preferences, learners are asked to evaluate a resource by assigning a value in the range $[0, 1]$ to it. Preferences and resource evaluations are maintained in each user's profile.

As suggested by Fig. 4.1, the community architecture linked through group agents supports the interaction between a set of user agents that might be geographically dispersed on a network. User agents (see Fig. 4.2) and group agents (see Fig. 4.3) provide the necessary functionality to interact with each other and with real persons.

To support *ubiquity*, we have developed a special recommender scenario that runs on mobile devices taking into account their hardware limitations including screen size, computing power, and lack of input devices like mouse or keyboard, which slows down text input speed. This scenario assumes that students rarely want to access comprehensive learning resources on their mobile devices but rather may want to push or pull recommendations about certain resources, provide information upon their learning status, and perform pre- and post-tests by selecting predefined options with the joystick of their device. They can also communicate with co-students by sending instant text messages. The additional code required on a mobile device for this service is extremely small. It consumes around 200 kBytes on mobile telephones that support J2ME (such as the Nokia 6600 and many others). The code was built with the Light Extensible Agent Platform (LEAP), an extension to the Java Agent Development Framework (JADE) [1] for implementing agents on mobile devices. Another application for mobile devices we provide is a "mobile English test system". It was particularly designed for our empirical evaluation with Chinese students of an English course. The mobile telephone side just uses the embedded Web Browser of the mobile telephone, while the server side was build with the ASP.NET Mobile Application, which supports an interface suitable for all kinds of mobile telephones with different resolutions.

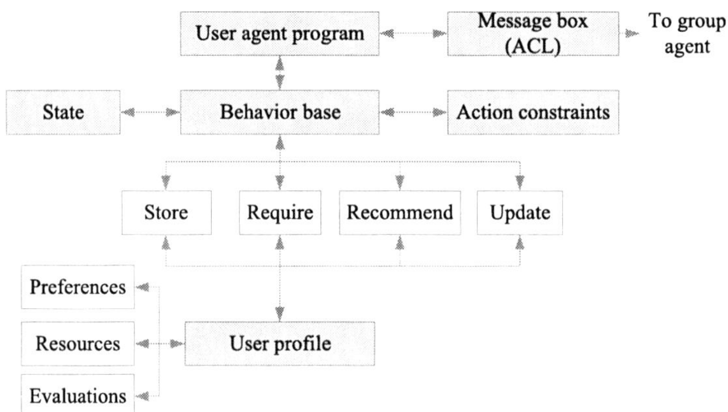

Fig. 4.2: Architecture of a user agent

The components of a user agent as depicted in Fig. 4.2 are explained below.

- ***User agent program*** (**UAP**) is the brain of a user agent. It defines an agent's behaviour in terms of the actions the agent may perform, must perform, or must not perform. Each agent must have at least one action to perform.
- ***Behaviour base*** includes implementations of the actions defined in the **UAP**. They typically have an effect on the agent's state and can be implemented through procedural code. The Java Agent Development Framework (JADE), which we used to implement our application, comes with a few predefined behaviours such as `CyclicBehaviour` or `Ticker-Behaviour`. These predefined behaviours can be inherited from application-oriented behaviours. Action `ReceiveRequest`, for example, which

regularly checks the agent's message box for incoming requests, inherits from `CyclicBehaviour`. Other actions we defined include `Store` or `Require` access to a resource, `Recommend` a resource to other user agents, and `Update` the user profile after the associated user has performed new actions. Each action must have a description of preconditions and effects. Preconditions are expressed in terms of the agent's state and must be satisfied for the action to be executable. Effects specify how the agent's state is changed as a result of the action execution.

- ***Message box*** lets multiple agents exchange messages asynchronously. Each agent manipulates its own message box autonomously. The message box consists of a set of data structures for message storage, with a set of operations for inserting, deleting, reading, and modifying messages. The message box is an optional component; an agent can use a different messaging program instead. A user agent's operating communications are specified by means of the FIPA[3] *agent communication language* (ACL).
- ***State*** comprises all the user agent's data objects (where each data object must belong to one of the data types associated with the agent), and all the messages in the agent's message box.
- ***User profile*** models the filtered data of a user, including the user's preferences, resources and evaluations.
- ***Action constraints*** are conditions depending on the agent's state. They prevent certain actions to execute concurrently.

4.2.2 Design of Group Agents

A group agent serves as a broker for requests sent out from a smaller community of user agents linked to that group agent. A group agent has two main functions:

1. **User agent management**
   The group agents of a community are distributed over a larger network and manage a sub-community of users with similar interests. Compared to a centralised community management approach this distributed solution can enhance the whole system's robustness, because there is still some probability of component or network failure. In our implementation we apply the usual remedy to cope with failures such as replication, redundancy management, and re-distribution of functionality and data. For the rest of this chapter we shall ignore, however, faults and failures.
2. **Similar user matching**
   A group agent is responsible for locating users with similar interests and managing the association of users into sub-communities. A group agent can interact with both the local user agents in its management domain and other group agents. If a request is raised from within a group of

[3] Foundation for Intelligent Physical Agents (`http://www.fipa.org/index.html`, last visited 18.02.2006).

user agents, it is first forwarded to the other members of that group. If no match can be found locally, the group agent passes the request to its peers.

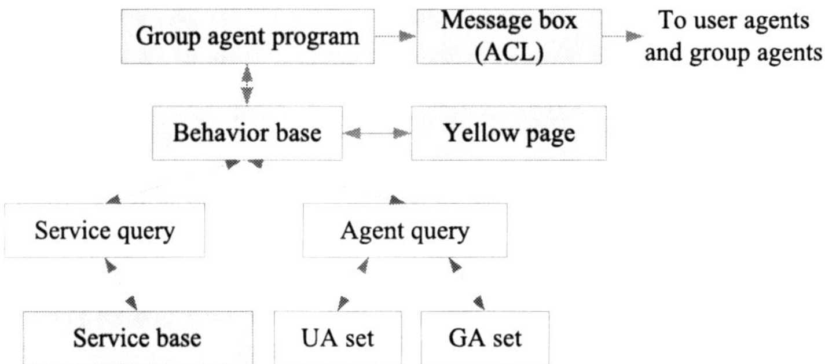

Fig. 4.3: Architecture of a group agent

Fig. 4.3 illustrates the architecture of a group agent. It consists of several components that behave similarly as related components of a user agent, including the group agent program, behaviour base, and message box. In addition, it contains special components, such as a yellow page and service base. The details of the new components are explained below:

- ***Behaviour base*** is a set of actions that can affect the state of a user agent and manipulate functions of the group agent. Examples include `localRecommend`, `globalRecommend`, `receiveRecommend`, `insert-Member`, `deleteMember`, and `deleteSelf`.
- ***Yellow page*** is the directory facilitator of a group agent. It serves to locate and manage all group members and their relationships. The yellow page component enables the management of dynamically changing relationships between user agents and group agents. Local user agents can also register their personalised information through this service.
- ***Service base*** is a set of services each group can offer. Here, we primarily consider what kind of resources users have and what kind of resources users want to recommend. In addition, a group agent can query other user or group agents' information or provide services via the service base.

4.3 Conceptual Model

In this section we provide a few formal definitions to make our concept of recommendation system more precise. We shall also rely on these definitions in the specification of our group formation algorithm in Section 4.4. For the sake of simplicity, we ignore the possibility that group agents can be generated dynamically, when an existing group agents splits off a group agent and delegates some of its user agents to it.

Let **G** and **U** be disjoint countable sets of group and user names, respectively, with typical elements g, g', g'' and u, u', respectively. Further, let **P** and **R** be sets of all possible user preferences and resources.

For $P \subseteq \mathbf{P}$ and $R \subseteq \mathbf{R}$ the mapping $s : R \rightarrow P$ models the fact that preference $p \in P$ is *supported by* resource $r \in R$ if $s(r) = p$. By $R_p = \{r \in R \mid s(r) = p\}$ we denote the set of *all resources in R supporting preference p*.

Now we can define a user agent by combining the user name with the user's preferences and the resources supporting these preferences. A *user agent* A^u acting on behalf of user u is a structure $A^u = (R^u, RV^u, ma^u)$ with

- $R^u \subseteq \mathbf{R}$ the set of resources of u;
- $RV^u = \{w^u_{r_1}, \ldots, w^u_{r_m}\}$ is the rating vector of u on the resources $r_1, \ldots, r_m$ held by user u, where $w_{r_i} \in [0, 1]$;
- $ma^u \in \{0, 1, \ldots, \}$ models the group-membership award of user agent u with respect to the sub-community it belongs to; it is a measure for the strength of the relationship between user agent u and its corresponding group agent.

As discussed above, the motivation of this model is to group users with similar preferences into the same sub-community. To solve this problem, an award mechanism is used to evaluate user preference similarity. The basic idea is that the agents of users who exhibit a strong congruence in their preferences with others will obtain high award marks (in terms of ma). These users usually have either recommended useful resources to others, or they requested some resources that others hold. In contrast, users who have a weak congruence with others will have low awards or, in case they have no congruence with others at all, their award will be zero.

From R^u we can compute the set P^u of preferences of u as follows:

$$P^u = \bigcup_{r \in R^u} s(r) \subseteq \mathbf{P} \tag{4.1}$$

A *self-organising community* **C** consisting of a set U of users $u_1, \ldots, u_n$ and a set G of group agents $g_1, \ldots, g_l$ with $l \leq m$ is a structure $(G, (A^u)_{u \in U}, m)$ where $m : U \rightarrow G$ is a mapping associating users (and thus user agents) with their managing group agents. Thus, if $m(u) = g$, user agent A^u is a *member of the group* managed by g. The set U^g of all user agents *managed by group agent* g is then defined by equation 4.2:

$$U^g = \{A^u \in U | m(u) = g\} \tag{4.2}$$

the set of resources R^g maintained by the sub-community of user agents managed by g, short *community* g, is defined by equation 4.3:

$$R^g = \bigcup_{A^u \in U^g} R^u \tag{4.3}$$

and the set of resources of community g supporting preference p is defined by:

$$R^g_p = \{r \in R^u \mid s(r) = p\}. \tag{4.4}$$

4.4 Self-organisation Algorithm H3C-SORC

As discussed above, the objectives of community self-organisation are to find users who share similar interests and re-organise them into smaller reciprocal communities. To achieve the first objective, we compare resource recommendations of different users. Our hypothesis is that, if a user agent A^u always recommends resources that satisfy another user agent $A^{u'}$, or if A^u is in a state of always receiving interesting resources recommended from another user agent $A^{u'}$, we deduce that u and u' must have a similar interest and we establish a strong relationship with the same group. Fig. 4.4 presents the work flow of a recommendation match.

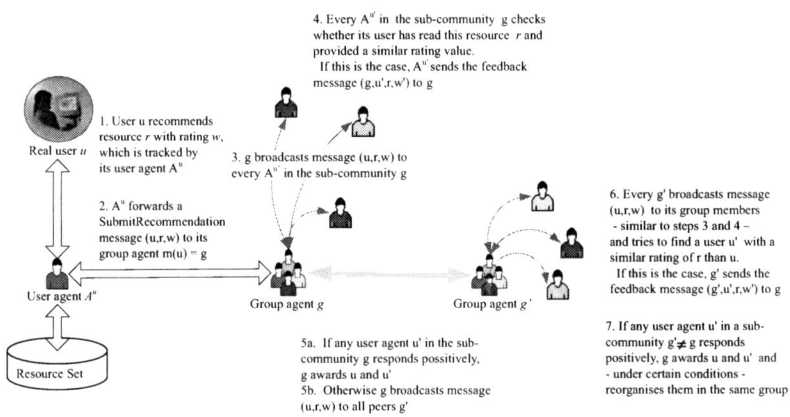

Fig. 4.4: Work flow of community self-organisation

In the self-organisation algorithm defined below we distinguish between two types of relevant events that trigger a change in the system, namely executions of the actions `RequestRecommendation` and `SubmitRecommendation`, which implement a pull and push side, respectively, of a recommendation process. The scenario depicted in Fig. 4.4 illustrates a recommendation push. If a user u wants to receive a recommendation for resource r, she will first issue a `SubmitRecommendation` message to its user agent A^u including the title of this resource and u's rating value w_r^u for it. User agent A^u then tracks this behaviour and forwards the message to its group agent g. When receiving this message, g will help to find users in the community also interested in resource r. To do so, group agent g first broadcasts the message to all other user agents in the sub-community it manages. In return, each user agent in that sub-community different from A^u checks if its user, say u', has read resource r and provided a similar rating value for it. If so, $A^{u'}$ will feed back a corresponding message to its group agent g including the user name (u'), the resource name r, and the rating of u' for r. If there is no positive feedback from within the sub-community managed by g, group agent g will forward u's request to its peers $g', g'', \ldots$. Each group agent g' that receives this message will check if there is any local user u' managed by group agent g' that is interested in this resource r and rated it. If this is the case, g' will feed back both the identifications u' and g' together with the rating of resource r to the requesting group agent g. To simplify the algorithm implementing this scenario, we assume that the requesting group agent g ignores any other feedback coming in after the first one was received. If a matching user was found, group agent g will award the recommender u and the responding user u', and will re-organise them into the same group if the award adjustment and member exchange rules apply.

4.4.1 Pseudo-Code

The heuristic community self-organising algorithm R3H-SORC (Resource Recommendation Request based Heuristic Self-Organisation Reciprocal Community) is abstractly defined in Fig. 4.5. R3H-SORC takes as input two non-negative integers n and l (with $l \leq n$) denoting the number of user agents and group agents to be generated, respectively, and a threshold value d in the range $[0, 1]$, which denotes the maximal difference between two rating values considered similar.

Initialisation. In the initialisation process, n user agents $A^{u_1}, \ldots, A^{u_n}$ are created and assigned to the community members. Based on a user u's individual profile, the associated user agent A^u will assign initial values to R^u, RV^u, and s, the mapping of u's resources to its preferences. The group-membership award ma^u of each user will be set to 0. In addition, a set G of group agents $g_1, \ldots, g_l$ will be generated, and the membership association m will be initially defined randomly in such a way that the number of user agents is approximately equally distributed among the group agents, and each group agent will include at least one user agent. Variable e denotes the name of any of the two event types `RequestRecommendation` and `SubmitRecommendation` or the termination event `stop`, which models the termination of the self-organisation process.

Assumptions. To make the algorithm work properly, we make the following assumptions: each user defines at least one preference, each of its preferences is supported by at least one resource, and the user is willing to provide a rating of these resources upon request by other users, or she takes initiative and submits a recommendation rating by herself. Further, we assume that the number of users is much larger than the number of preferences known in the community.

Lines 2–23 of the pseudo-code in Fig. 4.5 model the similar-user matching process, while Lines 24–44 model the award adjustment and dynamic exchange process. These steps will be explained in the following subsections. We use an empty feedback message $fb = ()$ to model the absence of a positive response, which could, e.g., be determined by a time-out in the implementation.

4.4.2 Implementation Design

System Initialisation

In the initialisation process, the implemented system employs three procedures: a user agent is automatically generated for every user based on the architecture defined in Fig. 4.2; user agents deal with system and user events according to the user preferences and requirements maintained in the user profile; and one or more group agents are generated based on the architecture defined in Fig. 4.3, together with an initial multi-agent structure specifying which user agents are managed by which group agent. In this step, every group agent builds up a yellow page to maintain the information of local user agents and group membership, respectively, which implement the mappings s and m introduced in Section 4.3. Each user agent should register randomly with one group agent before it seeks services from the system. The registration information may be summarised information (e.g., keywords or titles) of the resources the user holds. Group agents will maintain the registration information. In the initial community, it is likely that each group includes user agents with different interests. Furthermore, user agents can also de-register with a group agent and change to another when required (cf. Lines 36 and 38 in Fig. 4.5), which enable the exchange rule and community re-organisation.

User Matching

The heuristic algorithm we have devised to match users with similar interests relies primarily on the similarity of resource ratings. The recommendation scheme consists of the following steps. Once a user wants to recommend a new resource, its user agent will track this behaviour and send a `SubmitRecommendation` message to its group agent (Line 3 in Fig. 4.5). Upon receiving this enquiry, group agent g will first try to achieve a local match (thus strengthening the ties within its own group), and forward this message to its members, one by one (Lines 4 and 5). Each user agent of g's

Algorithm *R3H-SORC*

```
1  let $m(u) := g, fb := ()$; \\ $A^u$'s group agent, feedback message set to "no match"
2  while $e \neq$ stop {
3      $u$ sends message $msg := (e, u, r, w_r^u)$ to $g$;
4      upon receipt of $msg$, $g$ broadcasts $msg$ to all $u' \in C = \{u' \mid m(u') = g \wedge u' \neq u\}$;
5      for all $u' \in C$ {
6          if exists rating $w_r^{u'}$
7              then {
8                  if $\mid w_r^{u'} - w_r^u \mid \leq d$ \\ rating difference below threshold
9                      then {
10                         send message $fb := (u', g, r, w_r^{u'})$ to $g$;
11                         exit } \\ exit from loop
12                 endif }
13             else {
14                 $R^{u'} := R^{u'} \cup \{r\}$ }
15         endif }
16     endfor
17     if $fb = ()$ \\ no match found locally
18         then {
19             $g$ broadcasts message $(e, u, r, w_r^u)$ to all $g' \neq g \in G$;
20             for all $g' \neq g \in G$ {
21                 perform Step 4 to 14 by replacing $g$ by $g'$ }
22             endfor }
23     endif
24     if $fb = (u', g', r, w_r^{u'})$
25         then {
26             if $g = g'$ \\ $u, u'$ belong to the same group
27                 then {
28                     $ma^u := ma^u + 1$; $ma^{u'} := ma^{u'} + 1$ }
29                 else {
30                     if $ma^u = ma^{u'}$
31                         then {
32                             $ma^u := ma^{u'} := ma^u + 1$ }
33                         else {
34                             if $ma^u < ma^{u'}$
35                                 then{ \\ move $u$ to the group which $u'$ is a member of
36                                     $m := m - \{(u, g)\} \cup \{(u, g')\}$;
37                                     $ma^{u'} := ma^{u'} + 1$ }
38                                 else{ \\ move $u'$ to the group which $u$ is a member of
39                                     $m := m - \{(u', g')\} \cup \{(u', g)\}$;
40                                     $ma^u := ma^u + 1$ }
41                             endif }
42                     endif }
43             endif }
44     endif }
45 endwhile
```

Fig. 4.5: Community self-organisation algorithm R3H-SORC group, upon receiving that message, will check if it has read and rated this resource (Line 6). It will calculate the rating difference and test whether it passes the threshold (Line 8). If this test succeeds, the matched user agent will return its recommendation feedback to g (Line 10) and, in return, g will stop forwarding u's message to other user agents in its group (Line 11); otherwise this user agent, say $A^{u'}$, will signal a mismatch so that g can interrogate another user in its group. In addition, u' will add the resource named in the request into its unread resource set (Line 14), and eventually read and evaluate it at a later time. If there is no user in g's group interested in this resource (Line 17), g seeks help by forwarding the message to other group agents (Line 19). (In the prototype system we have introduced another input parameter `SearchLimit`, which specifies the maximum number of contacts to other group agents for one request to prevent endless searches in a large and distributed recommendation environment, but in the abstract algorithm we disregarded these details.) These other group agents then perform a local search in their community, much similar to the process g performed it its group before (Lines 20, 21), and they deliver any positive feedback to the requesting group agent g, which passes the feedback directly on to the original recommender u.

Award Adjustment and Member Exchange

Since every user agent is randomly registered with a group agent at the beginning, a group may have users with different interests, while the outcome expected from this heuristic algorithm is that only those user agents with similar interests are grouped in the same community. To solve this problem, we propose *award adjustment* and *member exchange* rules aiming to evaluate users and then group them accordingly.

When a group agent g relays search results to the active user agent u, it examines (a) whether the recommendation was successful, and (b) whether u and the matched user u' are in the same group. If the answer is positive, group agent g will increase the group-membership award ma of these two user agents by 1 (Lines 26–28). Otherwise, the requesting group agent g will compare the ma values of recommender u and the matching user u'. If both have the same membership award in there own group (Line 30), their award will be increased but no exchange takes place (Line 32), because we found it difficult to say which group has more users with the same interest. (If we performed a group switch here, we could only adopt a random strategy that would cause a random effect on community formation. Actually, we experimented with different combinations of rules, and this rule yielded the best experimental results.) If the awards of u and u' are different, the user agent with the lower group-membership award will be moved to the group agent managing the user agent with the higher award, and the award of the initiating user agent is increased by 1, while the award of the responding user agent is kept unchanged (Lines 34–40). This rule is based on the belief that a user with a high award is representative of its group's interest. It will be called the *group authority* and help attract users with similar interests. Hence, response to an authority user agent can drive other user agents with similar interest to join the same group quickly.

By means of the award adjustment rule, it is possible for group agents to differentiate between users who are contributing actively to their community and those who are passive. This information plays an important rôle in generating authority users and in forming sub-communities collecting users with similar interests. By using the member exchange rule, group agents can quickly cluster the user agents with the same interest, which is essential for a quick decrease of the matching time and an increase of the success rate.

4.4.3 Observations about the Algorithm's Convergence

Based on our former assumption, each user agent maintains a particular set of supporting resources in accordance to her preferences. In an arbitrary iteration step k of our algorithm, let $R^g_{p,k}$ be the set of resources supporting preference p maintained by some group agent g, and let $R^{u'}_{p,k}$ be the set of resources supporting p held by some user agent u' associated with another group agent g'. Further, let $award^g_{p,k}$ be the average award value of user agents with preference p in sub-community g in iteration k. Now, if in iteration k user agent $A^{u'}$ has been switched from group agent g' to g, then, based on the initialisation and the assumptions defined in Subsection 4.4.1 and Lines 30–42 of our algorithm in Fig. 4.5, we can conclude that the number of resources supporting p of sub-community g and its average award may have increased, i.e.,

$$R^g_{p_k} \subseteq R^g_{p,k+1}$$

and

$$award^g_{p_k} \leq award^g_{p,k+1}.$$

In other words, the award adjustment and member exchange rules proposed in our algorithm guarantee implicitly that the resources supporting a particular category will tend to flow to a group agent with higher award. This leads to the additional effect that this sub-community holds more and more resources supporting a particular preference, so that the probability increases that a response to a submitted or requested recommendation will occur. Gradually, the group will so attract more user agents looking for resources supporting their common preferences. The gradual assembly of users with the same preferences and their supporting resources ensures that the algorithm converges, which was also demonstrated by our experiments reported in the next section.

4.5 Experiments

The experiments presented in this section are based on the actual learning resources and rating values obtained by observation from the learners of the Network Education College at Shanghai Jiaotong University. We focus on the evaluation of the efficiency and scalability of this self-organisation algorithm. For simplicity of the experiment, we make the assumption that every user has one preference, only, and maintains one or more supporting documents.

4.5.1 ABSORB Simulation System

To visualise simulated operation and efficiency of the algorithm described in the previous section, we have developed a simulation platform, named **ABSORB** (**A**gent **B**ased **S**elf **O**rganisation **R**esearch **B**enchmark), which includes a user profile initialisation module and a simulation platform for community self-organisation.

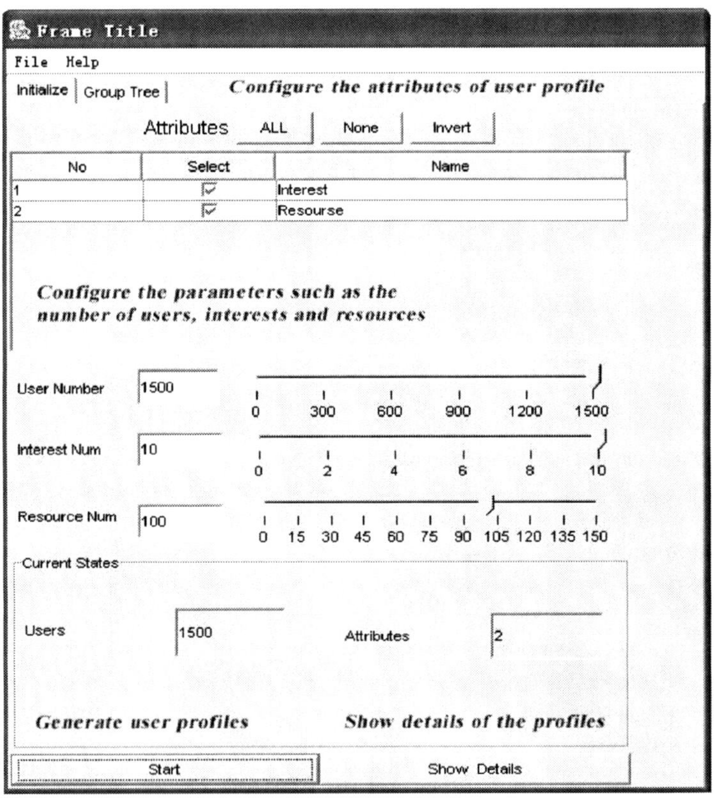

Fig. 4.6: Interface of the user profile initialisation module of ABSORB

Fig. 4.6 shows the interface of the user profile initialisation module. Through this window we can configure the attributes of users and generate corresponding user profiles. This module implements three types of functions. The "Attributes" panel is designed to specify the elements that should be included in the user profile; here the attributes "interest" and "resource" are of concern. We can also configure the initial numbers of user agents and interests (modeled as preferences), and the maximum number of resources supporting each interest. When clicking the "start" button, the system will collect the learned resources and the corresponding rating values from the learning database. Based on the filtered information, it can generate the user profiles, which can be checked by the user by pressing the "Show Details" button. They are also saved as .xml files.

Fig. 4.7 illustrates the simulation platform for community self-organisation. In the top panel, we can define the number of group agents to be generated and the maximum number of requests per user. The user distribution graph in the left panel at the bottom shows the user distribution in each group, while the right panel shows the statistic analysis of the recommendation accuracy. Here, the success ratio of a request can be defined as the percentage that the request could gain feedbacks from its own group, after each user agent has submitted or requested a recommendation.

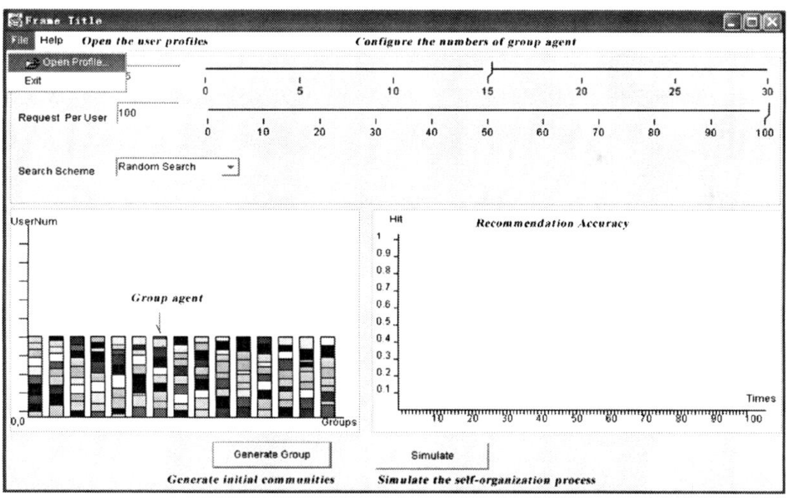

Fig. 4.7: Initial situation of the community

When the experiments start, the system first opens a user profile as discussed along Fig. 4.6, and then generates 1500 user agents and 15 group agents. The user agents randomly register each with one group agent, and transfer their summarised registration information to it. The group agents keep all information of users in their group. Fig. 4.7 shows the initial situation in the left lower panel. Each column represents a group agent, and differently coloured regions within each column represent different interests. The height of each coloured region in a column illustrates the percentage of users with corresponding interests belonging to the same group. Since each user agent is registered at random with a group agent, the colours of each column are mixed and the distribution is almost uniform as a result of the initialisation process.

After we press the "simulate" button, the user agents will simulate recommendation requests of the different users according to the **R3H-SORC** algorithm, which will gradually self-organise those users with the same interest into the same sub-community according to the award adjustment and member exchange rules. In the experiments, the user agents choose read resources from the user profiles and generate recommendation requests periodically. The request generation period is distributed normally around a mean Q.

4.5.2 Analysis of Recommendation Accuracy

After 20 recommendation requests per user, we observe an extension of coloured regions and a reduction of colour variation in each column as shown in Fig. 4.8, and the recommendation accuracy increased quickly from 0.31 to 0.79.

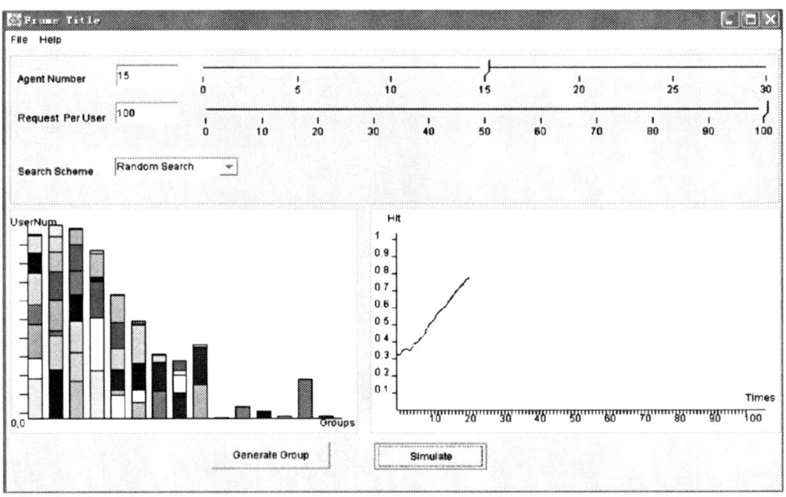

Fig. 4.8: Self-organising community situation after 20 requests per user

In Fig. 4.9 we can see the situation after 50 requests per user. The trend is towards fewer colours per column and larger coloured regions. Also we can see that some columns became shorter and some even disappeared. This means that some group agents lost all of their users during community formation. This result is consistent with the intended scenario of this experiment, because we only have 10 interests, while we generated 15 group agents. We observe an obvious trend of users being attracted by the authority user agent of a related group and migrating to that group, while some group agents lost all members, such as group agents 10, 11, 12 and 14 in Fig. 4.9.

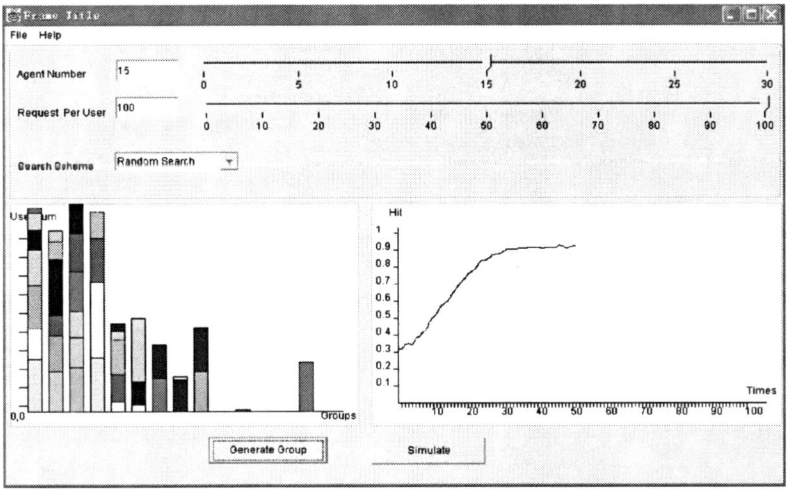

Fig. 4.9: Self-organising community situation after 50 requests per user

After more and more requests per user were submitted, users having the same interest migrated into one group and more group agents have finally lost all their members. Fig. 4.10 shows the situation after 100 requests per user agent. The formation of learner communities was quite successful. It is settled into a stable state with user agents that are interested in the same interest level being clustered into the same group.

From the recommendation accuracy graph, we can see that the accuracy is low when users are not well organised. This is because group agents often cannot find matching users locally, and need to send requests to other group agents. Since we limited the number of contacted group agents per request, the accuracy is lower during the first ten requests per user. Once user communities have started to form, however, the system exhibits an obviously improved accuracy and greater efficiency. As users with a common interest in resources are gradually grouped together, group agents can more easily get positive feedback in their own groups. As a result, there is an increase in the accuracy of recommendation.

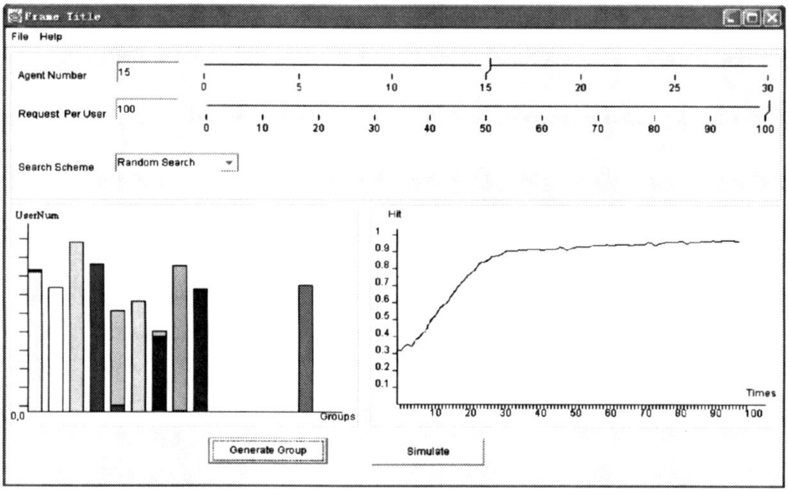

Fig. 4.10: Self-organising community situation after 100 requests per user agent

4.5.3 Scalability Analysis

Scalability is also an important issue in large-scale community organisation systems [17]. The system should function properly even when increasing numbers of users join it. In this experiment, we primarily investigated the scalability of self-organising communities with increased numbers of users. Scalability related to other issues such as number of documents and number of topics will be tackled in our future work.

Along with more and more users participating in the system, there will be more and more workload (e.g., communication and request processing) imposed on the system. In consequence, group agents will need more time to re-organise communities. However, the formation of communities is scalable if the required formation time exhibits a polynomial function. During simulation of our algorithm, we found that there is an important parameter that can illustrate the efficiency of community self-organisation. We name this parameter "recommendation request number". This number illustrates a critical tuning point of average request times. After this point, the community self-organization status will change from unstable to dynamically stable.

Self-organising communities have been tested with varying numbers of users in order to examine their scalability. The number of users ranges from 100 to 5000. Because requests are continuously generated in a certain interval Q, we use the increased number of recommendation requests to indicate the increased time required to form communities.

The experimental results depicted in Fig. 4.11 show that the time required by group agents to form communities (i.e., the recommendation request number per user) increased almost linearly in relation to the number of users. The whole system constructed correct communities even when the number of users reached 5000. This means that the self-organising communities we designed scaled quite effectively with increasing number of users.

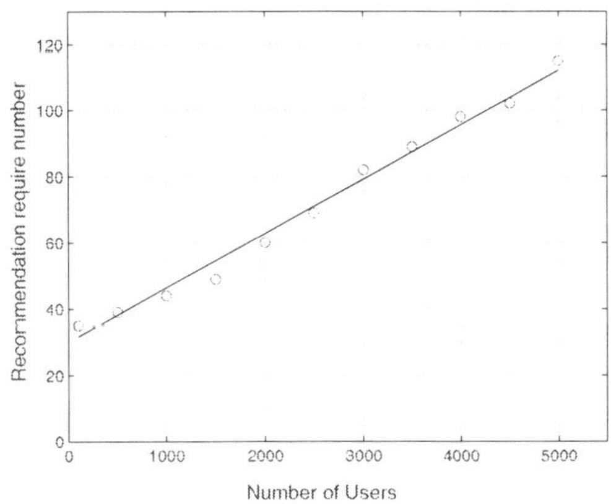

Fig. 4.11: Scalability of self-organising community

4.6 Conclusion

This chapter introduced a new method to self-organise users with similar interest by means of mutual recommendation into sub-communities that are managed by group agents. We demonstrated that this method can facilitate the appropriate organisation between users, and between users and group agents, according to the users' interests and behaviours. In particular, our self-organising communities utilised group agents to trace user behaviour and group users that have matching interest. When recommending mutually suitable resources, group agents also rewarded users that had contributed to the community and, based on the awards of users, they exchanged users with other group agents so as to organise users sensibly. This resulted in well-formed user communities, each of which had a common interest.

The formation of communities proved to be generic to different information resources and user interests. In addition, because user communities are organised based on the group agents' local knowledge and local interaction, self-organising communities are suitable to work in open and distributed environments. By organising users according to their interests and behaviours, the resulting recommendation system has an enhanced ability to provide efficient services for users. Experiments presented in this chapter support this assertion. In these experiments, the group agents were responsible for locating matched users with similar interest in response to user recommendations. Accordingly, the group agents re-organised users appropriately. As the user communities were homogenising, the group agents exhibited increased recommendation efficiency and accuracy. The experiments have also demonstrated that the formation of communities was scalable to increased numbers of users.

The method proposed in this chapter is just a heuristic approach that can be improved in various ways. For instance, we built the similarity relationship between users by referring to a resource with its title and, thus, viewed a resource as a single-featured entity. However, each resource may address many different content items that can be captured by different feature keywords. In that case, two users may give the same rating value for a resource, although they are interested in different content items. In such cases, the matching method based on resource names sometimes leads to invalid results. Therefore, we need to find a method to evaluate users' interests more precisely and include it in our algorithm. Inspired by resource similarity calculation methods used in content-based filtering technology [2], the interest of users can be evaluated based on interest feature vectors. This method would allow us to model user interest with increased precision and granularity.

Another challenge to recommendation systems lies in the fact that users are normally occupied by multiple interests, such as their programming, mathematics and other courses, for each of which they may want to match with different peers in different groups. If we would try to cope with this demand with the community construction scheme as proposed above, the multiple interests of the individual user may incur an increase of user switching and cause instability of the community structure as a consequence. Furthermore, when privacy protection issues become important, on-line users would like to share their personal information with their trusted neighbours, only. This is hard to implement in the method based on group agents described in this chapter, because of its broadcast communication used. Our idea to address this challenge is to drop the concept of group agent and let the user agents co-construct their trusted recommendation networks in a peer-to-peer (P2P) architecturepeer-to-peer architecture. In our future work, we shall pursue research on an architecture for a P2P recommendation system based on a trust weight between users. Each user agent will then be modeled as a peer in the P2P network, and it will maintain feature information and neighbour relationship on its own. A corresponding discovery scheme for users with similar interest, an adapted trust-weight learning rule, and a weight-normalisation strategy will be proposed.

References

1. Bellifemine F, Poggi A, Rimassa G (2001) JADE: a FIPA2000 Compliant Agent Development Environment. In: Proc 5th Int Conf on Autonomous Agents, ACM Press, New York, pp 216–217
2. Breese J, Heckerman D, Kadie C (1998) Empirical Analysis of Predictive Algorithms for Collaborative Filtering. In: Proc 14th Conf on Uncertainty in Artificial Intelligence, Morgan Kaufmann Publishers, Madison, WI, pp 43–52
3. Chan HK (2003) The Study of Different Grouping Arrangement in ICT supported Cooperative Learning. PhD Thesis, Hong Kong University
4. Chang CK (2001) Refining Collaborative Learning Strategies for Reducing the Technical Requirements of Web-based Classroom Management. Innovations in Education & Teaching Int 38 (2): 133–143
5. Carrier SI, Moulds LD (2003) Pedagogy, Andragogy, and Cybergogy: Exploring Best-practice Paradigm for Online Teaching and Learning. In: Proc 9th Annual SloanC/Asynchronous Learning Networks Conf. `http://www.sloan-c.org/conference/Proc/2003/track2.asp`
6. Dorneich MC, Jones PM (2000) The Design and Implementation of a Learning Collaboratory. In: Proc IEEE Int Conf on Systems, Man, and Cybernetics, IEEE Computer Society Press, pp 146–151
7. Krämer BJ (1997) New Possibilities for Distance Learning. Invited contribution to the round table discussion: Keeping Pace with an Information Society, IEEE Computer, November
8. Lave JA, Wenger E (1991) Situated Learning: Legitimate Peripheral Participation. Cambridge University Press
9. Preece J (2000) Online Communities: Designing Usability, Supporting Sociability. John Wiley and Sons, Chichester
10. Putnam RD (2000) Bowling Alone: The Collapse and Revival of American Community. Simon & Schuster, New York
11. Rigou M (2004) On the Development of Adaptive Web-based Learning Communities. In: Proc IASTED Int Conf Web-Based Education, ACTA Press, Calgary, pp 625–630
12. Rothmeier GR, Mandl H, Prenzel M (2000) Computer-supported Learning Environments. Planning, Formation and Assessment. Wiley-VCH, Munich
13. Seufert S (2002) Design and Management of Online Learning Communities. In: Proc 2nd Annual Conf on Innovative Research in Management, Springer, pp 10–29
14. Suthers, D (1999) Effects of Alternate Representations of Evidential Relations on Collaborative Learning Discourse. In: Proc 3rd Conf on Computer Supported Collaborative Learning, Lawrence Erlbaum Associates, Inc., Palo Alto, CA, pp 14–17
15. Suthers D (2001) Collaborative Representations: Supporting Face-to-Face and Online Knowledge Building Discourse. In: Proc 34th Hawaii Int Conf on the System Sciences, IEEE Computer Society Press, pp 32–37
16. Tang TY, Chan KCC (2002) Feature Construction for Student Group Forming Based on Their Browsing Behaviors in an E-Learning System. In: Proc 7th Pacific Rim Int Conf on Artificial Intelligence, Springer, pp 512–521
17. Wang F (2002) Self-organising Communities Formed by Middle Agents. In: Proc 1st Int Conf on Autonomous Agents and Multi-Agent Systems, IEEE Computer Society Press, pp 1333–1339

18. Wang MJ, Poole M (2004) Nurturing a Dynamic Online Learning Community Among Teens. The Int J of Learning, pp 859–870
19. Wang MJ, Kang J (2005) Cybergogy of Engaged Learning through Information and Communication Technology: A Framework for Creating Learner Engagement. Engaged Learning, ActiveInk pp 120–131
20. Yang F (2006) Analysis, Design and Implementation of Personalized Recommendation Algorithms Supporting Self-organized Communities. VDI-Verlag, Düsseldorf, Reihe 22, No 23

5

A Service Component Architecture to Federate E-Universities: A Case Study in Virtual Mobility

Bernd J. Krämer

Faculty of Electrical and Computer Engineering
FernUniversität in Hagen, 58084 Hagen, Germany
bernd.kraemer@fernuni-hagen.de

Summary. Global networking, digital knowledge media, and novel e-learning technologies challenge a fundamental renovation of contemporary universities as these factors strongly affect the way educational institutions create and impart knowledge. Universities need to co-operate and aggregate their learning resources and core competencies to reduce costs, and to be able to rapidly respond to changing conditions on an emerging international education market. It is argued that service-oriented computing offers appropriate computational constructs enabling inter-operation of today's heterogeneous university infrastructures to agile networks of collaborating applications. For a case study about virtual mobility of students, in the sense that students can custom-design their curriculum by combining e-learning offers from different providers, a service component architecture (SCA) is proposed. SCA is a novel engineering model created by an on-going industry-led standardisation consortium aiming to facilitate the development of coarse-grained service-oriented applications. The work builds on the outcome of a European project, CUBER, in which a personalised curriculum builder was produced, but offered no possibility to execute cross-institutional study programmes based on automated processes.

Key words: E-learning, service-oriented computing, web service, service component architecture, curriculum broker, virtual mobility, university federation.

5.1 Introduction

The electronic marketplace of virtual enterprises and mobile customers acting worldwide is quickly becoming a reality. Today, products and services are designed, produced, marketed, sold, installed, trained for, maintained, and resold in a networked fashion. This development has also touched universities. The creation of a European space for higher education [8] is on the political agenda for quite a while, and so-called e-universities are being planned or built all over the globe. E-universities are characterised by a comprehensive Internet portal acting as a one-stop, personalised access point to

B.J. Krämer: *A Service Component Architecture to Federate E-Universities: A Case Study in Virtual Mobility*, Studies in Computational Intelligence (SCI) **42**, 95–119 (2007)
www.springerlink.com © Springer-Verlag Berlin Heidelberg 2007

learning environments[1] and all kinds of student-centred and customisable services such as learning, library, administrative, community, and financial ones. Driving forces behind various governmental and institutional initiatives to evolve e-universities and university networks include the challenges caused by the transformative power of the Internet [30], rapid innovations in e-learning technologies, and the political intention to increase a region's or institution's agility and international competitiveness in higher education.

Similar to virtual enterprises, which typically involve many trading partners in joint value and supply chains, e-universities could be connected to realise the vision of inter-institutional co-operation and integrated programmes of study, training, and research [8]. But the extremely heterogeneous landscape of e-university implementations and a lack of appropriate conceptual and technology standards prohibit the seamless integration of university portal services to enable cross-institutional co-operation including the possibility of virtual mobility of students and academic staff.

5.1.1 E-Learning Technology Standards

The growing number of e-learning technology standards emerging at increasing speed is a must, but not sufficient to enable university service integration, because most of these standards focus on learning environments and digital content [24], but largely ignore integration of student administrative services. The Sharable Content Object Reference Model (SCORM) [1], the learning object metadata standard (LOM) [10], and the Question & Test Interoperability Specification (QTI) of the IMS Global Learning Consortium [21] aim at re-usability and accessibility of web-based learning content and assessment items, respectively, and the interoperability of related execution environments. Interoperability ensures the exchange of components among different contexts and their use in environments from different vendors. Re-usability is the ability to re-use content or assessment items with little or no modification and combine components from different providers. Interoperability and re-usability are prerequisites for the economic viability of content that is expensive to produce and maintain. Interoperability also avoids vendor lock-in. Accessibility guarantees ubiquitous access to content and its presentation on the device of the learner's choice.

Other standards address a reference architecture for learning systems [9], the exchange of metadata of digital content repositories [17] and digital repository interoperation [14], content packaging [18], competencies [12], learner profiles [20] and e-portfolios [19], intellectual property rights [11],

[1] In contrast to the traditional concept, in which a learning environment is viewed as a physical location like a class-room in which learning processes take place [29], a meaning of learning environment adequate to e-learning refers to a technical infrastructure consisting of computers, mobile devices, communication facilities, and appropriate software that allows students to deal with digital content, and supports social and pedagogical aspects of learning.

collaboration [7], and learning activity modeling languages like IMS Learning Design [15], which aims to specify educational scenarios in terms of learning objectives and activities, rôles, constraints, and resources.

5.1.2 Our Focus and Motivation

Our focus is different in that we aim at the federation of universities to a virtual university providing the illusion of an integrated institution — from the perspective of student administration services. With virtual we do not mean "unreal" but "simulated by software". In a networked virtual university, each node of the network is really existent, but student record systems, enrollment and admission functions, library systems, learning management systems, learning environments (VLE) directories, and other components of a university are exposed as software services.

In this chapter we present a service-oriented approach to enable the networking of e-universities. In particular, we propose a service component architecture that will enable universities willing to network and co-operate to exchange and combine educational offers, and allow their students to custom-design their curricula and move virtually between different institutions. The objective of our approach is not to homogenise education, but to evolve a hybrid approach in which individual institutions can develop their own strengths and competence profiles.

This chapter is organised as follows. In the next section, earlier work on a study module and curriculum broker is briefly reviewed and a running example is introduced. In Section 5.3, on-going standardisation efforts to design integration frameworks relying on service technology are discussed, and a conceptual architecture for institutional co-operation based on these findings is sketched. In Section 5.4, we introduce our approach based on a service component architecture whose behaviour is orchestrated by collaborative workflow models. In Section 5.5, we briefly report on related work, before we conclude and outline some ideas for future work.

5.2 E-university Collaboration from a Student Administrative Perspective

In the introduction we listed various efforts to enable the interoperation of learning (content) management systems and digital resource repositories. We spoke about re-use, sharing, and accessibility of educational content including assessment tasks, and about the exchange of personal information including learner profiles, learner portfolios and competencies. We also addressed learning and teaching strategies. All these endeavours largely focus on delivery and management of learning experiences.

In this section, we shall take a different perspective and consider the integration of student administration functions into intentional workflow processes that may span several educational institutions. Student administration functions typically include career planning, admission, enrollment, or financial services. In our case study, we shall particularly pick on a cross-institutional module award acquisition process. The process will be designed from the perspective of a student. Administration of the organisational and operational infrastructure are outside of our concern.

The core idea is to integrate learning and administrative systems of different institutions by exposing their application logic as services via portals providing a single systems' view (see Fig. 5.1). This service-based co-operation of universities supplemented by proper legal and organisational precautions would enable virtual mobility of students in the sense that students could design their personalised study programmes through integration of study offers from partner universities, and perform study and assessment tasks remotely.

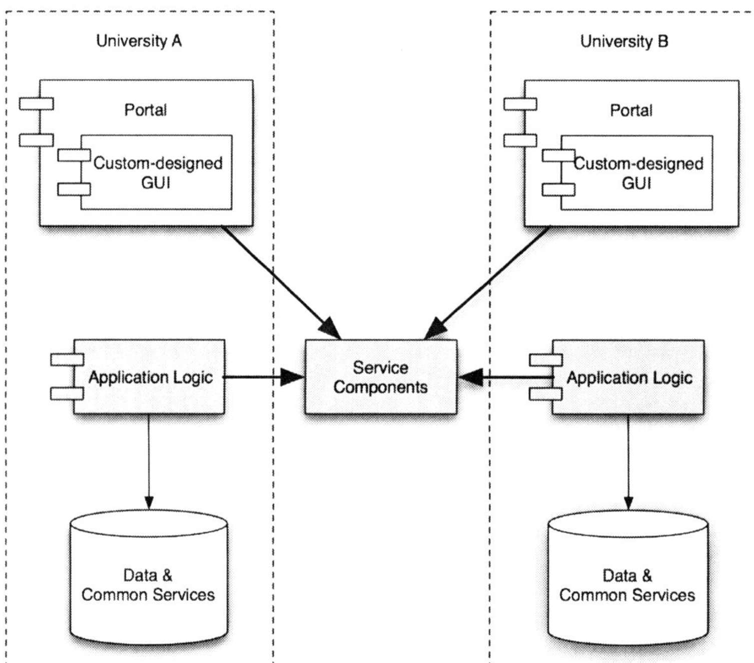

Fig. 5.1: Integration of heterogeneous e-university implementations through service abstraction

5.2.1 CUBER Course Broker

As a first step towards a federation of universities, the European project CUBER[2] has contributed an Internet-based curriculum broker [22, 23]. The broker's catalogue currently comprises hundreds of standardised descriptions of study modules and programmes offered by five European (distance) universities and three university networks in the topic areas computer science and computer engineering. CUBER's search engine allows students to browse the catalogue or search for specific modules and study programmes that match their personal needs, including individual learning objectives, learning style and study constraints. The broker also allows students to personalise their study programmes by selecting appropriate modules from different providers that are compatible with their actual study programmes.

The CUBER broker comprises three technical components [23]: knowledge base, search engine, and authoring interface (see Fig. 5.2). The CUBER broker insulates the learner from the heterogeneity of the different course information systems underlying a given university federation. The broker provides a single-system view on the sum of modules and study programmes offered despite their distribution across multiple universities, and the disparity of languages in which the modules are taught. To enable educational institutions to maintain and update their entries in the knowledge base, the system provides an authoring interface, which comprises all mechanisms and tools to facilitate module and programme data entry and establish relationships between different educational objects, e.g., a study programme and its constituent modules. The authoring interface represents the single point of update for all module and study offers of consortium members. It shields the knowledge base from direct access by content providers, and it performs plausibility and consistency checks upon data entry. The search engine provides a customisable search and exploration interface to the knowledge base. It supports browsing and search according to individual educational and learning demands. Free text search, keyword search, and conceptual content analysis relying on expert-defined classification schemes are supported. Unsuccessful search expressions can be saved to act as filters upon knowledge base updates. The filter engine may eventually issue a notification if an update matches the filter.

CUBER also provides a decision model for mutual recognition of credits among universities based on the European Credit Transfer System (ECTS). The methodology developed in the CUBER project allows users to check the equivalence of two modules. The result is obtained by applying a weighted equivalence measure including the following criteria: ECTS credits, placement of a module in a study programme, examination method, and module

---

[2] CUBER was financially supported by the European Commission under project number IST-1999-10737 in the Fifth Framework Programme in Action Line "Flexible Universities".

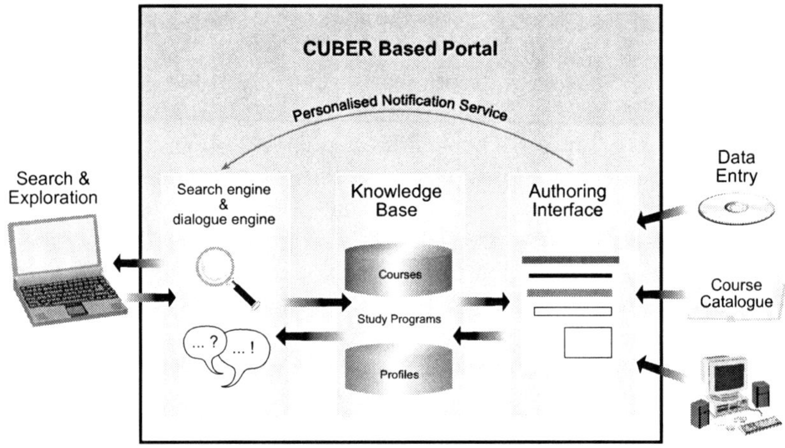

Fig. 5.2: Course information and brokerage system

contents. The method was derived from best practices as they are currently performed manually [5].

CUBER aimed to contribute to the following objectives listed in the Bologna Declaration:

Educational mobility The CUBER Knowledge Base embraces different levels of creditable study objects: modules, packages, and study programmes. The flexibility in substituting modules increases the mobility of students as they can choose the best match according to their learning objectives and individual needs.

Co-operation between universities on the curriculum level Compatibility of modules as a result of their equivalence stimulates the co-operation between universities. When compiling new study programmes, universities can use existing modules of other universities instead of developing their own. In addition, universities will also be able to find audiences for specific topics that would not have enough students in a single university.

Collaboration and competition among EU countries The broker system is a platform for universities to make their modules and study programmes internationally visible. This enhanced visibility can also be utilised to recruit students at a much larger scale than before.

Unfortunately, the project partners neither succeeded in launching an application project to demonstrate the virtue of virtual mobility, nor in approaching the larger vision underlying the CUBER project, viz., to build the technological ground for a federated European virtual university [22].

5.2.2 Running Example: A Scenario of Virtual Mobility

In the present situation, a user of CUBER may search for modules fitting her topic interests, personal learning preferences, financial constraints, or other conditions. If a match exists, the user receives a detailed description of a selected module or study programme with web links to further details on the offer and the offering institution. If the student selected a particular module she is interested in, she would need to contact the student office and ask for permission to enroll into the selected module. Normally, this request will only be granted if there exists a collaboration contract between the home and the foreign university. If the request is granted, the student needs to learn how to obtain access to and use the proper learning environment at the foreign university. It may also be necessary that she has to take care of obtaining a certificate of successful completion after having passed the module examination, and forwarding it to the administrative channels of her home university for acknowledgement of the credits acquired.

If a recognition agreement does not exist between the two universities, the request will be rejected with a high probability. The reason is that in many universities it is impossible to enroll in modules of regular study programmes without being registered as student in one of the schools or departments of those universities. But even given the fact that more and more universities make many information and administration services accessible over the web, these services are not programmatically composable, and students must interact separately with each module provider.

With a service-oriented solution, we envision that design and execution of the administrative facets of a personalised study programme can be achieved just by a few mouse clicks. As an example, consider the virtual mobility scenario depicted in Fig. 5.3. It assumes that a student has activated CUBER to search, for example, for an introductory database module that best fits her interest in a problem-based approach, and includes a thorough introduction to the Structured Query Language (SQL) as she is supposed to work on a database project in her company, soon. Assume further that the student found a reasonable candidate module in the match set CUBER displayed, and she tasks CUBER to check whether the selected module is equivalent to the corresponding one of her study programme at her home university. In case of a positive result, she signs into her learning environment to register on-line. If authenticated, she selects the preferred database module from the CUBER match and activates the module enrollment function of her web client. The enrollment service would recognise that the module is offered by a partner university, and transfer the registration application together with the necessary excerpt of the student record to the respective partner university. Although the documents handled in these administrative processess are subject to data protection, the transfer would take place, because the remote server is known to be trustable. Based on the information received, the remote registration server will check the university's admission condition

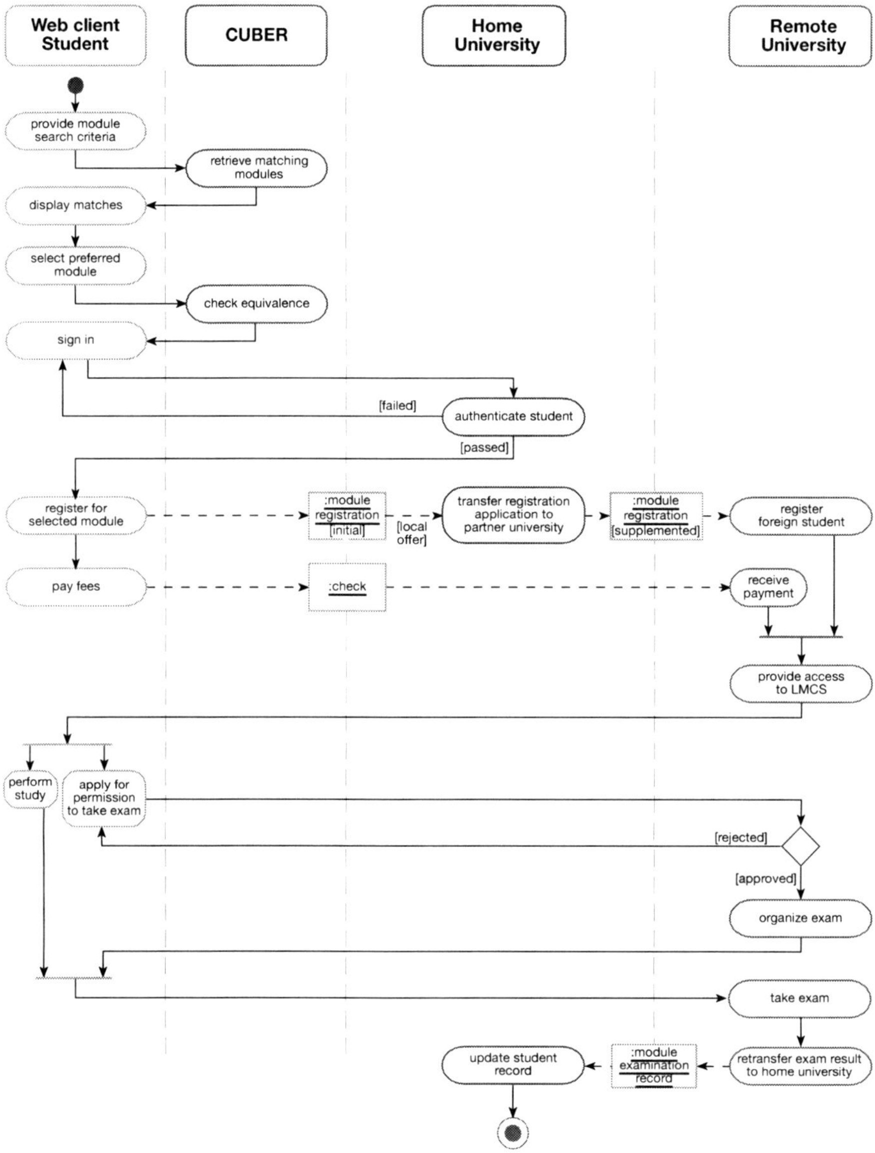

Fig. 5.3: Virtual mobility scenario

and, in case of success, provide access to the selected module in the module database and to module-specific collaboration and communication services. Once the semester starts, a learning service would be activated to track the student's progress, trigger appropriate learning activities, and provide suitable learning services. In the meantime, the student may want to register for a formal examination after the module is concluded. If the application is granted and after the student took the examination, the student's module examination record would be returned to the student's home university, where the student portfolio would be updated appropriately.

5.3 Integration Frameworks

Before developing our service-oriented federation solution, we review two different integration frameworks that are closely related to our problem, and provide a conceptual basis on which we can build.

5.3.1 IMS Abstract Learning Framework

After the IMS Global Learning Consortium has been involved for years in numerous interoperability specifications for e-learning technology, it published in 2003 the so-called IMS Abstract Framework (IAF, see Fig. 5.4) to establish a conceptual context and a set of interfaces for which interoperability specifications will be needed in the future [13]. IAF is an abstract representation of the services occurring in e-learning systems in the broadest sense. It is considered as a living document consisting of a glossary and application, service and component descriptions.

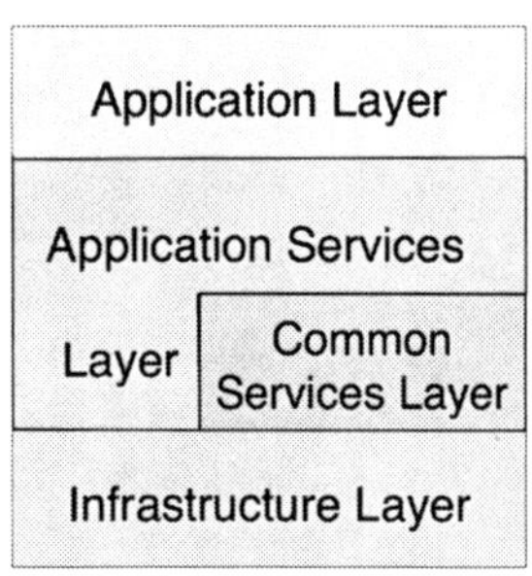

Fig. 5.4: Layered model of the IMS abstract learning framework

In Fig. 5.5, we have mapped into IAF typical advisory and administrative service components involved in the information flows caused by administrative processes like the one presented in Fig. 5.3. As the strict layering indicated in Fig. 5.4 is not enforced in concrete implementations of IAF, Fig. 5.5

shows an object-oriented presentation of service components. The *application layer* in Fig. 5.5 provides access to the module information system allowing learners to search and browse for proper content modules, to the learning (content) management system, to examination office services, and to the assessment system.

The *application services layer* comprises all services supporting e-learning-specific workflows like content, enrollment, and e-portfolio management, learner advisory and class management services, or human resources management. The latter is left unconnected for the sake of readability of the diagram. The brokerage application, for instance, would interact with the learner advisory and enrolment management service in this layer when checking admission constraints and enrolling a student, respectively.

Common services like authentication, authorisation, digital rights management, or access management are assigned to the *common services layer*. All these services are involved, either locally or remotely, through the application services as a student requests access to proper learning and formal assessment environments. Both application and common services have access to the *infrastructure layer* maintaining physical data stores, communication and messaging infrastructure, and financial management systems. Information flows across different applications, services, and infrastructure components triggered by learner interactions are illustrated in simplified form in Fig. 5.5.

In Fig. 5.5, we make no distinction between different service providers and owners of infrastructures, because all these applications, services, and proper infrastructure components would be implemented by each university participating in the federation following the same set of standards including domain ontology, service components, protocols, and data models. By and large, the heterogeneity of the underlying institutions would be transparent to students, tutors, and lecturers although, in detail, the services are likely to be realised in different ways.

5.3.2 Enterprise Integration Framework

The idea of using service technology as a backbone of systems and process integration within educational organisations is not new. Governmental committees and organisations like the Centre for Educational Technology Interoperability Standards (CETIS) in Britain or the Australian Department of Education have, for instance, teamed up with major U.S. universities to promote service-oriented framework concepts and standards in e-learning systems [33]. In [32], Wilson and his co-authors discuss several solutions to connect heterogeneous applications through web service abstraction. (The integration architecture presented in Fig. 5.1 was inspired by one of their sketches.)

The IMS Enterprise Specification aims at administrative applications and services that need to share data about learners, modules, performance, or credits across institutional infrastructures [16]. It defines a common data

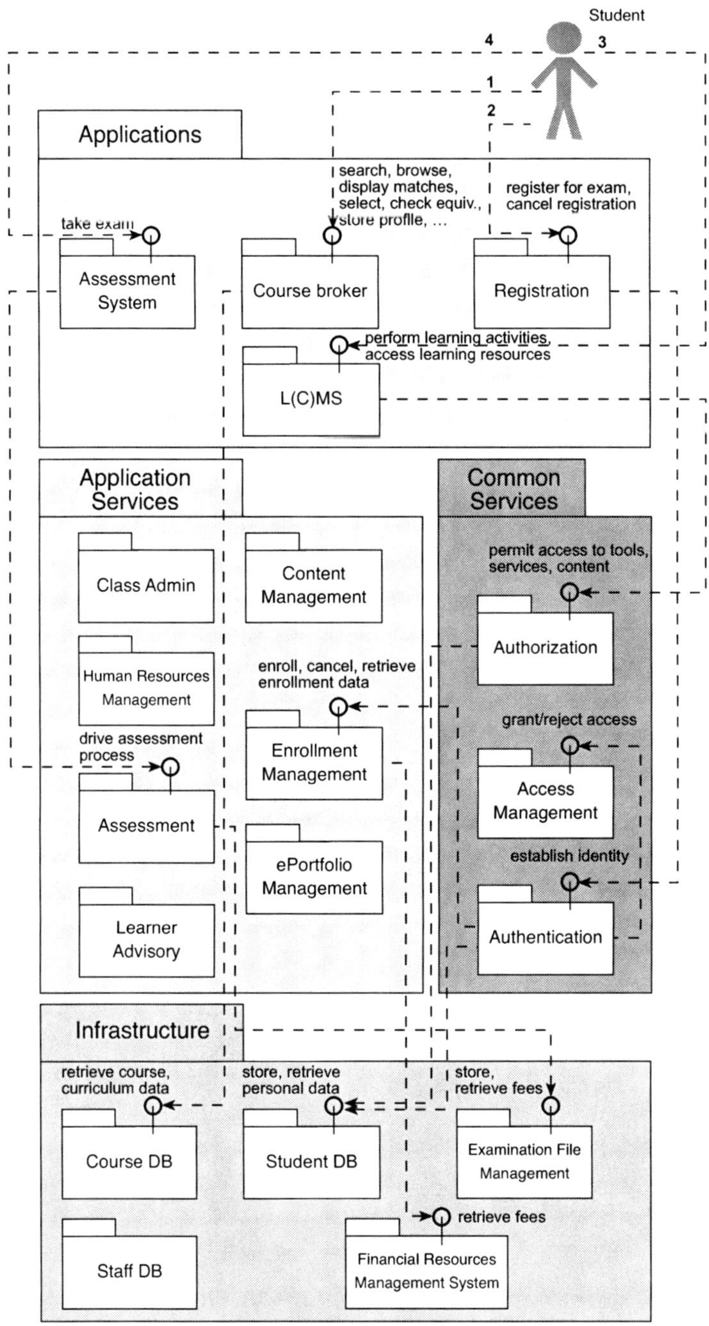

Fig. 5.5: Information flow among service components involved in the virtual mobility scenario

model to enable the exchange of data between learning management systems, human resources systems, student, module, and examination administration applications, or library management systems. The standardised data model allows for a diversity of implementations of the data model and instructional management processes. The latter can be coupled across institutional boundaries due to standardised interfaces. The IMS Enterprise Services bindings are expressed in the form of WSDL[3] files. The services are defined at a very fine level of granularity including operations like "createPerson", "deletePerson", or "replaceGroup". These operations may be useful to implement business processes, but do not provide the right level of abstraction to find agreements between business partners. In the next section, we shall propose a workflow-oriented integration approach relying on business process models and more coarse-grained service components.

5.4 Service Components for E-university Collaboration

The software applications in different organisational units of a university often evolved in isolation and focused primarily on tasks specific for the actual unit. This lead to redundant workflows and inconsistencies among overlapping data maintained in different units, which impedes the exchange of data and the seamless integration of individual workflows. Typical approaches to integrate this landscape of isolated data management systems through coherent administrative processes covering all relevant areas of responsibility include data integration by means of relational database management systems and the use of middleware like CORBA[4].

These solutions may work well from an intra-institutional perspective, but are less suited for inter-institutional co-operation. No university would want to set up shared databases. Moreover, different CORBA or other middleware implementations raise interoperability problems among different implementations, and their RPC[5] communication is blocked by common firewall security measures.

5.4.1 Service-Oriented Computing

Service-Oriented Computing (SOC) utilises services as constructs to support the development of rapid, low-cost, and easy composition of distributed applications [27]. Services are autonomous, platform-independent units of functionality that can be requested in a platform-independent way. A service

[3] Web Service Description Language, an XML format for describing Web services, which is promoted by the World Wide Web Consortium (W3C).

[4] Common Object Request Broker Architecture, an object-oriented middleware specification developed by the Object Management Group (OMG).

[5] Remote procedure call

performs one or more operations, typically observing business rules and often accessing databases. Services can be described, published, discovered, and dynamically assembled to develop massively distributed, interoperable, evolvable systems [34]. Any piece of code and any application component deployed on a system can be re-used and transformed into a network-available service [26]. Services use open Internet-based standards, such as the Simple Object Access Protocol (SOAP) to establish communication between them, the Web Services Description Language (WSDL) to define services, and the Business Process Execution Language (BPEL) to describe business processes and to co-ordinate the services they consist of. SOAP, WSDL, BPEL, and other related standards all use the Extensible Mark-up Language (XML) as encoding schema. SOC also uses the pervasive Internet concept of URLs to address document and service identification.

A key design concept of SOC is service-oriented architecture (SOA), which is an architectural style enabling loose coupling of services provided to either end-user applications or to other services distributed in a network, via published and discoverable interfaces.

5.4.2 Service Component Architecture

Service component architecture (SCA) is a recently developed model for building SOA applications by configuring and connecting components that provide service implementations [2]. The SCA assembly model allows tight coupling of functionally coherent services within modules, and loose coupling of service-oriented systems to business solutions. Service implementations are the elementary building blocks of the assembly model. A module is a unit of deployment, and the service components it provides share a common address space. SCA systems comprise services of a particular business area that is controlled by a single organisation. Next, we consider tight coupling, in Section 5.4.4 we also address loose coupling.

A service implementation provides a service consisting of a non-empty set of operations that are defined by an *interface*. This interface can be used by other service implementations, which *reference* the services they require. Service interfaces and references can be defined as Java interfaces or as WSDL port types (i.e., named sets of abstract operations and messages involved). A service implementation may have properties that are expressed in terms of data values. Examples for properties of a service supporting the virtual mobility scenario discussed in Section 5.2.2 could be the course fee requested (which may vary from institution to institution), currencies in which fees are paid, or deadlines for enrollment, examination registration, or fee payment.

Properties and referenced services can be configured externally to produce a *service component*. One or more service components are packaged into *modules*, which are the units of deployment in SCA. The services provided by the components of a module may be referenced by other components of that module as local services, or they are made available for use outside of the module through *entry points*. A module may rely on *external services* that are provided by components outside of that module.

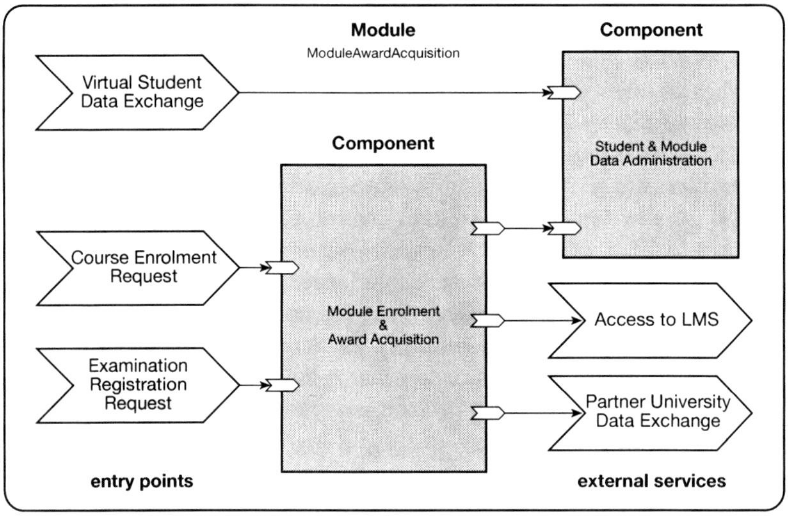

Fig. 5.6: SCA module supporting virtual mobility

Fig. 5.6 shows an SCA module named `Module Award Acquisition`[6], which contains two components `Module Enrollment & Award Acquisition` and `Student Portfolio Update`. The former provides a remotely accessible service including operations to enroll in a study module and register for a module examination, independent of whether it is offered by the student's home university or a partner university. The second component might provide a service adapter to the legacy system in which the university that deploys module `Module Award Acquisition` is maintaining its student portfolios. Our module offers one entry point and depends on two external services providing access to student portfolios maintained by partner universities, and to services of the university's learning management system (LMS). Fig. 5.6 is just an abstraction from a hierarchically structured XML file that represents these artefacts in all details in a portable format.

5.4.3 Implementing a Service Component through a BPEL Process

An implementation of a service component can be a BPEL process, a collection of Java or C++ classes, Enterprise JavaBeans (EJB), CORBA components, or XML queries and transformations (XLST). This flexibility simplifies the integration of legacy applications, and provides freedom in developing new

[6] Concepts referring to names used in a figure are written in teletype font.

service components. In the following, we present a BPEL implementation of the module in Fig. 5.6. Then, in Subsection 5.4.6, we address other implementation technologies like Java and web services.

The virtual scenario introduced in Section 5.2.2 involves services, operations, and data that are usually processed and maintained by different administrative and academic units. A student administration office typically handles student and module enrollment data. A central or departmental examination office controls examination regulations, study programme data, examiner data, and student performance data, and is responsible to organise examinations. The financial department collects student fees, and the computing and media centre manages the university's e-learning management system (LMS).

Fig. 5.7 illustrates these connections and details our sample scenario in terms of a workflow process diagram. The diagram serves two functions: (a) it models a cross-institutional module enrollment and award acquisition process from a student perspective, and (b), the BPEL process to be derived from it implements the SCA module depicted in Fig. 5.6. The diagram uses the Business Process Modeling Notation (BPMN), which aims to provide a visual notation that is readily understandable by domain experts and business analysts. Given proper XML-based definitions of the services, messages, and documents that are referred to in a BPMN model and are maintained underneath the surface by the BPMN modeling tool used, another objective behind the notation is to automatically generate executable BPEL processes from a BPMN model. A growing number of BPMN tools[7] is available on the market, and some of them support translation into BPEL. But due to some conceptual mismatches between BPMN and BPEL [28], this translation process still requires human intervention to refine and supplement the XML code produced by those tools.

The rounded-corner rectangles in our BPMN process diagram represent tasks or subprocesses, empty diamond-shaped symbols denote divergence and convergence of sequence flow, those with a cross inside the diamond shape denote forking and joining of flows enabling concurrent activities, circles represent different types of events (e.g., time out, send or termination events), solid edges denote sequence flow, dashed lines represent message flow, and the rectangular areas to the left and right of the flow model (e.g., `Partner University` or `Examiner`) represent participants in the process whose behaviour is not further detailed in this context.

Tasks provide some sort of service, which can be implemented by a web service, a programmed application, or a manual activity. Tasks are marked with different icons representing the different types of tasks between which BPMN and BPEL distinguish, including receive, send, invoke, wait,

[7] The model shown in Fig. 5.7 and 5.8 were designed with the Business Process Visual Architect 1.0 offered by Visual Paradigm; see, e.g., http://www.visual-paradigm.com.

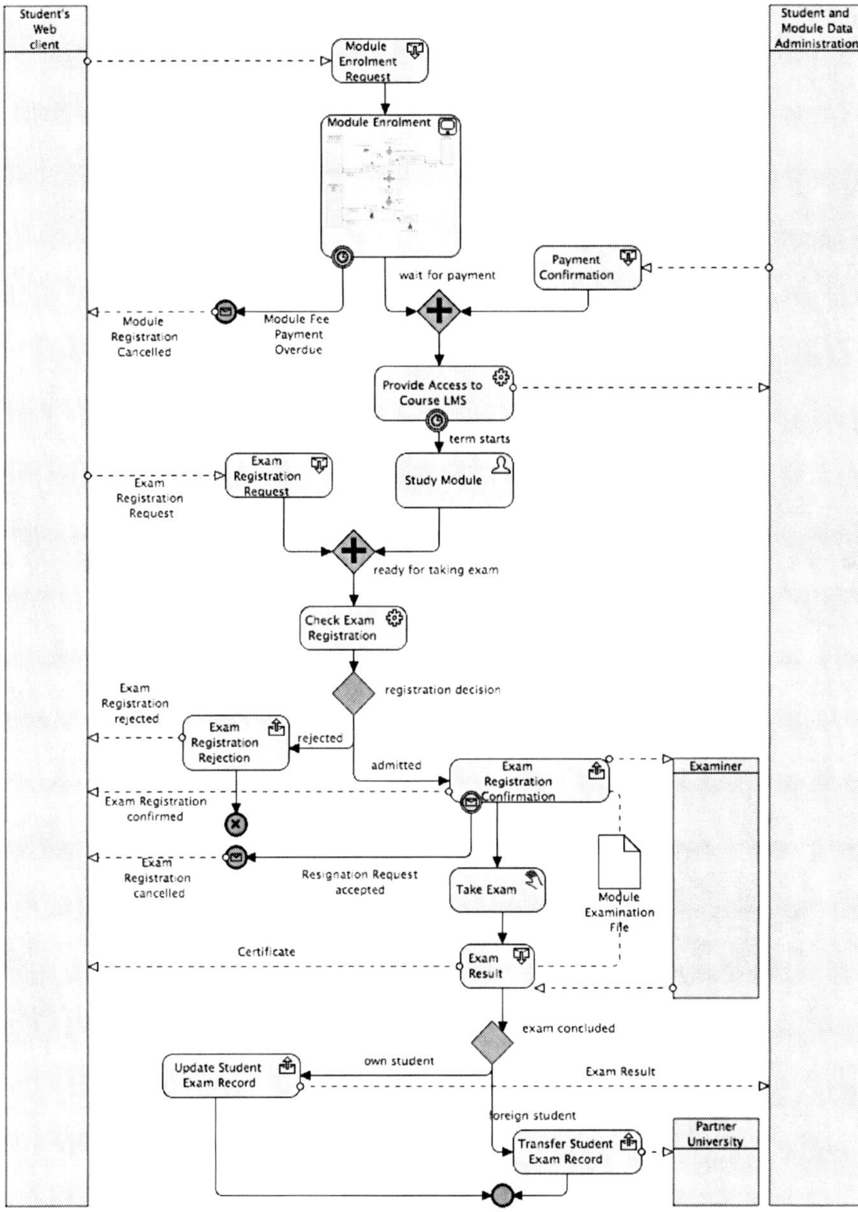

Fig. 5.7: BPMN model of the module enrollment and award acquisition process

and manual tasks. The process begins with a *receive task* named `Module Enrollment Request`. This task waits for a message from an external participant, in our example from a student who wants to communicate a module enrollment request through its web client. The receive task is completed once the message has been received. In the associated textual specification, which is not presented here, we have stated that every activation of this receive task will instantiate a new process. After receipt of the enrollment request, a *subprocess* named `Module Enrollment` is activated. Its details are described in Fig. 5.8. The subprocess performs several tasks: it checks the correctness of the request by invoking a service or otherwise implemented checking procedure; it forms an enrollment document needed for administrative purposes by either extracting the necessary information from the local student database through a method invocation, or requesting appropriate data from a remote server managed by a partner university by invoking an external service; it sends the enrollment document to the enrollment department and, depending on whether the application is granted or not, a confirmation is submitted to the requesting student including a fee payment notice, or a rejection is communicated and the whole process is cancelled.

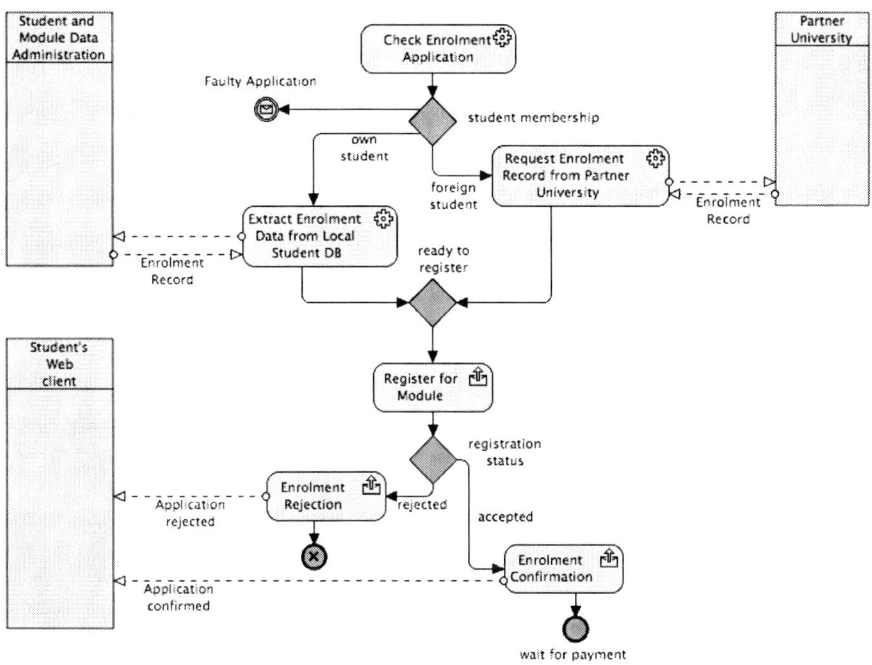

Fig. 5.8: Details of the composite activity in Fig. 5.7

Both the positive conclusion of the enrollment subprocess and the due payment of the module fee are prerequisites to provide access to the LMS as specified in Fig. 5.7. Once the student has submitted an `Examination Registration Request` and finished to `Study` the `Module`, the request is granted, and when the examination has taken place, the result is used to send an `Update Student Examination Record` message to the local `Student and Module Data Administration`, or the student examination record is transferred to the partner university the student is registered with. (Note that `Study Module` is marked as a manual task.)

In the process we have modeled a few exceptions like the situation that a payment is overdue and, in consequence, an enrollment process is cancelled, or that a student may resign from an examination after having been registered. Other deviations from the normal flow that may occur in daily practice have been omitted for the sake of readability.

A BPMN is not just a graphical notation for business processes, each object in a diagram must be supplemented by additional specifications of WSDL files for web services implementing certain activities, messages communicated, variables to maintain intermediate data or documents, participants of the web services, and properties like implementation language, rôles, and other information.

5.4.4 Loose Coupling of Service Modules to a Cross-Institutional Enrollment and Module Award Acquisition System

Loosely connected coarse-grained components are assembled in the SCA engineering model within a *system*. A system has many similarities with a module but rather than containing service components it includes *module components*. They represent configured instances of modules, i.e., the external services and property values of a module have been set in a module component. A system can be divided into subsystems to facilitate separate updates and subsystem deployment.

Fig. 5.9 shows an SCA system view of our example. We have assigned several subsystems to each university participating in a federation. The module components of the different subsystems have been grouped according to the distribution of responsibilities among different administrative and academic units in an institution. Subsystem `Provided Service Components` just provides two entry points offering services to partner universities, while subsystem `Required Service Components` comprises two external services provided by partner universities implementing student data exchange functionality and services to access a remote study module database. Subsystem `Module Enrollment & Award Acquisition` comprises a configured instance of module `Module Award Acquisition` introduced earlier, and a module component `Web Client` that is implemented by a configured instance of another module not further discussed here. This module would provide browser-based functionality to sign into

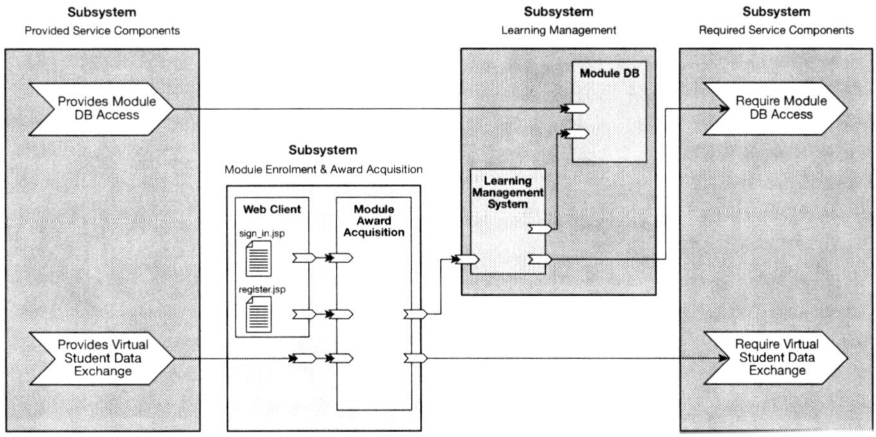

Fig. 5.9: SCA subsystems controlled by a single university

the LMS, select study modules, submit a module enrollment, an examination registration request or a resignation message, and perform other administrative services. An example how such browser functionality can be implemented in terms of Java server pages or other technology is given in [3]. Module `Web Client` relies on services provided by module `Module Award Acquisition`. These services are invoked when the user makes a selection or enters personal data through her browser. Subsystem `Learning Management` comprises two module components providing LMS services and access functions to the database maintaining the university's study modules. The database's functionality is both used by the local LMS and by LMSs of partner universities in case one of their students needs access to a foreign study module.

5.4.5 Security Issues and Policies

Service component architecture addresses security requirements such as authentication, authorisation, auditing, or privacy through policies that are associated with entry points and external services of service components. An example for a security requirement in our sample application would be that a student wants to submit a module enrollment request and needs to be authenticated as a registered student of the university providing the enrollment service. In addition, an application for examination registration must be digitally signed by the applicant, because an examination appointment not exercised counts as "failed". Similarly, the `Student and Module Data Administration` component would require the authentication of a partner university submitting a request to `Transfer` an `Enrollment Record` or a `Student Examination Record`. Encryption of the digital documents communicated would also be required.

SCA builds on the Web Services Security (WS-Security) specification 1.1 that has been passed by OASIS[8]. This standard includes mechanisms supporting message encryption, digital signatures, authentication, authorisation, and key exchange. The WS-Security specification includes various parameters that can be set through web service deployment and configuration files. The first open-source implemention of WS-Security is Apache WSS4J[9], which basically consists of a Java class library. WSS4J is still under development, but is to become part of the Apache Axis[10] distribution.

According to SCA developers, a web service environment can initially be developed and tested regardless of security concerns. With WSS4J, security features can be added to a tested environment, while obviating the need to change existing code. To impose security requirements, application developers need to modify the deployment files on the server side, and apply appropriate XML configuration files on the client side of a web service binding. For instance, if a digital signature is required, the configuration needs to define the certification infrastructure to be used for issuing and checking certificates.

WSS4J uses Axis handlers to instrument security specifications. The security handlers act like filters or interceptors attached to data input and output streams. WSS4J offers two Java classes for this purpose: `WSDoAllSender` and `WSDoAllReceiver`. The receiver is always the first instance on the input stream, while the sender is attached as last instance to the output stream. On the server side, the input stream is the request flow, and the output stream is the response flow, and vice versa on the client side.

For an implementation of the proposed university federation, security measures would be indispensable. In our case study we omitted security mechanisms.

5.4.6 Experimental Implementation

To test core components of our design, we have set up a simulation and test environment including a web service server with Apache Axis version 1.2 and Apache Tomcat[11] version 5.5. The web service specifications we needed to interface with external services referred to in our BPMN process were defined with the tool Altova XML Spy. We used the BPMN modeling tool

[8] Organisation for the Advancement of Structured Information Standards, see `http://www.oasis-open.org/committees/wss/`

[9] `http://ws.apache.org/wss4j/`

[10] Apache Axis is an open-source implementation of the Simple Object Access Protocol (SOAP).

[11] Apache Tomcat is a Java servlet container that supports Java server pages and executes Java code on web servers.

ActiveBPEL[12] to simulate the award acquisition process depicted in Fig. 5.7. All component services were designed with XML Spy and imported into ActiveBPEL to enable the simulation of the BPEL process.

Besides BPMN and BPEL tools, no development environment is available yet to support effective and comfortable development of SCA systems. Basis for SCA developers is the SCA XML schema defined by a standardisation group and an associated semantic description. The schema definition can be used in combination with a generic XML tool like XML Spy.

An associated standard, BPELJ, enables Java and BPEL to co-operate by allowing sections of Java code, called Java snippets, to be included in BPEL process definitions. Snippets can be used to implement loop conditions, branching conditions, variable initialisation, and web service message preparation, to calculate values to be inserted into a document, to construct or deconstruct a document to be sent to or received from a web service, or to create side-effects without the need to develop a web service. BPMN and BPEL are rather preferred when describing the logic of business processes. They ensure transactional properties of processes, and facilitate to design the flow of activity sets that should occur at designated times and in pre-defined order. Finally, one of the primary differences between a Java interface and a web service port type is that the methods of a Java interface can accept arbitrary Java objects as parameters, whereas a web service is defined to accept XML code.

5.5 Related Work

Several other researchers have proposed service-oriented e-learning architectures. But they all focus on services related to individual learning experiences and ignore federation aspects aiming at an integrated educational institution. Vossen and Westerkamp [31] suggest to partition e-learning systems into separate applications and realise them as web services. They also advocate the idea of evolving learning objects from content packages into web services that include presentation information. But their work particularly aims at the support of learning activities, while we take an organisational point of view. In [25], a service-oriented functional architecture for building interoperable e-learning systems is introduced. Again, this work mainly focuses on the functional partitioning of learning management systems (LMSs) and learning content management systems (LCMSs). Blackmon and Rehak present a web service solution implementing a range of customisation functions for learning content, and its creation, delivery and access [4]. In [6], an open web-service-based architecture to connect learning and teaching services is introduced.

[12] ActiveBPEL generates and executes BPEL processes from a BPMN model that was supplemented with proper service definitions and stubs (see `http://www.activebpel.org`).

5.6 Conclusion

In this chapter, we presented a case study on virtual student mobility in the context of a fictitious e-university federation relying on an IT infrastructure built on top of a service component architecture. We argued that many attempts have already been made to enable sharing and re-use of learning content in terms of information and learning objects and assessment tasks. Also the interoperability of learning environments and co-operation and communication tools is on the agenda of different standardisation bodies for a while. But little effort has been spent so far on the programmatic integration of student administration services that would allow students to take study modules and acquire the credits they need to obtain their degrees from different universities. We briefly reported on an earlier research project, CUBER, that developed the first milestone on the pathway to enable software-supported university collaboration in the form of a module information and brokerage system covering the educational offers of different European universities and university networks.

The solution we developed in this chapter extends this work by designing a cross-institutional student administrative software infrastructure. Our assumption was that all participants of a federation agree on the service components and process specifications, but preserve the freedom to implement the components in the preferred way including the possibility to integrate individual legacy software. The process specifications and security constraints agreed upon need to reflect the most strict processes and constraints of the federation participants. On the implementation side, this still leaves the freedom to implement a service through a null operation if a single institution does not insist on this service element. An example would be that some university allows students to register for an examination without having been enrolled in the corresponding module.

The design of the infrastructure we introduced shares the advantages of service-oriented computing with the benefits of component architectures. The former provides interoperability between various software applications running on disparate platforms [27]. These applications can be loosely coupled based on open specifications, interoperability standards and protocols. Data formats and protocols are widely based on XML, which allows the application of standard XML tools to edit, understand, and validate service, document, and protocol specifications. Relying on the HTTP protocol, web services can pass firewall security measures without requiring changes to common firewall filtering rules. Although services can be composed to composite ones, service components provide a better abstraction mechanism and better structuring capabilities. A component-based system has a coarse-grained structure, which allows application developers to map structures of the application domain into corresponding software concepts. Component-based software also facilitates extensions that avoid unexpected interactions with existing components.

Finally, we argued that a range of emerging standards including Web Service Policy, Web Service Security, or Web Service Reliability and supporting implementations provide the supplements necessary for realistic applications relying on strict security and reliability measures.

With our case study we have shown that software-supported university federation is technically feasible. We are fully aware, however, that legal and organisational barriers and objections might be much more difficult to overcome than the technical challenges we addressed. But we share the optimism of Tsichritzis when he argued [30]: "To date, universities dispense modules only regionally to their students. The global market, however, offers the possibility for export" – and collaboration, we would like to add.

Acknowledgments

Much of this work was achieved during the author's sabbatical, which he partly spent in Saas-Fee, Switzerland, and San Diego, California. This work was supported in part by Deutsche Forschungsgemeinschaft (German Research Foundation) under code number 44200719 and by the European Union under contract number EC-US U8041. Ms. Xia Wang developed the experimental implementation and Volker Winkler perfected some of the illustrations.

References

1. Advanced Distributed Learning (ADL) Initiative (2004) Sharable Content Object Reference Model. http://www.adlnet.gov/index.cfm
2. Beisiegel M et al. (2005) SCA Service Component Architecture: Assembly Model Specification, Version 09. http://www.siebel.com/soa/docs/SCA_AssemblyModel.pdf 28.03.2006
3. Beisiegel M et al. (2005) SCA Service Component Architecture: Building your First Application — Simplified BigBank, Version 0.9, http://www.oracle.com/technology/tech/webservices/standards/sca/pdf/SCA_BuildingYourFirstApplication_V09.pdf
4. Blackmon WH, Rehak DR (2003) Customized Learning: A Web Services Approach. In: Proc Ed-Media, AACE Digital Library. http://www.aace.org/dl/index.cfm/fuseaction/toc/paperid/12694
5. Boursas L, Keller J (2002) A Method to Automatically Recommend Course Alternatives in the Web-Based Coure Broker CUBER. In: Proc 2nd EDEN Research Workshop, Hildesheim, pp 141–143. http://www.eden-online.org/contents/publications/Hheim_02/2002-hildesheim.pdf
6. Bry F, Eisinger N, Schneemeyer G (2003) Web Services for Teaching: A Case Study. In: Proc 1st Int Conf on Web Services, CSREA Press, pp 402–408
7. CEN Learning Technology Standards Observatory (2006) Collaboration Overview. http://www.cen-ltso.org/Users/main.aspx?put=203

8. Confederation of EU Rectors' Conf (1999) The Bologna Declaration. http://europa.eu.int/comm/education/policies/educ/bologna/bologna.pdf
9. Farance F, Tonkel J (2001) LTSA Specification — Learning Technology Systems Architecture, Draft 9. http://edutool.com/ltsa/09/IEEE_1484_01_D09_LTSA.pdf
10. IEEE Learning Technology Standards Committee (2002) Draft Standard for Learning Object Metadata. IEEE Publication P1484.12.1/D6.4. http://ltsc.ieee.org/wg12/files/LOM_1484_12_1_v1_Final_Draft.pdf
11. IEEE Learning Technology Standards Committee (2006) Digital Rights Expression Languages. http://www.ieeeltsc.org/wg4DREL/
12. IMS Global Learning Consortium (2002) IMS Reusable Definition of Competency or Educational Objective, Version 1. http://www.imsglobal.org/competencies/
13. IMS Global Learning Consortium (2003) IMS Abstract Framework, Version 1.0. http://www.imsglobal.org/af/
14. IMS Global Learning Consortium (2003) Digital Repositories Specification, Version 1.0. http://www.imsglobal.org/digitalrepositories/index.html
15. IMS Global Learning Consortium (2003) IMS Learning Design. Information Model, Best Practice and Implementation Guide, Binding Document, Schemas. http://www.imsglobal.org/learningdesign/index.html
16. IMS Global Learning Consortium (2004) IMS Enterprise Services Specification, Version 1. http://www.imsglobal.org/es/
17. IMS Global Learning Consortium (2004) IMS Resource List Interoperability — Best Practice and Implementation Guide, Version 1.0, http://www.imsglobal.org/rli/rliv1p0/imsrli_bestv1p0.html
18. IMS Global Learning Consortium (2005) Content Packaging Specification, Version 1.2. http://www.imsglobal.org/content/packaging/
19. IMS Global Learning Consortium (2005) IMS ePortfolio Specification, Version 1.0. http://www.imsglobal.org/ep/index.html
20. IMS Global Learning Consortium (2005) IMS Learner Information Package Specification, Version 1.0.1. http://www.imsglobal.org/profiles/index.html
21. IMS Global Learning Consortium (2006) IMS Question and Test Interoperability Overview, Version 2.1. http://www.imsglobal.org/question/qtiv2p1pd/imsqti_oviewv2p1pd.html
22. Krämer BJ (2000) Forming a Federated Virtual University Through Course Broker Middleware. In: LearnTEC 2000, Schriftenreihe der Karlsruher Kongress-und Ausstellungs-GmbH I, pp 137–148
23. Krämer BJ (2006) CUBER oder Ausbildung à la carte. In: Lernort Virtuelle Universität, Leske und Budrich
24. Krämer BJ, Schmidt HW (2001) Components and Tools for Online Education. European J of Education 36 (2): 195–222
25. Liu X, El Saddik A, Georganas ND (2003): An Implementable Architecture of an E-learning System. In: Proc IEEE Canadian Conf Electrical and Computer Engineering, vol 2, pp 717–720
26. Papazoglou MP, Georgakapoulos G (2003) Service-Oriented Computing. Comm ACM 46 (10): 25–28

27. Papazoglou, MP (2006) Principles and Foundations of Web Services: An Holistic View (Technologies, Business Drivers, Models, Architectures and Standards). Addison-Wesley
28. Recker J, Mendling J (2006) On the Translation between BPMN and BPEL: Conceptual Mismatch between Process Modeling Languages. In: Proc 11th Int Workshop on Exploring Modeling Methods in Systems Analysis and Design
29. Tessmer M, Harris D (1992) Analyzing the Instructional Setting. Kogan Page, London
30. Tsichritzis D (1999): Reengineering the university. Comm ACM 42 (6): 93–100
31. Vossen G, Westerkamp P (2003) E-Learning as a Web Service. In: Proc 7th Int Database Engineering and Applications Symp, pp 242–246
32. Wilson S, Olivier B, Jeyes S, Powell A, Franklin T (2004) A Technical Framework to Support e-Learning. http://www.jisc.ac.uk/uploaded_documents/Technical
33. Wilson S, Blinco K, Rehak D (2005) Service-Oriented Frameworks: Modeling the Infrastructure for the Next Generation of e-Learning Systems. http://www.jisc.ac.uk/uploaded_documents/AltilabServiceOrientedFrameworks.pdf
34. Yang Y, Papazoglou MP, Krämer BJ (2003) A Publish/Subscribe Scheme for Peer-to-Peer Database Networks. In: On the Move to Meaningful Internet Systems, LNCS 2888, Springer, pp 244–262

Part II

Components and Solutions for Business and Industry

6

Wireless Asset Management

Peter Neumann

Institut für Automation und Kommunikation Magdeburg e.V.
39179 Barleben, Germany
peter.neumann@ifak-md.de

Summary. In process automation systems, there is a need for unified access mechanisms to the objects within the devices to use these objects in different applications such as maintenance despite of the diverse tools and infrastructures employed by the stakeholders involved. Based on a detailed description of the activities required, a platform for maintenance application integration is presented. The status information distributed in the field, but needed for maintenance applications is collected and pre-processed by a special device, the Asset Management Box. This data server has application modules for reverse documentation, field device state representation, support for vendor-independent engineering, notification on maintenance and repair measures needed, and for engineering-free device exchange. To improve its flexibility and to extend its use to harsh environments, a wireless version is desirable. This was prototypically built and validated, and the radio transmission technologies available for its implementation are compared.

Key words: Industrial automation, maintenance application integration, web-based on-line maintenance, asset management, wireless.

6.1 Introduction

In installations for contemporary process automation, the installed communication infrastructure consists of fieldbus systems, Ethernet-based industrial communication systems, and network components used as gateways between different communication protocols. Additionally, smart field devices can be connected via a digital parameter channel with fieldbus devices to incorporate the variable or parameter values into the data stock of the fieldbus system and, via its attached components, into all interested levels of an automation hierarchy. Fig. 6.1 depicts a typical architecture within the field area of a Distributed Computer Control System (DCS) including two different fieldbus systems and the HART interface.

These different fieldbus systems use different communication protocols and object models. Thus, there is a need for unified access mechanisms to

P. Neumann: *Wireless Asset Management*, Studies in Computational Intelligence (SCI) **42**, 123–138 (2007)
www.springerlink.com © Springer-Verlag Berlin Heidelberg 2007

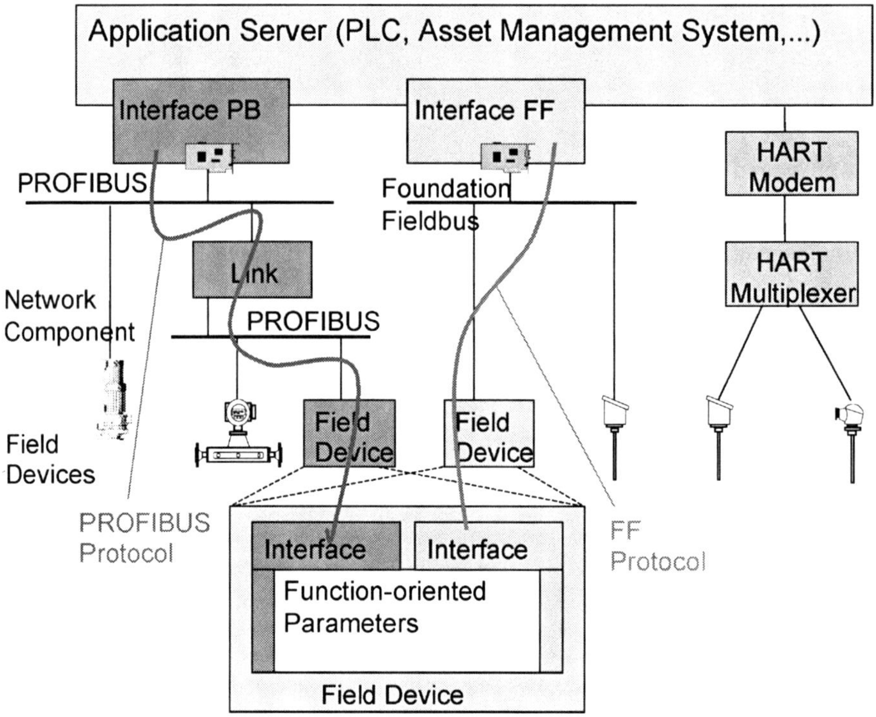

Fig. 6.1: Field area of a distributed computer control system

the objects within the devices to use these objects in different applications of the enterprise hierarchy levels [4, 16]. Besides pure process control tasks, other on-line applications for the overall management of a production plant are of high interest, e.g.,

- business administration (comparison of similar plants, delivering and visualisation of important data),
- visualisation of performance indicators,
- quality assurance, or
- workflow management (automation, optimisation).

For this purpose, the right information has to be at the right location at the right time. In this chapter, the focus will be on the maintenance process, which is a time- and resource-consuming activity [8, 19, 1]. Maintenance management includes design, control, and optimisation of maintenance workflow and the resources used. It means that plant assets, material as well as personnel have to be managed. To support these tasks, a maintenance management system has to offer the following functions,

- documentation (state, history),
- processing and visualisation of process data,
- visualisation of necessary business data,
- interfaces and access to external data,
- unified platform for different investigations,
- global access to different media,
- individual view on the offered data (information) stock, and
- unified view on data, delivered by different sources,

which have to be implemented on various platforms such as personal computers, personal digital assistants, barcode readers, and other mobile devices.

Nowadays, these platforms are used in combination, distributed over the partners involved (enterprise operating the plants/assets; service organisations providing maintenance work). These partners are using different resources, workflow control methods and tools. Thus, there is a co-existence of partners and their methods/tools with a lack of information and efficiency. Therefore, as carried out in the next section, the activities required have to be described more in detail. The data, which are needed for maintenance applications and which describe the current state of plant assets including the automation devices installed, have to be collected and pre-processed by a special device, the Asset Management Box (AMBox). The functions and a prototypical implementation of an AMBox are discussed in Section 6.3. To improve the flexibility of an AMBox and to extend its use to harsh environments, a wireless version of an AMBox should be suitable. Section 6.4 details the desired wireless version of an AMBox and compares the technologies available for its implementation. The mentioned investigations were carried out in course of completed research projects [2, 9].

6.2 Infrastructure for Web-based On-line Maintenance

Caused by the distribution of the maintenance work on non-unified infrastructures, there is a trend towards integrating maintenance applications. This means, that in a first step the infrastructure for maintenance has to be unified. The second step focuses on the unification of the business logic and on advanced maintenance functions. Fig. 6.2 depicts the scenario of so-called Maintenance Application Integration.

Maintenance-related status information is indicated in Fig. 6.3, and the types of maintenance-related information according to [20] is depicted in Fig. 6.4. This information is spread over distributed data storages

- in field devices,
- in application servers (fieldbus, controllers, maintenance servers),
- in local system manager databases, and
- at referenced locations.

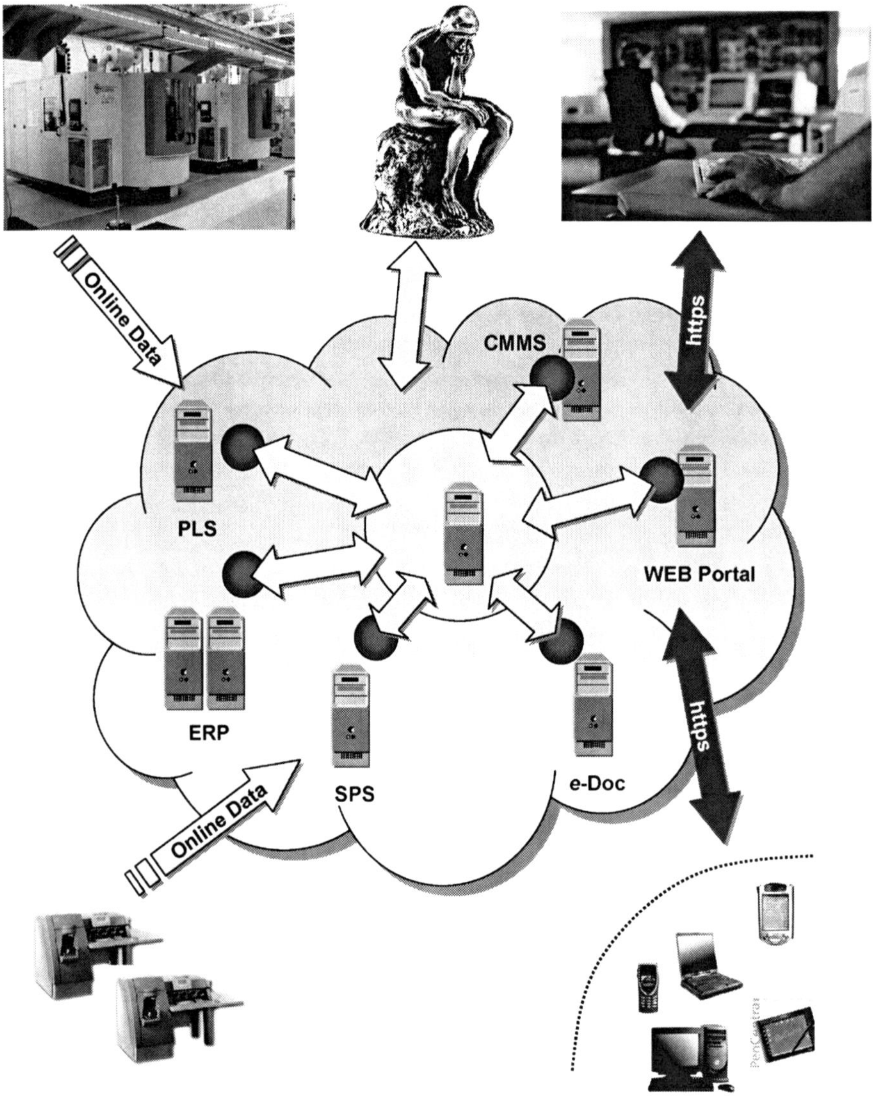

Fig. 6.2: Distributed maintenance systems [3]

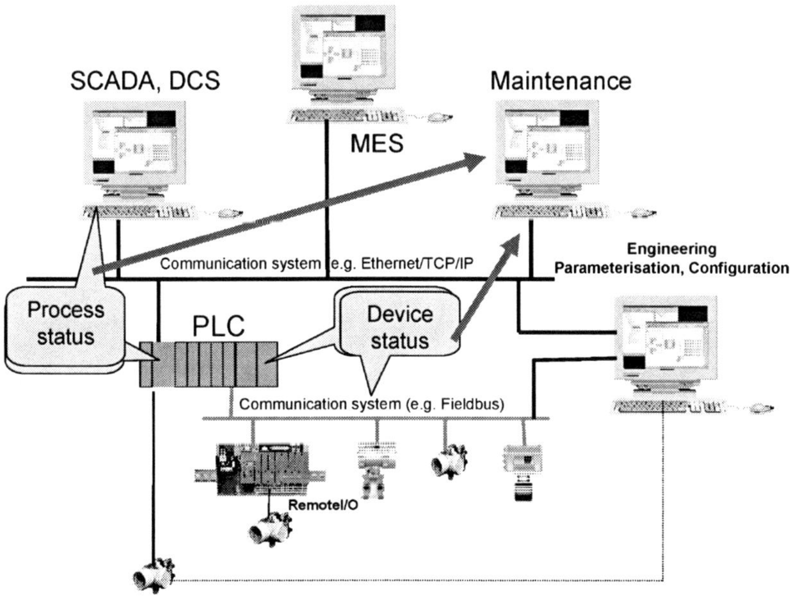

Fig. 6.3: Maintenance-related status information

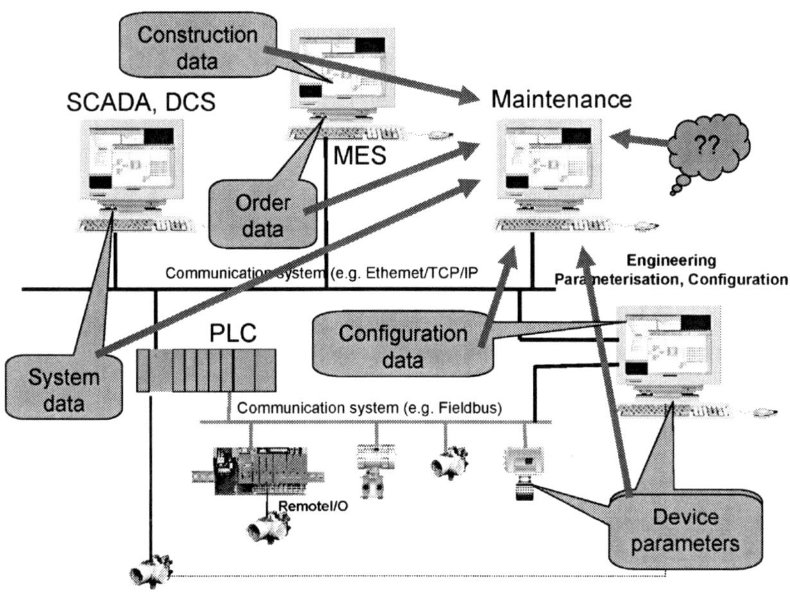

Fig. 6.4: Types of maintenance-related information

Fig. 6.5 depicts the access of the integration platform, as shown in Fig. 6.2, to the data of the installed automation devices using direct web-based access or via an asset management box [3, 18]. The data server maps device-specific data to business logic objects, which can be used by the maintenance application functions.

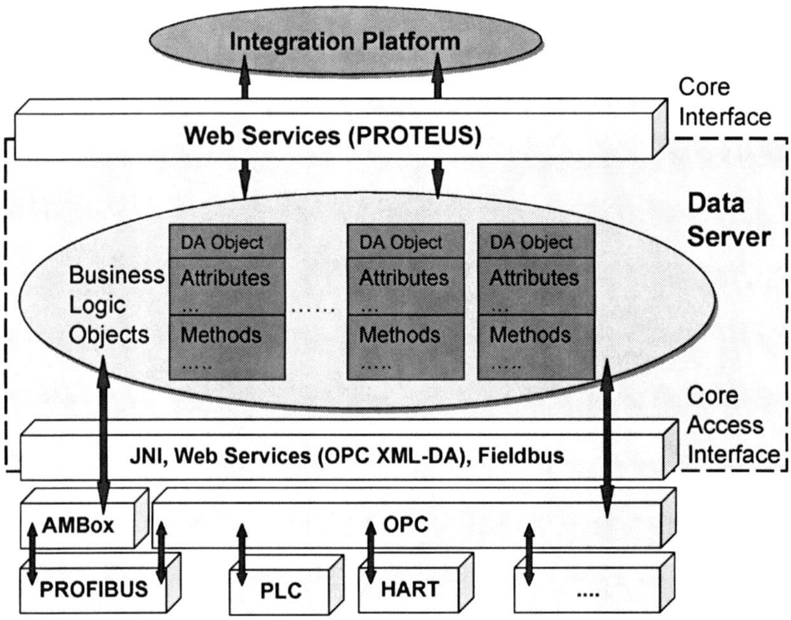

Fig. 6.5: Data access of the integration platform

The Core Access Interface handles the various data, produced in the underlying automation devices or asset management boxes, and supports the various maintenance applications using web services:

- application registration,
- browsing objects,
- access rights' management,
- event distribution, and
- data handling (create, find tag, get/set attribute, read value, archive, snapshot, event notification etc.)

The central services, located in the common central maintenance management infrastructure, are:

Global Access Protection realised by a Central Authorisation and Authentication Server (CAAS), and

Central Object Relation Database (CORD) containing the relation between various objects describing differents view to selected assets.

The decentral services, located in the partner-dependent maintenance management infrastructure, are:

- access to automation systems (data acquisition server),
- knowledge management,
- documentation,
- logistics and specific maintenance management services
- business applications (SAP applications, ERP), and
- access to the Internet (web portal).

Fig. 6.6 shows the installed platform prototype of a maintenance application integration, which was realised within the EUREKA/ITEA project PROTEUS [2]. It comprises a number of computers and automation devices at various locations (as mentioned in parentheses) as well as different software tools, transfer protocols, and application programs.

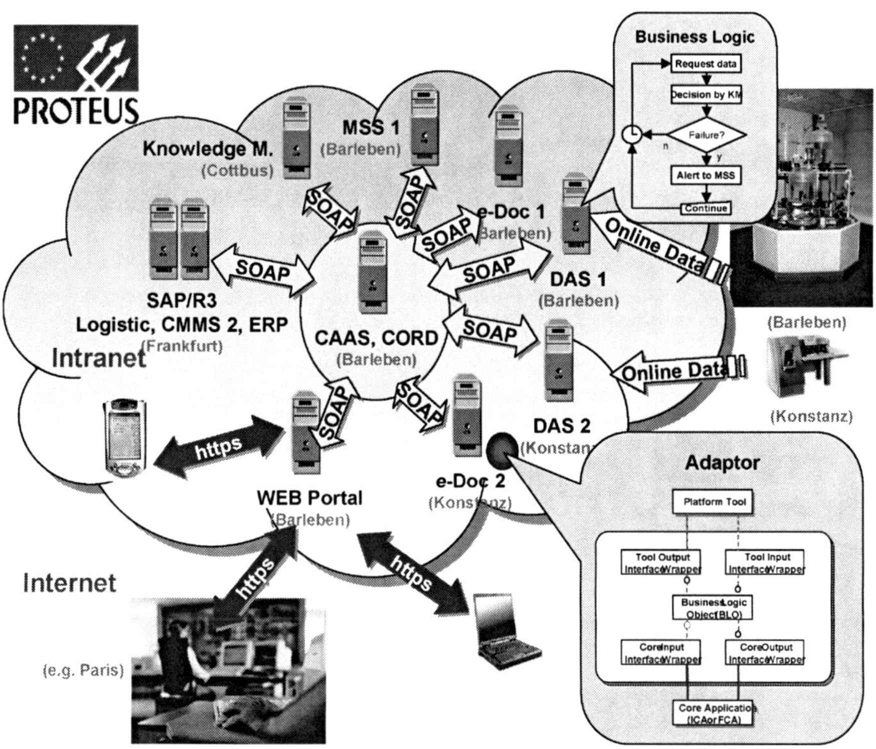

Fig. 6.6: PROTEUS platform for maintenance application integration

6.3 Asset Management Box

Besides pure process control tasks, applications of on-line asset management are increasingly gaining importance in distributed computer control (cp., e.g., [17]). In general, asset management is understood as the utilisation of plant and device data for the economical optimisation of technical plant operation. This comprises, for instance, optimising planned maintenance, improving process diagnostics and monitoring, as well as increasing production throughput by exploiting the reserves still available in the field devices [18, 1, 10, 5]. To implement these applications, it is necessary to install an additional information channel between the control levels for the field and for operations [7, 6]. Into this information channel the tool for asset management, the so-called Asset Management Box (AMBox), is incorporated (Fig. 6.7).

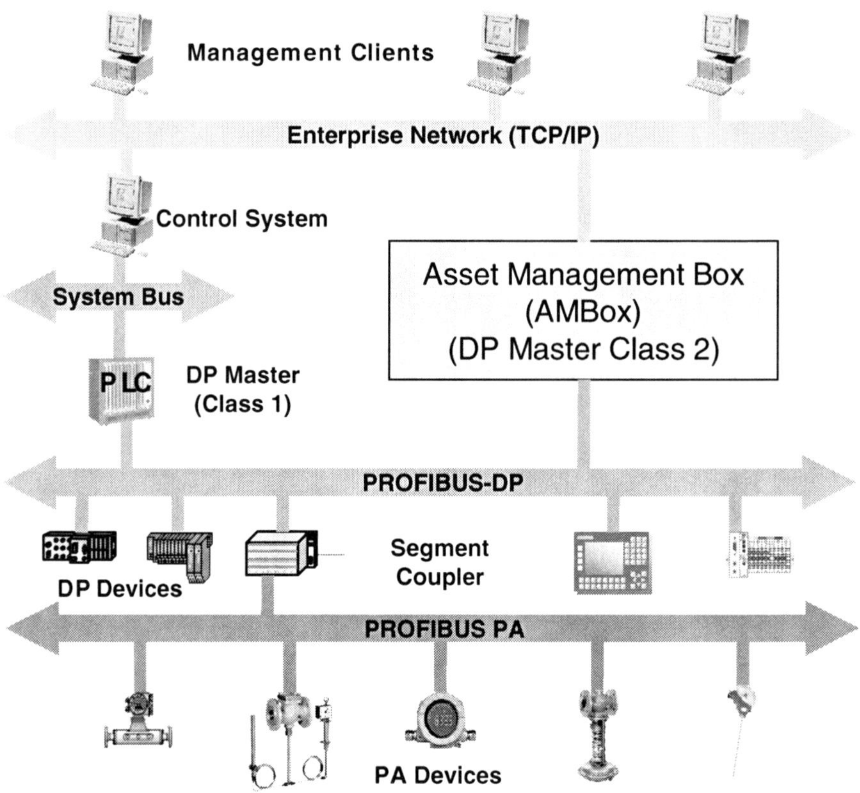

Fig. 6.7: Asset management box as an information server

The scenario in Fig. 6.7 shows how an AMBox is integrated into an automation system according to the present state of wired technology. Without requiring any special engineering provisions, the AMBox makes available information from a fieldbus installation. An outstanding feature of the easy access to the information relevant for on-line asset management is the automatic configuration of the AMBox. At run-time, a complete set of information is read out of the devices attached to the fieldbus without the need to carry out additional parameterisations. With the help of the automatic configuration, the AMBox can be connected to a distributed process control system at any time without influencing the system's software implementation. When the AMBox has been connected to the fieldbus installation, the connected fieldbus segments will be explored without any additional engineering using object libraries. The related instances are automatically generated during the start-up phase of the AMBox. The installed field devices are identified without any necessary a priori knowledge and, therefore, without engineering effort. At run-time, the fieldbus segments are periodically checked. Thus, changes of the fieldbus installation are recognised automatically.

The AMBox acts as an information server. The installed field devices can be identified by their bus addresses and represented by proxy objects within the object management of the AMBox. A proxy offers the data content of the fieldbus device within the environment of the DCS. The data structure of a proxy corresponds with the internal data structure of the linked field devices. The data structure uses different types of blocks. The blocks, attributes and parameters are offered to the various applications as periodically updated data objects. The collected data can be transmitted via the Internet, and be visualised by management clients. At run-time, dynamically loadable libraries, representing different asset management applications, can be added at any time. The libraries realise the different universal or plant-specific application modules of the maintenance management [10]. They have access to the information stock of the field devices, represented by the various proxies. The application modules offer their results openly and in a self-describing fashion to other maintenance applications, office applications etc. Therefore, the object management system enables access to the objects of the database.

There are various application modules:

Reverse documentation module realising automatic reverse documentation of the underlying fieldbus systems without any necessary engineering.

Field device state representation enabling the adaptation of the actual to the desired operation mode of a device as well as meeting the needs of diagnosis and maintenance. The actual status of the installed field device can be represented as an XML file or an Excel sheet and, therefore, be processed by common office application programs.

Support for vendor-independent engineering of the installed field devices, as limited to the possibilities specified by the field device profiles [17]. The blocks and parameters, defined by the profiles and identified using suitable clients, can be taken over to the AMBox object model by related application modules and transmitted to the connected fieldbus devices.

Notification on maintenance and repair measures needed The diagnostic information provided as well as the internal device parameters can be used directly. In addition, specific information for complex maintenance models can be generated. Since the generated information is accessible using standard Internet technology without previous system configuration, remote diagnosis and maintenance are possible.

Application module for engineering-free device exchange and device persecution which describes the rôle behaviour of the field devices within a plant and device environment.

Further properties of the AMBox are the completely automatic on-line documentation of the underlying fieldbus segments, and the possibility to exchange devices transparently without any engineering measures. The mentioned proxies, representing the field devices within a plant environment, remain persistent within the system after removal of field devices. Thus, important information remains within the information stock of the DCS.

6.4 Wireless Asset Management Box

The scenario in Fig. 6.7 showed how an AMBox is integrated into an automation system according to the present state of wired technology. Both for the attachment to the fieldbus system as well as for the connection to the higher-level enterprise network with the asset management clients the hardware of the existing communication system can be used, but the architecture has to be physically extended by the asset management box. According to practical experiences, this may cause failures or, under unfavourable conditions, even stoppages of communication and, thus, of operation of the automated plant.

Therefore, it appears feasible to create a radio-based communication platform, which reduces the effort for the integration of the AMBox and the corresponding risk of failures for the existing communication system. In doing so, the mentioned advantages and functionalities of the AMBox must be maintained completely. Moreover, new possibilities for the mobile and flexible deployability of tools for asset management are opened up. To attain these goals, the AMBox is incorporated into an existing control system by wireless radio-based communication, using the the functions of the wired AMBox, but extending it to a "Wireless Asset Management Box" (WAMBox). Besides reducing installation expenses and the risk of disruptions of communication, the utilisation of radio transmission in this context yields further advantages. Thus, the WAMBox does not need to be installed in direct vicinity to the fieldbus, which is especially useful for applications of on-line asset management in environments characterised by the risk of explosions.

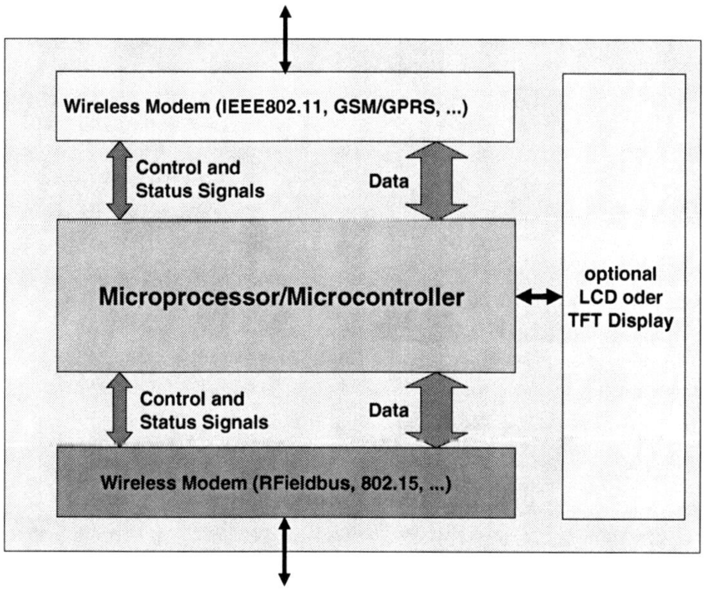

Fig. 6.8: Simplified device hardware architecture of a WAMBox

Radio-based communication creates an opportunity for mobile asset management, too. If additional possibilities for visualisation are given in the WAMBox, a powerful tool for maintenance operations of any kind is available. Besides the parameters directly acquired from the devices, also pertaining manuals and guidelines for repair can be consulted on-site. Thus, the contribution of the travel costs to the total maintenance costs, which varies between 10 and 20% depending on plant structure and application area, can be reduced considerably. These costs are mainly due to not having available on-site all information required and, therefore, the need to search for it in different locations of an enterprise.

A further advantage of the WAMBox consists in increasing flexibility. Given an appropriate software implementation, heterogeneous communication systems widely used in industrial process automation such as Foundation Fieldbus, HART, or Profibus-PA can be accessed in a common way without having to modify the hardware of the WAMBox. Here, the radio link and the corresponding implementation serve as a unified software and hardware interface. Fig. 6.8 shows the simplified hardware architecture of a WAMBox with its essential hardware elements. The central element is a microcontroller with the drivers for the radio components, the communication stack, the AMBox functionality, and the implementations for Foundation Fieldbus (conceptually), HART and Profibus-PA. Fig. 6.9 depicts a simplified software architecture of a WAMBox.

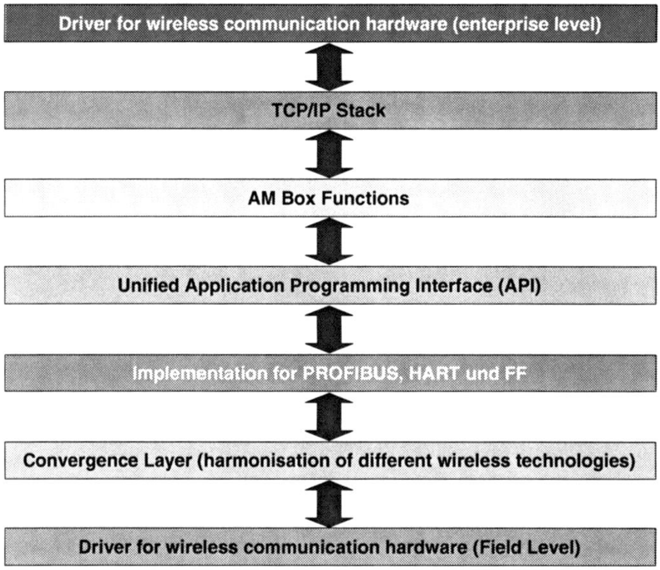

Fig. 6.9: Simplified software architecture of a WAMBox

The essential results presented in this chapter are specifications of new, effective scenarios for on-line asset management, of the system architectures correspondingly required in plants with communication based on Foundation Fieldbus, HART, and Profibus-PA, as well as of the communication architecture of the network components (bridge, base station) and of the Wireless AMBox. Also essential is the mapping of the abstract specifications on implementation specifications for the hardware and the software, including a common Application Programming Interface (API) for the on-line asset management applications in the WAMBox.

Prototypically, the implementation of the hardware and the software for the radio access and the integration of the AMBox functionality was carried out for the systems HART and Profibus-PA. With the prototypes of wireless fieldbus bridges or gateways and of the Wireless AMBox combined operation was validated in a semi-professional plant, and behavioural investigations were performed. In the course of this, the experiences gained in and the results of the European project RFieldbus [13, 11, 12] could directly be utilised, e.g., when specifying the radio access to Profibus-PA.

In plant structures, which correspond to the structure with a wired AMBox (Fig. 6.7), the wireless link can be realised by radio techniques like RFieldbus or solutions according to IEEE 802.15 and their additional application layers (ZigBee, Bluetooth). From all segments underlying Profibus-DP (Profibus-PA, Foundation Fieldbus) the device information can be read out. This variant is particularly advantageous if several segments and/or fieldbus segments of different types are subordinated to Profibus-DP. Then, only one radio component needs to be installed at Profibus-DP. The direct connection to Profibus-DP imposes, however, high requirements on the radio technology. In particular, the high data rates can presently be achieved with correspondingly high expense, only. Radio components meeting these requirements are considerably more expensive than the variants mentioned below. A further variant to integrate a WAMBox into an existing industrial communication system consists in the immediate linkage to a Profibus-PA segment (Fig. 6.10). Advantageous is that the radio component can be realised easily and cheaply. As radio technologies RFieldbus or ones according to IEEE 802.15 can be used here, too, with the latter appearing more suitable for this application mainly due to lower energy consumption. Disadvantages can arise for very many Profibus-PA segments, as a separate radio component has to be installed for each segment.

Fig. 6.11 shows the same concept used for the Foundation Fieldbus.

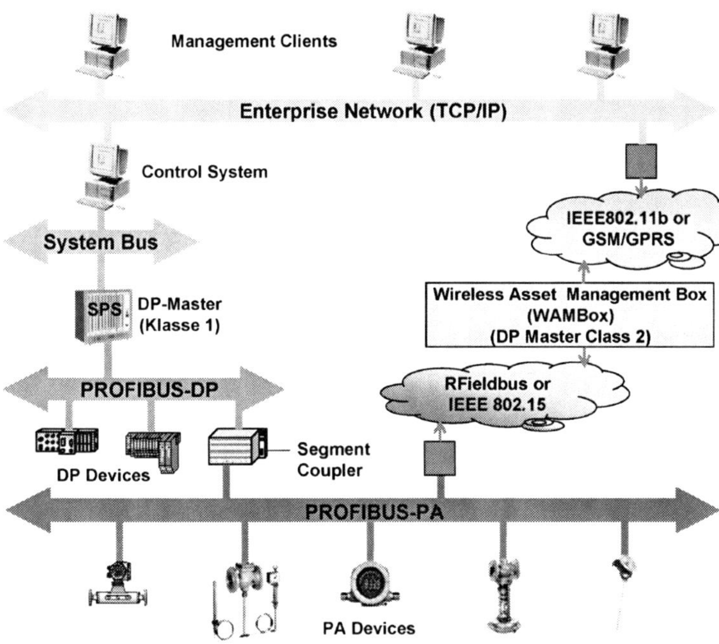

Fig. 6.10: WAMBox as information server at Profibus-PA

A third variant is the direct radio-based attachment to HART devices (Fig. 6.12). Here it needs to be considered whether the radio component can be miniaturised to enable its direct integration into HART devices, e.g., via the connector, or if an external separate device is necessary.

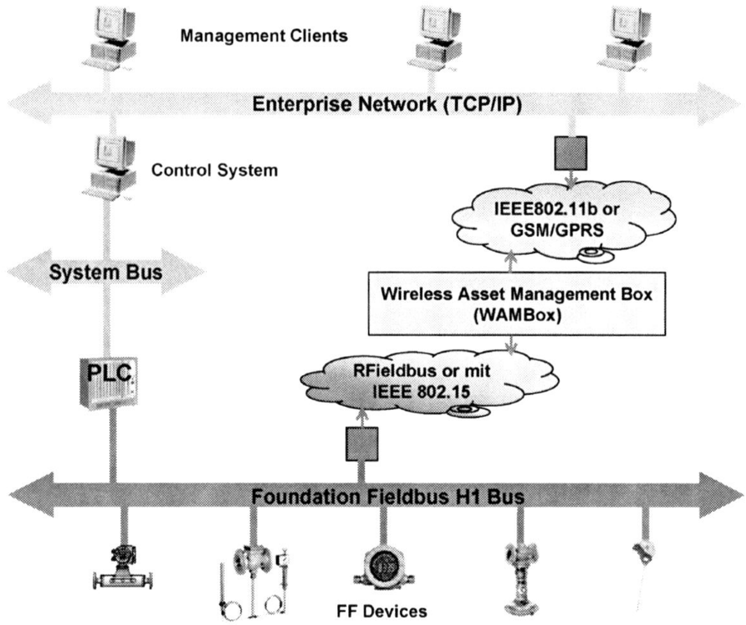

Fig. 6.11: WAMBox as information server at Foundation Fieldbus

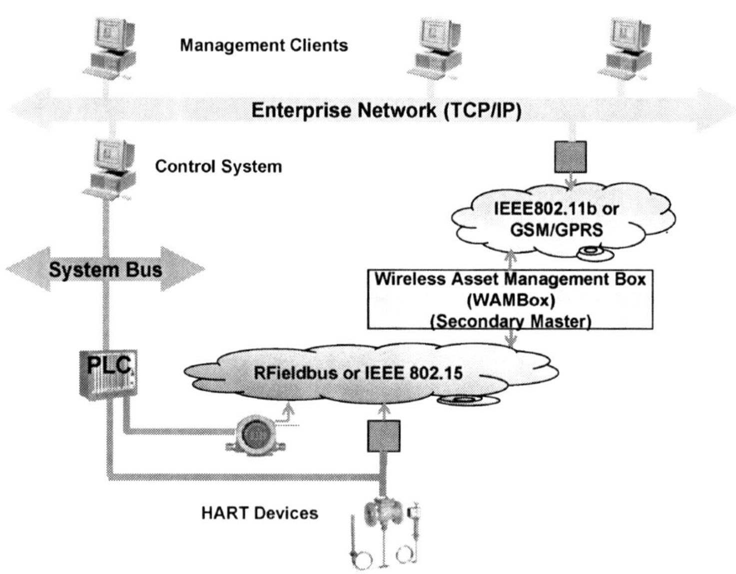

Fig. 6.12: WAMBox as information server at HART

To summarise, there are different ways to integrate an asset management box into a networked control system using different concepts as well as different technologies of wireless transmission. The latter could not all be described in this chapter. For details, the reader is referred to [15].

References

1. Bangemann T, Wollschlaeger M (2002) IT-based Management zur Unterstützung von Instandhaltungsprozessen in der Prozessindustrie. In: Proc IT & Automation
2. Bangemann T, Schmidt D, Thron M (2005) PROTEUS Generic e-Maintenance Platform. Final report of BMBF project 01ISC01A, Institut für Automation und Kommunikation, Magdeburg
3. Bangemann T, Schulze A (2005) Globale Asset Management Dienste auf der Basis von Maintenance Application Integration. Presentation for ZVEI
4. Epple U, Uecker F (1999) Integration of Maintenance Functions into the Process Control Environment. In: Proc 3rd Int Exhibition and Conf on Plant Maintenance
5. Koschel J, Müller U, Nicklaus E (2000) Automatisierungstechnische Praxis 42 (12): 41–47
6. Müller J, Uecker F, Wollschlaeger M, Diedrich C, Epple U (2003) Computing & Control Engineering J, 34–39
7. Müller J, Wollschlaeger M, Epple U, Diedrich C (2003) Asset Management Applikationen auf der Asset Management Box. In: VDI-Berichte, No 1756, VDI-Verlag, Düsseldorf
8. Müller J, Wollschlaeger M, Enste U, Epple U, Diedrich C (2004) Opening Up the Information Sets of Modern Field Devices for Various Asset Management Applications. In: Proc 1st European Conf on Electrical and Instrumentation Applications in the Petroleum and Chemical Industry
9. Müller J, Wollschlaeger M (2005) Simplification and Automation of the Engineering of On-line Asset Management Applications — Asset Management Box. Final report of project 13321 BG/1, Institut für Automation und Kommunikation, Magdeburg, and Rheinisch-Westfälische Technische Hochschule Aachen
10. Nicklaus E, Fuß H-P (2000) Automatisierungstechnische Praxis 42 (5): 30–39
11. Rauchhaupt L (2002) System and Device Architecture of a Radio Based Fieldbus — The RFieldbus system. In: Proc 4th IEEE Int Workshop on Factory Communication Systems
12. Rauchhaupt L (2003) RFieldbus — a Survey. In: Proc 5th IFAC Int Conf on Fieldbus Systems and their Applications
13. RFieldbus (1999) High Performance Wireless Fieldbus in Industrial Related Multi-Media Environment. EU IST 1999 11316. `http://www.RFieldbus.de`
14. Strohbach R (2003) Anwendungsmodule für das Online Asset Management von PROFIBUS PA-Systemen. Otto von Guericke-Universität, Magdeburg
15. Virtual Automation Networks (2005) European Integrated Project FP6/2004/IST/NMP/2-016696, Work Package 1: Requirements and Trend Screening, Deliverable 01-1 State of the Art
16. Wollschlaeger M (1999) Mapping of Fieldbus Components to WWW-Based Management Solutions. In: Proc IFAC Conf on Fieldbus Technology

17. Wollschlaeger M, Müller J, Epple U, Diedrich C (2001) Asset Management Solution based on PROFIBUS-PA Profiles. In: Proc 8th IEEE Int Conf on Emerging Technologies and Factory Automation
18. Wollschlaeger M, Neumann P, Bangemann T (2002) Web Services for Remote Maintenance of Fieldbus Based Automation Systems. In: Proc IEEE AFRICON
19. Wollschlaeger M, Bangemann T (2003) IT-based Management of Maintenance Processes in Fieldbus-based Automation Systems. In: Proc IFAC Conf on Fieldbus Technology
20. Wollschlaeger M, Bangemann T (2004) Requirements, Structures and Content Model for Web-Based Maintenance Portals in Automation. In: Proc INCOM

7

Knowledge Management for E-Maintenance of Industrial Automation Systems

Wei Zhang

Faculty of Electrical and Computer Engineering
FernUniversität in Hagen, 58084 Hagen, Germany
wei.zhang@fernuni-hagen.de

Summary. Knowledge management is now considered as a method to improve the global competitive capacity of an enterprise. The evolution of industrial automation systems makes the maintenance tasks more and more important. These tasks imply to deliver the right knowledge/information in a right format at the right time to the users. Some problems such as knowledge/information exchange and system complexity make it difficult to achieve this goal. The relationship between knowledge and information is elucidated. Hence, knowledge management is thought to be facilitated by information management. A framework of multi-agent systems is introduced to make information/knowledge exchange available. All information sources in industrial automation systems are analysed and modeled by agents. The relationships between these sources are emphasised in the context of maintenance, especially of proactive maintenance. They are actually the knowledge to be created in order to facilitate proactive maintenance.

Key words: Knowledge management, maintenance, proactive maintenance, agents, multi-agent systems.

7.1 Introduction

Nowadays, an enterprise, especially a large industrial one, is equipped with various systems from hardware to software, from workshop operation to enterprise management, in order to improve its competitive capacity. For instance, Distributed Control Systems (DCS), Supervisory Control and Data Acquisition (SCADA) systems, Manufacturing Execution Systems (MES), Computerised Maintenance Management Systems (CMMS), or Enterprise Resource Planning (ERP) systems are used to achieve this goal. These systems may be connected through industrial bus systems or Ethernet, as illustrated in Fig. 7.1. This is known as the automation pyramid. However, the reality of maintenance of distributed control systems and corresponding devices remains non-satisfactory.

W. Zhang: *Knowledge Management for E-Maintenance of Industrial Automation Systems*, Studies in Computational Intelligence (SCI) **42**, 139–155 (2007)
www.springerlink.com © Springer-Verlag Berlin Heidelberg 2007

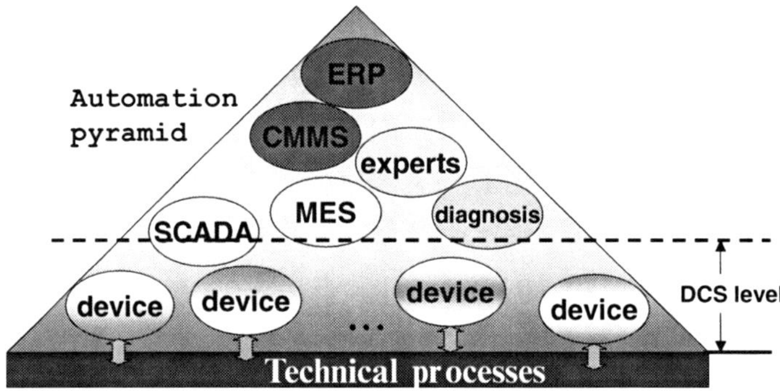

Fig. 7.1: Automation pyramid

Maintenance is becoming extremely important for companies to sustain their manufacturing productivity and customer satisfaction at the highest possible level in the global market [10]. Nowadays, the degree of automation is rather high in industrial areas. One unavoidable consequence is that the cost of maintaining devices and equipment is increasing due to using complex automation systems and intelligent field devices to realise advanced processes. That is also the reason why considerable research effort is devoted to reliability and safety-related issues. Reliability means that systems function correctly as long as they are within their lifecycles, whereas safety means that persons are not hurt nor economic loss occurs in cases of failure. However, after devices or systems are manufactured or released, the reliability- and safety-related tasks have to be transferred to maintenance. What maintenance can provide is to guarantee and support the service times of devices and systems as long as possible. Therefore, entire automation processes should keep running stably without unnecessary shutdowns.

Two types of problems are still open. One of them is how to acquire suitable and sufficient information and knowledge, the other is how to manage the information/knowledge for certain special applications. For instance, predictive maintenance necessitates detailed runtime information on automation processes and knowledge of device behaviour modes (including failure modes). Generally, a running system represents a certain domain knowledge or information source. Co-operation among these sources has to answer the question of how to make them understand each other. Also social knowledge is required. Therefore, one of the urgent problems is to realise knowledge management (KM) for enterprises. In addition, current maintenance engineers create their maintenance worksheets manually and organise maintenance operations themselves, i.e., this part of maintenance-related knowledge exists in the heads of maintenance engineers, only. When facing the increasing requirements for maintenance, maintenance engineers find it a problem to attain correct information and knowledge at the right time. On the other hand, enterprises have to carry the high costs caused by shutdown times due to unnecessary or not well-arranged maintenance operations.

E-maintenance is a concept for collecting, processing, and providing information from various individual sources to work out predictive and corrective maintenance. Obviously, collaboration has to be in terms of information exchange and knowledge sharing. In this chapter, a platform is introduced to solve (but not limited to that) the problem of information integration and KM for e-maintenance. In this work, a multi-agent system (MAS) is used to model the platform, due to some inherent features of agents:

autonomous having computational entities for independent problem solving;
reactive having the ability to perceive and respond to changes occurring in their environment in a timely fashion to satisfy their design objectives;
proactive being able to exhibit goal-directed behaviour by taking the initiative to satisfy their design objectives;
social being capable of interacting with other agents to satisfy their design objectives;
decision-making having the ability to select suitable actions on perception by reasoning on the base of their knowledge;
mobile being able to move from one environment to another to satisfy their design objectives.

Of course it is not always necessary to consider them all in an application of e-maintenance. We do not use the mobility feature in our model.

In the following Section 7.2, the concept of proactive maintenance is discussed. Then, in Section 7.3 some concepts about KM and special requirements of industrial automation systems are pointed out. Based on an analysis of information available from automation processes in Section 7.4, the agent-based platform is introduced in Section 7.5. Then, knowledge construction for proactive maintenance is discussed in Section 7.6, before some conclusions are drawn and future work is identified in Section 7.7.

7.2 Proactive Maintenance

Maintenance can be classified to be corrective, planned, predictive, or proactive. Corrective maintenance is to execute the maintenance operations after failure, while planned, predictive maintenance (PdM), and proactive maintenance (PaM) seek to execute the maintenance operations before failure. Maintenance operations are checking, calibrating, cleaning, repairing, and replacing failed components or equipment. Planned maintenance executes pre-set maintenance operations on all components before some minimum expected failure time. However, it may cause a maintenance operation to be too early or too late. PdM consists in deciding whether or not to maintain a device/equipment according to its state. Currently, PdM techniques are highly regarded as the most effective means assuring equipment/device reliability, efficiency, and cost effectiveness. Benefits include fully useful lifecycle for equipment and lower redundancy inventory [2].

PaM is sometimes regarded as equal to PdM, because they are so close. As discussed in [4], however, the key principle of the PaM strategy is to consider the integration of the duality between functioning and malfunctioning. PaM commissions corrective actions aimed at the sources of failure. For instance, for a valve or hydraulic system, its contamination is often considered the root cause of failure, which may not be mainly covered by PdM. As stated in [20], some features of the PdM technologies (the hardware and software tools) are not fully employed by users. That means PdM needs KM support to perform at its full potential. In automation systems, there are different views on system functioning oriented at, e.g., material flows, product flows, energy flows, and information flows. From the perspective of e-maintenance, we focus on functioning in terms of information management (IM) and KM. The emerging intelligent field devices and Electronic Device Description (EDD) technology [14] can support our idea. Therefore, in automation processes, we simply consider PaM to be equal to PdM plus KM. Associated with the corresponding IM, KM is the main part of our PaM solution, while technologies used by PdM provide possible additional data/information to PaM, such as vibration monitoring, infrared thermography, or oil quality analysis. In [9], modeling of system functioning is considered an approach to deal with analysis of functioning. The relationship of functioning and malfunctioning is mapped to the relationship between normal and abnormal (degradation) states of a system. Later in this chapter, we analyse functioning models (knowledge) of a device, and available information as well as all necessary information.

Based on PdM, PaM necessitates a high level of KM (efficiency and accuracy), which may depend on seamless combination of existing methods and information integration. From the simplified PaM scenario (in a graphic description similar to a UML sequence diagram) shown in Fig. 7.2, the requirements of KM in PaM are refined as:

- Integration of distributed domain knowledge (information) bases from the process control level (field) up to higher management level. (It is implied that the common knowledge representation method is needed to create a unified maintenance view.)
- Models of functioning devices
- Models of malfunctioning devices
- Tracking device behaviour, capturing abnormal information/data, and fault events in a timely manner.
- Since there are control view, management view as well as maintenance view on processes in the area of industrial automation, having different goals [9], maintenance-oriented KM has to support and to be supported by other views in an active way.
- Reliability and safety-related considerations

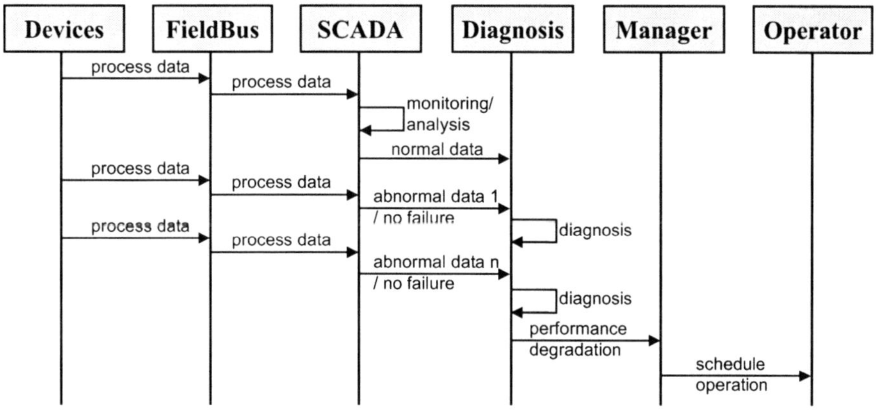

Fig. 7.2: A scenario for proactive maintenance

7.3 Knowledge Management in Industrial Automation

Here we consider KM for PaM in terms of models of device functioning and malfunctioning. To actualise PaM in industrial automation, the knowledge has to be ready before carrying out a PaM strategy.

An overview on KM is given in [6] from the design perspective. *"In this perspective, knowledge workers constantly create new knowledge as they work. The goal of KM is to enable innovative practice at an organisational (community) level by supporting collaboration and communication among knowledge workers in the same domain and across domains."* This concept emphasises the activeness of KM, since participants in KM are, in fact, human beings. However, the implementation of conventional views on KM is essentially limited to document management [23]. This means that knowledge is seen as a passive effector in organisational operations. Actually, as the design perspective implies, knowledge actively affects automation processes in terms of information acquisition, transformation, representation, and storage. This could be called information management (IM). Knowledge existing in the heads of experts directs further knowledge generation. This is one of the steps of the cyclic KM process, called knowledge acquisition. Additionally, knowledge representation, knowledge communication and maintenance are other activities in KM. From this point, KM can be mapped into IM.

Traditional document management could work as an alternative to KM if assumptions could be attained and satisfied according to our real-life experience, i.e., if participants in KM could understand each other precisely. This assumption, however, can never be completely satisfied. Therefore, we view documents as a sort of information. The information space of industrial automation systems is evolving with operation activities, which represent some kind of knowledge, and organisational memory [21]. With the support of information technology (IT), KM is the combination of IM and the application of artificial intelligence (AI) algorithms. This naturally leads to the question what distinguishes knowledge from information and data in industrial automation systems.

Stenmark [19] and Tuomi [21] discussed the relationships between data, information and knowledge, both based on Polanyi's knowledge theories [15], which stated that knowledge is tacit. Stenmark argued that all knowledge is tacit, and that information is what can be articulated and made tangible outside the human mind. It is generally accepted to view data as simple facts which become information when combined into meaningful structures which, in turn, becomes knowledge when put into a context and when useful to make predictions, i.e., data are considered as a prerequisite for information, and information as a prerequisite for knowledge. On the other hand, Tuomi proposed a reversed hierarchy of knowledge, showing that knowledge exists before information, and that information emerges before data. Namely, knowledge is necessary to capture information from data. These two points of view can both be found in the processes of KM, especially in industrial automation systems. In a temperature controlling process, for instance, temperature data captured by temperature sensors are time-stamped and fuzzilised into structured data, the information. In this step, fuzzy rules embodying associated knowledge are used to transform the raw data into information (such as connecting the logical concepts hot, warm, cool, and cold to contexts of the temperature controlling process). Therefore, the already existing knowledge supports the generation of information, i.e., the above hierarchy is reversed.

7.4 Analysis of Information in Automation Systems

As, on one hand, information is easy to transmit and store, but, on the other, knowledge is hard to distribute among different working groups, we approach knowledge management from information processing.

In industrial automation systems, information is processed and exchanged (flows) in the automation hierarchy. The characteristic structures of automation systems are analysed in [22]. In addition to the automation pyramid itself, up-stream and down-stream enterprises are combined into various chains of material and information flow due to the increasing demand for co-operation. This results in information exchanges (in term of information flow) between various components or systems. In this section, we analyse the information available through industrial communication networks, and all necessary information for PaM. There are three classified types of information in industrial processes [1]:

1. static information representing the documentation only modified during an evolution of the physical process or physical field components of a Computerised Maintenance Management/Intelligent Actuation and Measurement System (CMMS/IAM),

2. semi-static information representing the current configuration, parameterisation, or evolving history of a CMMS or IAM system, and
3. dynamic information representing the real-time behaviour of a system, i.e., its status, state, or model of variables.

They show us different features of the information and also imply different information sources. All information/data necessary for PaM in industrial automation can be listed:

- process signals (temperature, pressure, binary signals, etc.),
- control signals (current, voltage, binary signals, etc.),
- environmental data/information,
- device and communication system information/data, and
- relationships between these data/information.

On the other hand, the information available via industrial communication networks is:

Device Profiles: some characteristics of devices, using device description technology, i.e., EDD.

Fieldbus: real-time data, such as process signals, control signals, other data provided by the devices and transmitted over a fieldbus (e.g., Profibus).

SCADA: sensor data, historical data about devices and fieldbus, statistical data about devices, fieldbus, and fault/failure reports, operation and system performance reports, etc.

Operators and other sources: some may be represented in certain forms of knowledge, but others are difficult to acquire directly, e.g., operator experience.

These data/information come from different sources, which are as heterogenous as automation systems themselves. Given the characteristics and relationships of these information/data, the goals of device behaviour monitoring and performance assessment can be achieved by making use of the information available. Therefore, one task of KM is to build the knowledge to represent the relationships between these data/information. This knowledge could be represented at several levels, such as the epistemological level (connection), the logic level (conceptual), and the heuristic level (symbolic).

A solution to handle device configuration in process automation is to use a standard device desription language, e.g., Electronic Device Description Language (EDDL). The latter is already an industrial standard, used to describe configuration and operational behaviour of devices. It covers the following aspects [14, 3]:

- description of the device parameters, semantically defined by the field device model,
- support of parameter dependencies,
- logical grouping of the device parameters,
- selection and execution of supported device functions, and

- description of the device parameter access method.

A device profile specifies the information necessary such as parameters and value ranges of devices in a profile format. Based on the common format of device profiles, device integration can be achieved. Device integration in the communication environment of Profibus is shown in [3].

For conventional communication as based, for instance, on the HART protocol or 4–20 mA technology, is difficult to provide the large amount of diagnostic and measurement data required for purposes of predictive maintenance. All modern automation systems, however, use certain kinds of fieldbus networks, such as Profibus, Foundation Fieldbus, or CAN. Smart field devices, equipped with microprocessors, connected to such a fieldbus are able to provide these data. Since values of process variables and control signals as well as information on intelligent devices are transmitted via the fieldbus, directly to/from controllers and indirectly to SCADA system, some of these data/information are meaningful to PaM when their relationships are explicit. Additionally, intelligent field devices can provide statistical data about their operation behaviour.

The open, platform- and vendor-independent communication system ACPLT/KS was developed for use in process control and engineering [17]. It is kind of a "Process Control Engineering Internet", designed to interconnect varying information partners. It is suitable for purposes of process control, plant and process supervision, human-process communication, product tracking, and on-line asset management. Actually, ACPLT/KS is a "middleware" between users (clients) and providers (servers) of industrial process data/information. For example, the "Diagnosis" system/component shown in Fig. 7.2 is a client of data/information servers, which provides device information and process data. Via ACPLT/KS, the following can be accessed [17]:

- data (on-line, off-line, parameters),
- structure information and meta-information, and
- objects and structure management.

ACPLT/KS defines five basic object classes whose objects, so-called "Communication Objects", flexibly represent the specialised objects of respective domain-specific systems (DCS, PLC, etc.). The clients of ACPLT/KS can retrieve structural information over the network at any time [17]. The data/information available via ACPLT/KS are used as follows:

- variables provide access to data, may they be process data, parameters, model data, or other,
- domains structure hierarchies, e.g., when assigning technological objects to plant sections,
- histories provide access to archived data-like time series and message logs,
- likes represent typed relations between objects, and

- structures group data (variables).

The ACPLT server was implemented in hardware for Profibus-DP as a so-called AMBox (cp. Chapter II). This way, the data/information listed above can be accessed and used for purposes of proactive maintenance. In addition to device profiles, fieldbus, and AMBox, process data/information can directly be acquired from sensors. Normally, sensors are connected to SCADA systems, where useful historical data in form of statistical logs and reports are produced and stored in databases. On the other hand, the experience of operators and maintenance engineers may not be ignored. They have good understanding of cause-consequence relations of certain kinds of problems, which may be very simple and straightforward, but difficult to acquire and to represent in computer programs.

Communication networks of process automation systems may often be overloaded with data. That does not mean, however, that we have sufficient information. The inconsistency is that, at one hand, we can access amounts of data even bigger than we can handle, but, on the other hand, we may always need extra information to achieve our goals. For example, when we receive a sequence of data representing an abnormal output flow at a electro-hydraulic valve and the associated profile of the valve, it is still not clear whether the valve is wrong, or there is just an erroneous transmission. Diagnosis is often needed to find a problem. Therefore, all information available has to be integrated according to the maintenance view.

Based on the above discussion about industrial automation systems, just simply assembling information at the control and field levels does not enable the middle and management levels to satisfy the performance goals set. Their relationships will tell more. The relationships of information sources are very helpful to integrate information. Information integration is, however, inherently complex because of the following fundamental challenges:

1. Systems running on various operating platforms were developed differently. Therefore, information integration is often extended to enterprise integration which is another challenge, because the de facto standards behind heterogeneous devices and technologies have little in common in terms of data structures.
2. Hardware and software systems are deployed in different networks. Normally, the information in the field network has no direct meaning to systems at higher networks layers. Hence, information integration has to be network-independent.
3. There exists no common information format for various components/systems.

Normally, the aim of integration is usually achieved by using some "middleware". Therefore, file transfers, data sharing, remote procedure calls, and message passing could be supported. From the maintenance point of view, CMMS and SCADA, being located at the middle layer of the automation pyramid, are the pivots to information integration, which vertically forms both lower and higher layers. In addition to vertical integration, horizontal integration is to be considered by integrating the views of control, management as well as maintenance, called CMM concept [13]. Since IM systems are often overloaded with data, an important task of IM is to enable information users to access the right information in a right format at the right time (right context). On the basis of available information sources, we focus on the relationships between information transferred over fieldbuses. These relationships are actually the so-called social knowledge. They are the key to realise IM and KM, which is discussed in the following section.

7.5 Agent-based Infrastructure for Knowledge Management

The architecture of an agent-based platform presented in this chapter seeks to integrate diverse automation systems. Its basic idea is to map industrial automation systems onto an MAS-based platform applying a rôle-based view. Each system/component currently in use in an enterprise can be represented by at least one agent located in the e-maintenance platform. Some agent types will now be defined. Fig. 7.3 illustrates the agent platform, and some intelligent agents designed for maintenance.

Agent Management Core (AMC) is like the Agent Management System (AMS) specified by the Foundation for Intelligent Physical Agents (FIPA) [5], except that AMC provides all services required for integration, including yellow page, white page, addressing services, dynamic regulation service, agent registration as well as information flow control, by employing some web technologies. In the platform shown in Fig. 7.3, AMC is responsible for creation and deletion of agents.

Knowledge Agent (KA) comprises knowledge management functionalities, e.g., diagnosis in e-maintenance that might be supported by existing intelligent tools and social knowledge.

Field Agent (FA) communicates with programmable logic controllers and fieldbuses via corresponding interfaces. In the platform, FAs acquire real-time data, capture errors or failures, and perform first statistical analyses. The resulting information can be used further to form knowledge. For intelligent maintenance, the functionality of recording and analysing the controlled system's performance degradation could also be provided by an FA. Its results will help in developing strategies for predictive maintenance.

Configuration Agent (CA) communicates with software tools for configuration. Normally, a system is configured before being put in use. During its lifecycle, supervisory software tools are monitoring the field processes. To support timely intelligent maintenance and adaptation of control systems to any changes in the field, CAs search the existing case bases located locally, and match them with current (problem) cases.

Diagnosis Agent (DA) mainly deals with failure diagnosis as well as architectural changes at the field level. Both adding new devices and removing failed ones are considered architectural changes. Then, model-based and case-based diagnosis modules can be used for data mining and knowledge discovery.

Management Agent (MA) and ERP systems use the information generated by FAs, DAs, CAs, and KAs (if possible) for maintenance scheduling and to select maintenance teams. They may get in touch with various management systems like ERP systems.

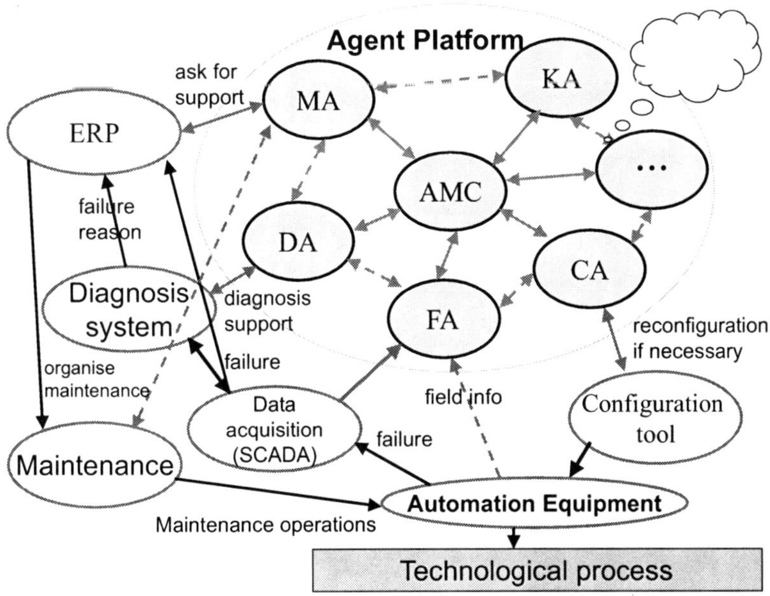

Fig. 7.3: Agent-based platform for e-maintenance

This way, industrial automation processes, business management process, and an enterprise's worldwide interactions can be integrated in a macroview. Internally, each agent is structured into five modules as illustrated by Fig. 7.4:

1. the local knowledge module (LKM) defines the meta-knowledge for the domain knowledge of the automation system/component that is represented by the agent;
2. the social knowledge module (SKM) defines the agent's community associations;
3. the decision-making module (DMM) defines algorithms and modes of decision-making providing, e.g., fuzzy logic, artificial neural networks,

or case-based reasoning; the classical Belief-Desire-Intension (BDI) [16] can be utilised as decision mode;
4. the communication module (CoM) defines communication mechanisms and protocols;
5. the interface module (ItfM) defines the specifications for interfaces and adapters.

Obviously, the two modules LKM and SKM in this agent structure are designed to serve for KM. The types Field Agent and Knowledge Agent are especially designed for field device maintenance. The knowledge about a type of a field device, e.g., its functioning and malfunctioning models, are to be stored in the LKM part of an FA. With the Knowledge Query and Manipulation Language (KQML) [5] the interaction between agents is specified at the language level. The Knowledge Interchange Format (KIF) [5] is used to construct knowledge bases.

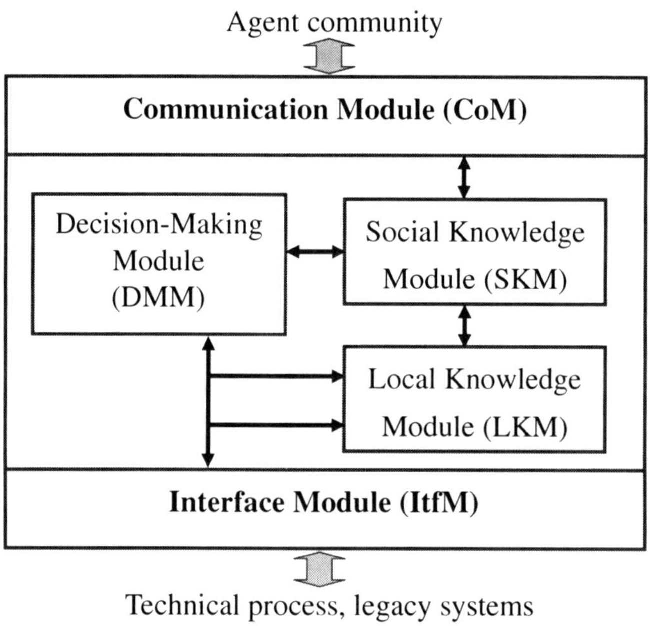

Fig. 7.4: Structure of an agent

Normally, in an agent platform, all agents form a community, in which they share a common name space, domain backgrounds, and get to know each other during information exchange. We express this social knowledge as:

$$K_s = K_1 \bigcap K_2 \bigcap \ldots \bigcap K_n,$$

with n the number of agents in the community. The social knowledge K_s is the common part of each individual knowledge space, by which agents can understand each other. On one hand, it exists in the knowledge bases of the community members, on the other, it supports members to communicate with each other by providing common knowledge to individual members lacking it. Since the interactions among agents in the community embody social behaviours, the social knowledge influences every member's actions to some extent. For instance, agents have to register themselves with the AMC by providing their name, URL, services (functionalities), languages accepted etc. If interactions, e.g., knowledge queries, are expressed in terms of the FIPA Agent Communication Language (ACL) [5], member agents have to communicate with each other by the message format specified in FIPA-ACL. Thus, social knowledge can be specified according to a PaM context. As we understand, the PaM strategy seeks a solution to reduce maintenance cost as well as to improve maintenance efficiency and accuracy. So, in the next section, a top-down approach is discussed, allocating social understanding of PaM into an MAS.

7.6 Knowledge Construction

As outlined in [9], models of system functioning and malfunctioning should be constructed for PaM. But, in fact, it is really hard to construct an exact model of a field device even in collaboration with the manufacturer early in the design phase of the device. Hence, no "exact model" exists. While saying "model", we mean a certain degree of approximation, which depends on our goal. One point is worth meaning, viz., that it is essential to make the functioning and malfunctioning models as simple as possible. As [7] suggested, simplicity is a fundamental principle to design safety-related systems. Here, we regard the functioning and malfunctioning models as black boxes with information/data inputs and outputs (e.g., maintenance requirements), as illustrated in Fig. 7.5.

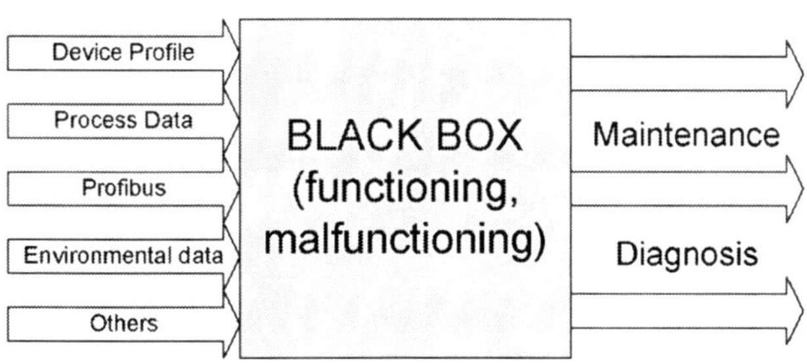

Fig. 7.5: Functioning and malfunctioning models as black boxes

To realise the full potential of PaM, we revisit some features of PdM. The "core" PdM technologies — hardware and software tools monitoring equipment/device condition — are continuously being enhanced by suppliers, but users often feel to be behind the development and not fully using the tools' extended features [20]. Therefore, the complex relations between available data/information should be interpreted in the framework of an explicit cause-consequence table, which constitutes the contents of the black box. At the other hand, the black box is overloaded by the information available on the fieldbus, which is not always suitable. This could be another reason why PdM programs may fall short of their full potential [20], raising the demand for efficient IM and KM to promote PdM to PaM.

Among the several agent types introduced above, FA and KA are the most important ones for KM of PaM. As FA is designed to capture information from the field layer (PLC systems, fieldbus), instances of type FA are essential to implement the integration of field device data, process and control signals. Thus, the corresponding knowledge should be decomposed. Some decomposed parts can be the LKM parts of FA instances. A watchdog agent was designed [8] and implemented on a chip to carry out prognostics [11, 12]. The watchdog agent is similar to our FA but somehow hard-wired, which limits its integrability with other information sources such as sensors, controllers, and human. Therefore, FA instances may be deployed in AM-Boxes. Then, they will have good connections with their working environments, e.g., Profibus communication networks, sensors, controllers, CMMS and SCADA systems.

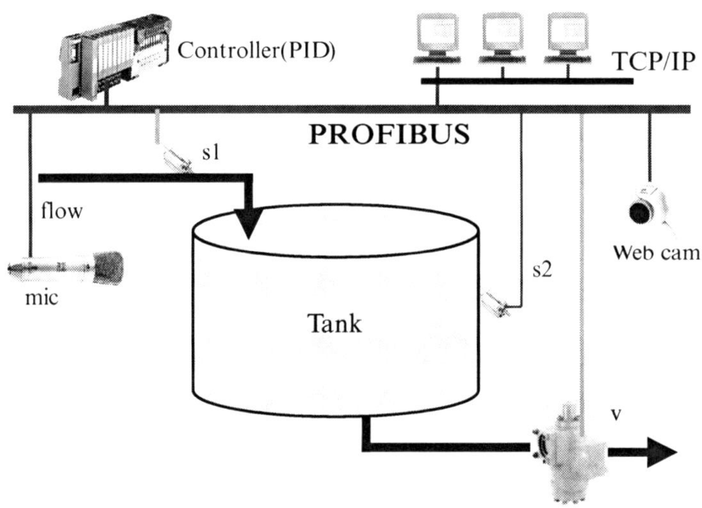

Fig. 7.6: Example

For illustration, we now consider a simplified application with some devices located on a Profibus network as depicted in Fig. 7.6. This small process includes a tank, two intelligent sensors and one intelligent electro-hydraulic valve controlled by a controller running the PID algorithm. The control objective is to keep the level of the liquid in the tank stable as given by the set-point. When the input flow changes, the valve "v" has to react by opening more or less. A camera and a sound sensor are installed to provide further information on the process. Our maintenance target is the valve "v". First, assume that the knowledge about the valve (malfunctioning) is known available and represented in a simple cause-effect table. Therefore, we have the following variables:

fin: input flow sensed by s1;
fout: output flow of v;
lev: liquid level sensed by s2;
ctl: control signal from the controller;
d: time delay of v1's action.

Some cause-effect sequences together with associated information/data are given in Table 7.1. Beyond this, a diagnostic system/component based on artificial intelligence algorithms is necessary to deal with more complex relationships.

Table 7.1: Malfunctions of a valve

Cause	Effects	Associated information/data
contamination	noise	noise captured by microphone; *d* is in normal range; but *lev*, *fin* and *fout* are not stable
	performance degraded	value of *d* increases; *lev* is not stable;
	...	...

7.7 Conclusion

Having the requirements of industrial automation in mind, we tried to give support to knowledge management for proactive maintenance. This chapter outlines an attempt of and some experiences with applying agent technology in industrial systems. Based on an agent-based platform for e-maintenance, we seek to realise KM in terms of IM. On the basis of clarifying the relationship between knowledge and information, this chapter discusses the importance of KM for PaM. By observing information flows in industrial automation, we analyse the requirements of KM for PaM, and both information necessary and available in automation systems. The local knowledge of field agents is represented by relationships between information (e.g., directly available on a fieldbus) and maintenance requirements. On the basis of current research work, we intend to further pursue the decomposition of general maintenance requirements in order to explicitly allocate them to FAs. The next step will then be the consideration of safety issues in diagnosis and performance assessment of automation systems.

Acknowledgement

This work was financially supported by a Matching Funds Scholarship of the German Academic Exchange Service (DAAD) and Institut für Automation und Kommunikation e.V. Magdeburg.

References

1. Antsaklis PJ, Passino KM (1993) Introduction to Intelligent Control Systems with High Degrees of Autonomy. In: An Introduction to Intelligent and Autonomous Control, Kluwer, Boston, pp 1–26
2. Bartos FJ (1999) Predictive Maintenance Maximizes Machinery Health. Control Engineering 46 (7): 51–52
3. Diedrich C, Riedl M, Neumann P, Simon R (2001) Field Device Integration. In: Proc 8th IEEE Int Conf on Emerging Technologies and Factory Automation
4. Ferro BA, Leger JB, Iung B, Pinoteau J (1999) An Innovative Approach for New Distributed Maintenance System: Application to Hydro Power Plants of the REMAFEX project. Computers in Industry 38 (2): 133–150
5. The Foundation for Intelligent Physical Agents (2002) Specifications of FIPA. `http://www.fipa.org`
6. Fischer G, Ostwald, J (2001) Knowledge Management: Problems, Promises, Realities, and Challenges. IEEE Intelligent Systems and Their Applications 16 (1): 60–72
7. Halang WA (2003) Software Dependability Considered as the Main Problem of Embedded Systems. In: Proc IEEE Int Conf on Industrial Technology, IEEE, Piscataway, pp 14–21
8. Intelligent Maintenance System Center (2003) `http://www.uwm.edu/Dept/IME/IMS`
9. Iung B (2003) From Remote Maintenance to MAS-based E-maintenance of an Industrial Process. J of Intelligent Manufacturing 14 (1): 59–82
10. Koc M, Lee J (2001) A System Framework for Next-Generation E-Maintenance Systems. Transaction of Chinese Mechanical Engineer 12 (5)
11. Lee J (1999) Machine Performance Assessment Methodology and Advanced Service Technologies. In: Proc 4th Annual Symp on Frontiers of Engineering. National Academy Press, Washington, DC
12. Lee J, Wang B (1999) Computer-aided Maintenance: Methodologies and Practices.

13. Morel G, Iung B, Galara D, Russo F (1994) Prototyping a Subconcept of Computer Integrated Manufacturing Engineering (CIME): The Integrated Control, Maintenance and Technical Management System (CMMS). Int J on Intelligent Actuation and Soft Computing-Trends in R&D and Applications 2: pp 25–30
14. Neumann P, Diedrich C, Simon R (2002) Engineering of Field Devices Using Device Descriptions. In: Proc IFAC World Congress
15. Polanyi M (1958) Personal Knowledge. Harper & Row, New York
16. Rao AS, Georgeff MP (1995) BDI Agent from Theory to Practice. In: Proc 1st Int Conf on Multi-Agent Systems
17. Rheinisch-Westfälische Technische Hochschule Aachen (2005) Aachener Prozessleittechnik. `http://www.plt.rwth-aachen.de/ks/english`
18. Russell S, Norvig P (2003) Artificial Intelligence: A Modern Approach. Prentice Hall
19. Stenmark D (2002) Information vs Knowledge: The Role of Intranets in Knowledge Management.
20. Swanekamp R (2001) Predictive Maintenance Technologies Come Together. Power Magazine 145 (6): 43–50
21. Tuomi I (1999) Data Is More Than Knowledge: Implications of the Reversed Knowledge Hierarchy for Knowledge Management and Organizational Memory. In: Proc 32nd Annual Hawaii Int Conf on System Sciences, vol 1, p 1071
22. Wagner T (2002) An Agent-oriented Approach to Industrial Automation Systems. In: Kowalczyk et al. (eds) Agent Technologies, Infrastructures, Tools and Applications for E-services, LNAI 2592, pp 314–328
23. Yu B, Singh M, Source P (2002) An Agent-based Approach to Knowledge Management. In: Proc ACM/SIGIR/SIGMIS 11th Int Conf on Information and Knowledge Management, ACM, pp 642–644

8

Using Formal Concept Analysis for Semantic Web Applications

Yi Zhao

Faculty of Electrical and Computer Engineering
FernUniversität in Hagen, 58084 Hagen, Germany
yi.zhao@fernuni-hagen.de

Summary. The explosive growth of information in the Internet and the widespread use of databases have created an immense need for semantic web mining methodologies. The semantic web is envisioned to bring structure to the content of the World Wide Web, and ontologies are seen as the key to realise the vision of the semantic web. The effect the theory of formal concept analysis (FCA) has on semantic web mining, and the contribution of semantic web mining techniques to building concept lattices are described. In more detail, semantic web mining techniques such as ontology building and visualisation based on FCA, and ontology interoperability techniques such as ontology merging and mapping based on FCA are reviewed. Particularly, our ontology mapping method employing rough set theory and FCA is provided to improve the mapping results. The challenge of information retrieval is to extract information from the large amount of data available. Semantic web techniques together with FCA can help to improve the quality of the results retrieved. Furthermore, semantic web techniques can be used for FCA applications. It is evident that there is high potential in studying the relations holding between concept lattice theory and semantic web mining techniques.

Key words: Concept lattices, formal concept analysis, semantic web mining, ontologies, RDF, XML topic map, ontology interoperability, information retrieval.

8.1 Introduction

The semantic web is expected to be the next step that will lead the World Wide Web to its full potential [6]. The explosive growth of information on the Internet and the widespread use of databases have created an immense need for semantic web mining methodologies. Concept lattices provide a natural and formal setting to discover and represent concept hierarchies, and can be considered as classification tools for knowledge units in concept hierarchies that can be used within a knowledge-based system.

Y. Zhao: *Using Formal Concept Analysis for Semantic Web Applications*, Studies in Computational Intelligence (SCI) **42**, 157–176 (2007)
www.springerlink.com © Springer-Verlag Berlin Heidelberg 2007

The semantic web is a vision, viz., the idea of having data on the web defined and linked in such a way that they can be used by machines not just for display purposes, but for automation, integration and re-use across various applications [6]. The main objective of the semantic web is to enable agents to interact with each other without having to specify exchange formats or application programming interfaces a priori, as it is necessary in today's distributed systems. Ontologies play a crucial rôle for the semantic web. They are defined as explicit conceptualisations of domains, organising information into taxonomies of concepts, attributes and the relationships between them, enabling knowledge to be shared and re-used. With the rapid development of the semantic web, it is likely that the number of ontologies will greatly increase over the next few years, and that many will describe similar or overlapping domains, providing a rich source of material. The differences between ontologies comprise mismatches at the syntactic level, for instance, using different representation languages, and mismatches at the semantic level, which include using the same linguistic terms to describe different concepts, or using different terms to describe the same concept.

Since ontologies are seen as the key to realise the vision of the semantic web, the ability to automatically and effectively perform mappings between ontologies is becoming an important requirement. Ontology mapping does not intend to unify ontologies and their data, but to establish correspondences among the source ontologies. A good mapping tool should find the maximum number of potential mapping pairs, and put ontologies in correspondence in such a way that both mathematical structure and ontological axioms are preserved. The problem of mapping different ontologies is well known in knowledge engineering.

The purpose of this chapter is to find mutual utilisations between concept lattices and semantic web mining. The state of the art in applying concept lattice theory and carrying out semantic web mining is given. In particular, we shall discuss how concept lattice theory is employed in semantic web techniques, how both concept lattices and semantic web techniques are applied to information retrieval, and how concept lattices are generated with semantic web techniques.

This chapter is organised as follows. In Section 8.2, the background of concept lattice theory and of semantic web mining is introduced. The application of concept lattices to semantic web mining is presented in Section 8.3. How concept lattice theory and the semantic web are utilised for information retrieval is described in Section 8.4, followed by showing how concept lattices are generated with semantic web techniques in Section 8.5.

8.2 Background

8.2.1 Concept Lattice Theory

Formal concept analysis is a mathematical approach to data analysis. It provides information with structure, and visualises hierarchies by line diagrams. In artificial intelligence, FCA is used as a knowledge representation mechanism and as a conceptual clustering method. In database theory, FCA has extensively been used for design and management of class hierarchies. Concept lattices constitute a mathematical framework which allows building embedded classes from a set of objects. As a machine learning technique, concept lattice theory (CLT) can be used to represent and discover knowledge. The characteristic of CLT lies in reasoning on the possible attributes of data sets. The Hasse diagram of a lattice represents a generalisation/specification relationship between the concepts (nodes). Therefore, lattice and Hasse diagram corresponding to a set of objects described by some properties can be used as an effective tool for symbolic data analysis and knowledge acquisition [39]. Concept or Galois lattices are used to represent conceptual hierarchies, in which each node represents a subset of objects (extent) with their common properties (intent). They are the core of the mathematical theory of Formal Concept Analysis (FCA) [54].

We recall some necessary basic notions about FCA. It starts with a formal context, which is a triple $K := (G, M, I)$, where G (set of objects) and M (set of attributes) are two finite sets, and I is a binary relation between G and M. In order to express that an object g is in a relation I with an attribute m, we write gIm or $(g, m) \in I$, and read "the object g has the attribute m", which is a unique ordered set that describes the inherent lattice structure defining natural groupings and relationships among the elements of G and M. This structure is known as concept lattice.

For a set $X \subseteq G$ of objects, X' can be defined as the set of attributes common to the objects in X, $X' = m \in M \mid gIm,$ for all $g \in X$. Correspondingly, for a set Y of attributes, Y' is the set of objects which have all attributes in Y, $Y' = g \in G \mid gIm,$ for all $m \in Y$. A pair (X,Y), $X \subseteq G$, $Y \subseteq M$ is called a formal concept of the context K, with $X' = Y$ and $Y' = X$. Then, X is called the extent and Y the intent of concept (X,Y).

A partial order relation ($<$) can be defined on the concept sets of the context K, and it can be used to generate the lattice graph: given two nodes H_1 and H_2, if $H_1 < H_2$ and there is no other element H_3 in the lattice such that $H_1 < H_3 < H_2$, then H_1 is called parent of H_2, and H_2 child of H_1. The graph is usually called Hasse diagram, and is a valid visualisation tool to data analysis and knowledge discovery. A Hasse diagram is a directed acyclic graph with an additional constraint: each pair of nodes in the graph has a unique nearest common descendant — or meet ($\wedge$) — and a unique nearest common ancestor — their join ($\vee$). These are referred to as greatest lower bounds, and least upper bounds, respectively. Meet and join of the

BudgetCarRental: service-peer

CarRental	<pickup-info,rate-qualifier, car-class-preference, car-availability, rate, coverage>

HappyTravelAgent: service-peer

AirlineBooking	<air-itinerary, customer-loyalty, price, ticketing, invoice-detail, confirm-itinerary, cancel-flight>
HotelBooking	<hotel-search, hotel-availability, room-price, rate-plan, rate-quote, room-profile, make-reservation, make-cancellation>
CarRental	<pickup-info, rate-qualifier, car-rental-preference, car-availability, rate, coverage, transmission-preference>

FlexiCarRental: service-peer

CarRental	<pickup-info,rate-qualifier, car-class-preference, car-availability, rate-qualifier, rate, coverage, car-make-preference, transmission-preference, air-conditioning-preference, loyalty-programme>

LeisureHotelGroup: service-peer

HotelBooking	<hotel-search, hotel-availability, room-price, rate-plan, rate-quote, room-profile, make-reservation, make-cancellation, shuttle-service, dining-service, health&fitness-service>
GolfCourseReservation	<course-search, course-availability, course-reservation, golf-trait, rate-qualifier, price>

GolfTeeTimes: service-peer

GolfCourseReservation	<course-search, course-availability, course-reservation, golf-trait, rate-qualifier, price>

AirlineTravel: service-peer

AirlineBooking	<air-itinerary, customer-loyalty, price, ticketing, invoice-detail, confirm-itinerary, cancel-flight, meals-&-special-requests, specific-flight-availability, specific-airline-availability>
CreditCheck	<customer-credit-enquiry, charge-customer-credit, process-credit-payment>
InsuranceCoverage	<plan-option, plan-type, insurance-type, obtain-quote, plan-cost>

Fig. 8.1: Sample publications in an e-travel service syndication

lattice are characterised by:

$$\wedge(X_t, Y_t) = (\cap X_t, \cup Y_t) \vee (X_t, Y_t) = (\cup X_t, \cap Y_t)$$

where (X_t, Y_t) is a formal concept.

As an example from [40], Fig. 8.1 illustrates possible publications made by airlines, car rental agencies, hotel corporations, and leisure operators that have published their services as part of an e-travel service syndication. For reasons of brevity we show only six service peers in this figure. Each publication consists of three parts: the peer's (service provider's) name, e.g., BudgetCarRental, its WSDL [58] port-type name, e.g., CarRental, and the name of the operations contained in it. In WSDL, port-types contain the abstract definition of operations. All service providers in the marketplace (and syndication) use standard (unique) names for their port-types and associated operations.

Fig. 8.2 illustrates the concept lattice of the complete sample publication. It was automatically produced from Table 8.1 and includes all missing port-types and operations found in Fig. 8.1.

Table 8.1: Example of a publication context for the travel syndication service depicted in Fig. 8.1

	AirlineBk(ai,cl,p,t,id,ci,cf)	AirlineBk(ai,cl,p,t,id,ci,cf,msr,sfa,saa)	CarRental(pi,rq,ccp,ca,r,c)	CarRental(pi,rq,crp,ca,r,c,tp)	CarRental(pi,rq,ccp,ca,r,c,tp,cmp,acp,lp)	CreditCheck(cce,ccc,pcp)	GolfCourseRes(cs,ca,cr,gt,rq,p)	HotelBk(hs,ha,rpr,rp,rq,rpf,mr,mc)	HotelBk(hs,ha,rpr,rp,rq,rpf,mr,mc,ss,ds,hfs)	InsuranceCov(po,pt,it,oq,pc)
AirlineTravel		1				1				1
BudgetCarRental			1							
FlexiCarRental					1					
GolfTeeTimes							1			
HappyTravelAgent	1			1				1		
LeisureHotelGroup							1		1	

8.2.2 Semantic Web

The great success of the current World Wide Web lead to a new challenge: a huge amount of data is interpretable by humans, only, as machine support is limited. As a possible remedy, Berners-Lee [7] created the idea of the semantic web, which is to annotate web resources with machine-processable meta-data. This extension of the current web is to give data well-defined meaning, better enabling computers and people to work in co-operation. For instance,

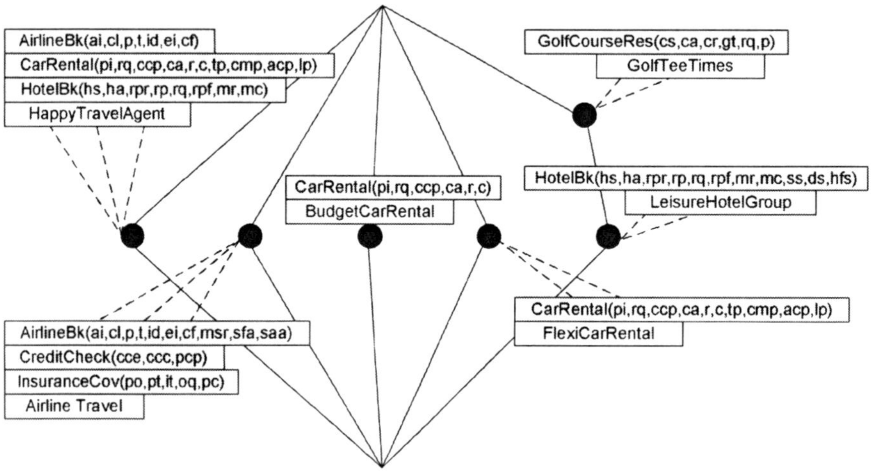

Fig. 8.2: Concept lattice including subtype relationships among published services

today's search engines are already quite powerful, but still return too often too large or inadequate lists of hits. Machine-processable metadata can point search engines to the relevant pages and are, thus, expected to improve both precision and recall.

Ontologies

Ontologies play an important rôle in the semantic web for providing semantic information to assist communication between heterogeneous information repositories, and are seen as the key to realise the vision of the semantic web. The term 'ontology' is borrowed from philosophy, where it refers to a systematic account of what can exist or 'be' in the world. In the fields of artificial intelligence and knowledge representation, the term refers to the construction of knowledge models that specify a set of concepts, their attributes, and the relationships between them, as is defined as a "explicit conceptualisation of a domain" [27].

A core ontology is a tuple $O := (C, is\_a, R, \sigma)$, where C is a set whose elements are called concepts, is_a is a partial order on C, R is a set whose elements are called relations, and $\sigma : R \to C^+$ is function which assigns to each relation its arity [48].

An ontology allows to explicitly specify a domain of knowledge, which permits to access and reason about agent knowledge [19], incorporating semantics into data, and promote its exchange in an explicit and understandable form. With the rapid development of the semantic web, it is likely that the number of ontologies will greatly boom over the next few years. Many formal languages for representing ontologies were proposed, such as XML (eXtensible Mark-up Language) Schema, RDF (Resource Description Framework), XSLT (EXtensible Stylesheet Language Transformations), RDFS (RDF Schema), DAML-ONT (DARPA Agent Mark-up Language-Ontology), OIL (Ontology Inference Layer) [20], or OWL (Web Ontology Language) [15].

Resource Description Framework

The Resource Description Framework (RDF) [55] is the emerging standard to annotate web documents with metadata. These annotations are related to ontologies, declared in RDF Schema (RDFS) [56], which is very close to the Simple Conceptual Graph Model, such that the work on one formalism can easily be generalised to the respective other. An RDF annotation of a web resource consists of a set of statements, each one specifying a value of a property of the resource. A statement is, thus, a triple (subject, predicate, object). The subject is the resource, the entity being described. The predicate is a trait or aspect about that resource, and often expresses a relationship between the subject and the object. The object is the object of the relationship or value of that trait.

XML Topic Maps

As defined in ISO/IEC 13250 [46], XML Topic Maps are used to organise information in a way that can be optimised for navigation. Topic maps were designed to solve the problem of managing large quantities of unorganised information. Information is not useful if it cannot be found or linked to. In publishing, there are several mechanisms to organise and index the information contained within a book or document. Indices allow readers to go directly to the portion of the document that is relevant to their information needs. Topic maps can be thought of as the on-line equivalent of printed indices. Topic maps are also a powerful way to manage link information, much as glossaries, cross-references, thesauri, and catalogues do in paper-based publishing. Topic maps allow users to create a large quantity of metadata and tightly interconnected data. They constitute a kind of semantic network above the data themselves. A new specification which aims to apply the topic map paradigm to the web is called XML Topic Maps (XTM) [53]. XML Topic Maps allow to structure data on the web and, therefore, make web mining more efficient.

8.3 Semantic Web Mining with FCA

A typical task for which concept lattices are useful is to unfold given data, making their conceptual structure visible and accessible in order to find patterns, regularities, exceptions, etc. This may help in ontology processing such as building, structuring, refining, merging, or mapping.

8.3.1 Ontology Building and Visualisation with FCA

Formal concept analysis is a mathematical tool that can express an ontology in a concept lattice. Owing to inference, "the ontology structure is more convenient for people to understand than a tree structure". Furthermore, it can help to find all possible abstractions that may be useful in refining the original classification. This is how FCA can contribute in ontology structuring [28].

In the semantic web, models of web access behaviour can be shared as ontologies. Agent software can then utilise them to provide personalised services such as recommendation and search. So the key problem is to transform web access activities into ontologies. In [60], an approach is proposed to incorporate fuzzy logic into FCA (Fuzzy Formal Concept Analysis, FFCA) to extract data on client-side web usage for automatic ontology generation.

Definition: A fuzzy formal context is a triple $K = (G, M, I = \varphi(G \times M))$, where G is a set of objects, M is a set of attributes, and I is a fuzzy set on domain $G \times M$. Each relation $(g, m) \in I$ has a membership value $\mu(g, m) \in [0, 1]$.

According to the FFCA approach, a web usage context is constructed with fuzzy logic and, then, the fuzzy concept lattice is constructed. After pruning the lattice, the final ontology expressed in OWL [57] is built.

A similar automatic method for ontology generation (Fig. 8.3) is introduced in [43] for the so-called scholarly semantic web. The scholarly ontology contains scholarly information as a hierarchy of research areas from an experimental citation database. A conceptual clustering technique that can cluster the fuzzy concept lattice to generate a concept hierarchy is proposed, then the concept hierarchy can be converted into the scholarly ontology. The proposed framework is used to construct the scholarly semantic web from a citation database. The authors declared that the FFCA framework has been evaluated and good performance has been achieved.

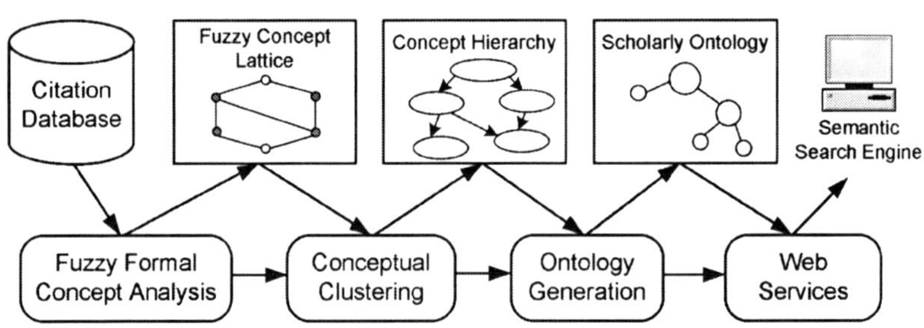

Fig. 8.3: Ontology generation with FFCA

Another automatic ontology generation method based on applying FCA in combination with natural language processing (NLP) is presented in [29].

In this approach, objects for FCA can be any domain-specific texts that use domain-specific vocabulary and describe domain-specific entities. Attributes of an object for FCA are noun-phrases present in the domain-specific texts describing a given domain-specific entity, which are discovered by NLP. Together with a binary relationship, the concept lattice constructed from a context can be considered a formal domain-specific ontology, the authors declared. Actually, a concept lattice is not totally equal to an ontology, since the relationship in ontologies may not be binary.

Table 8.2: Part of a movie database

Attributes/Objects	Genre	MPAA Rating	Awards
Brave Heart	Drama	R	Oscar
Lion King	Action/Adventure	PG	Golden Globe

An improved ontology construction method is put forward by Bashir [5]. It automatically extracts semantic information from a given database schema and structured textual data based on FCA. There are two distinct problems for an ontology-based model: one is the extraction of semantic concepts from a database schema and underlying data, and the other one is constituted by the relationships among these semantic concepts to generate the actual ontology. For a given database with objects and attributes (as shown in Table 8.2 from [5]), the attributes are specific terms, which means that the attributes present non-binary characteristics of objects. A non-binary attribute can be transformed into a binary relation by splitting it into atomic classes and using a new set of expanded attributes to create a cross table. For instance, the attribute 'Genre' can be split into 'Gen-D' and 'Gen-A', and 'MPAA Rating' and 'Awards' can be replaced by 'Rat-R', 'Rat-PG', 'Aw-Os', and 'Aw-GG', respectively (Table 8.3). Finally, an ontology can be generated with the line diagram from the concept lattice in a top-down fashion.

Table 8.3: Movie data with binary relations

Attributes/Objects	Gen-D	Gen-A	Rat-R	Rat-PG	Aw-Os	Aw-GG
Brave Heart	1	0	1	0	1	0
Lion King	0	1	0	1	0	1

In the project *Conceptual E-mail Manager* [11] an ontology, which consists of a hierarchy of concepts (catchwords) together with a lexicon, is used to store e-mail messages. From an FCA perspective, multiple instantiation can be seen as storing the messages in a formal context: the objects are the messages, and the attributes are the catchwords. So, FCA attributes and ontology concepts are considered equivalent. Multi-inheritance is modeled as an additional partial order on the set of attributes to allow the user to retrieve messages via a concept lattice.

OntEx (Ontology Exploration) [23] is an FCA-based ontology creation method. It relies on the knowledge acquisition technique of attribute exploration [22], which has been developed within the mathematical framework of FCA to execute the merging process. OntEx is based on an interactive, top-down knowledge acquisition process, which assures that the knowledge engineer considers all possible cases while avoiding redundant acquisition. In detail, using the concepts of the source ontology and all hierarchical relationships known, it interacts with the knowledge engineer to decide whether further hierarchical relationships hold. If the user agrees, then the relation between the concepts is added, if he denies, then he has to provide a counterexample.

For ontology building, especially for bottom-up building, concept lattices can serve as guideline for developers to carry out abstractions [28]. A concept lattice first classifies given concepts by abstracting common attributes. Once the attributes are determined, the concept lattice can classify concepts mathematically. These classifications are implied by the concepts' attributes objectively, and do not rely on any domain expert. As a guideline for ontology developers, these classifications can be chosen as formal ones. The author of [28] developed the FcaTab plug-in for the ontology editor Protege-2000 [47] to use concept lattices to analyse ontologies. FcaTab can generate an XML file that can be understood by the tool ConExp (Concept Explorer, a lattice drawing application of FCA) [13] to draw a lattice for an ontology that is more expressive and natural in representing the multi-inheritance relationship in a class hierarchy.

In an e-Learning environment, the ontology-based courseware Watchdog [51] supports the user in retrieving and organising distributed courseware resources. There are two kinds of relations in ontology structures: hierarchical and non-hierarchical ones. To display specific hierarchies, Watchdog employs FCA to support visualisation with its conceptual clustering technique, while relational browsing can be used to display non-hierarchies.

8.3.2 Ontology Interoperability with FCA

By themselves, ontologies do not provide semantic interoperability, since a single ontology cannot be used to represent all kinds of domains and applications. Ontology interoperability is, therefore, introduced to achieve knowledge sharing and semantic integration in an environment with different underlying ontologies. In this section, we give a short review on applying concept lattice theory in ontology merging and ontology mapping.

Ontology Merging with FCA

Ontology merging means to create a new ontology from two (possibly overlapping) source ontologies. This concept is closely related to that of integration in the database community. Several approaches have been presented to support the merging of ontologies based on FCA.

Stumme and Maedche have presented the FCA-Merge [48] method for semi-automatic ontology merging following a bottom-up approach. It extracts, for each source ontology, contexts by applying natural language processing techniques, where the objects are the documents, and the attributes are the ontology attributes. The contexts from both source ontologies are merged into a pruned concept lattice with algorithm Titanic [49]. The pruned concept lattice provides a hierarchical, conceptual clustering of the source ontologies' concepts, where concepts are suggested to be merged if and only if they have the same extent. Finally, the lattice is transformed back to an ontology by an ontology engineer.

In [26] an approach based on formal concept analysis and relational concept analysis for ontology merging is proposed to enhance collaboration between different organisations that use different ontologies. It consists of several steps including encoding and mapping. The aim of encoding is to represent each ontology by ontology concepts and ontology relations. The mapping step is designed to find candidate mappings by pre-defined heuristics such as name-similarity between concept names, definition-similarity between concept definitions, hierarchical similarity based on similarity between taxonomical hierarchies and is-a relationships, and relation-similarity matches based on similarity of other taxonomical and ontological relations between concepts. FCA algorithms are used to link two ontologies by producing a merged lattice that combines the two ontologies, thus allowing for a seamless exchange of information. Both concepts and relations are considered in this method, which can cause little loss of information when interoperability is required.

Rather than constructing a Conceptual Graph (CG) [52] to represent an ontology, the authors of [14] assert that the CG formalism is better exploited by using the concept type hierarchy to act as framework for a knowledge base. An unpopulated ontology is represented by the type hierarchy without specific individuals, while a populated ontology is represented by a hierarchy and specific conceptual graphs which instantiate individuals, constraints, situations, or concepts. Ontologies can be compared with concept type hierarchy.

Ontology Mapping with FCA

Given the decentralised nature of the semantic web's development, it is likely that the number of ontologies will greatly increase over the next few years

[18]. Many ontologies may describe similar domains but, with different syntaxes, their representations are not the same. For instance, some ontologies may define "address" with just one item "street name, house no., zip code, city name, country name", while others may define it with five items: "street name", "house no.", "zip code", "city name", and "country name". However, ontologies themselves do not overcome per se any interoperability problems, since it is hardly conceivable that a single ontology is applied in all kinds of domains and applications. As more applications exploit semantic interoperability by employing an increasing number of ontologies developed by diverse communities, the demand for rapid ontology mapping is arising. Ontology mapping defines equivalence relations between similar concepts or relations from different ontologies.

Based on the theory of formal concept analysis, a hybrid similarity measure combining structural and featural similarity measurement based on Tversky's similarity model model is proposed in [17] for the alignment of subdomain ontologies using an upper-ontology:

$$SIM(a,b) = \frac{|(a \vee b)^{\wedge}|}{|(a \vee b)^{\wedge}| + \alpha|(a-b)^{\wedge}| + (1-\alpha)|(b-a)^{\wedge}|} \tag{8.1}$$

This upper-ontology is represented by a lattice, over which some similarity measures can be performed to identify cross-ontology-related concepts. With the typical lattice (structural) operations join ($\vee$) and meet ($\wedge$), the set of common meet-irreducible elements is given by the meet-irreducible elements which are the intent of $a \vee b$, i.e., $(a \vee b)^{\wedge}$. This measure is both structural and featural at the same time, and is based on the structural elements that really contribute to the definition of the lattice.

To further improve the performance of FCA-based similarity measurement, an extension of this measure employing rough set theory is introduced in [59] as follows:

$$SIM(a,b) = \frac{|(a \vee b)_{LA}|}{|(a \vee b)_{LA}| + \alpha|a_{LA} - b_{LA}| + (1-\alpha)|b_{LA} - a_{LA}|} \tag{8.2}$$

In Eq. 8.2, the rough lower approximation is used to take the place of the meet-irreducible operation in Eq. 8.1. The function "| |" represents the cardinality of a set. This approach combines rough set theory and formal concept analysis to determine the similarity of concept nodes from two ontologies on the basis of Tversky's similarity model. The similarity measure is, therefore, both featural and structural in this sense.

Another kind of ontology mapping method, IF-Map [33], is based on information flow theory (IF) [4]. The notions of ontology, ontology morphism, and ontology mapping are formalised and linked to the formal notions of local logic and logic infomorphism stemming from IF theory and FCA. Then, they are applied to map a reference and various local ontologies automatically.

8.4 Information Retrieval with FCA and Semantic Web Techniques

The challenge of information retrieval is to extract information from the large amount of data available. FCA software appears to show a promise for applications in information retrieval. It is, however, not suited for direct manipulations of very large data sources [42]. To compensate for this limitation, the semantic structure can help to improve the results of information retrieval.

In [36], XML topic maps are used to characterise and clean website data by analysing a lattice generated by a classification algorithm — the Galois algorithm. Based on the concept lattice, the web data can be clustered, organised, and visualised in different ways to enhance the users' navigation and understanding of these documents (cp. Fig. 8.4).

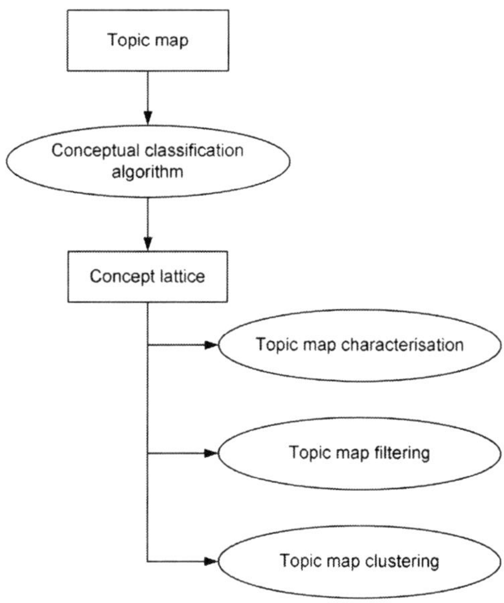

Fig. 8.4: Algorithm to analyse a website

To build a retrieval system incrementally and dynamically, an ontology browsing mechanism combining the incremental knowledge acquisition approach Ripple-Down Rule (RDR) [12] with FCA is introduced [35]. An overview of this system is shown in Fig. 8.5. One of its key feature is the incremental development of a knowledge base via a web-based interface. The formal context used here is (D, K, I), where D is a set of documents, K is a set of keywords, and I is a binary relation between D and K. This key extension to FCA is similar to the philosophy of RDR, where an expert is asked to gradually build up axes of differentiation between objects. The system suggests keywords that the document may have in common with other documents, and also indicates documents that have the same annotations, and which the author may wish to differentiate. These techniques are very simple, but their utility is well established in the other areas of incremental knowledge acquisition.

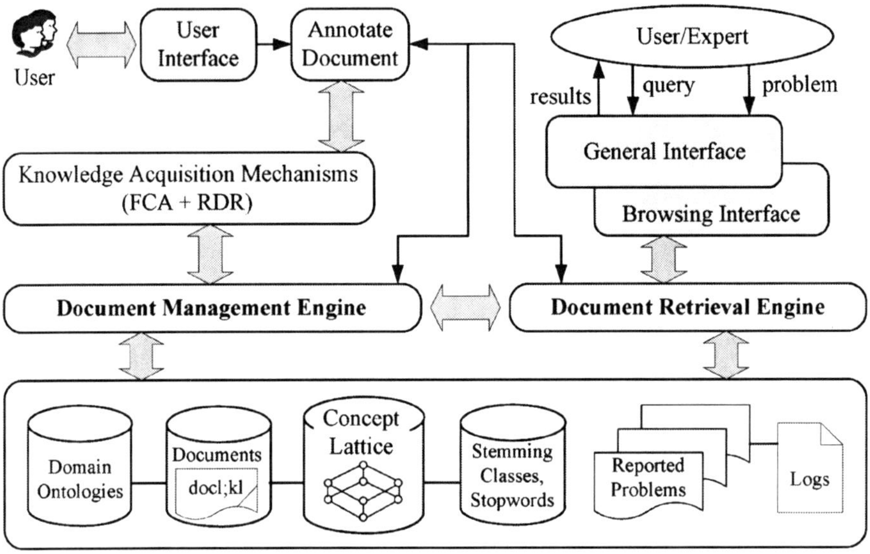

Fig. 8.5: Overview of a system combining RDR and FCA

The JBraindead information retrieval system [9] combines a free-text search engine with on-line FCA to organise the results of a query. Unlike most applications of conceptual clustering to information retrieval, JBraindead is not restricted to specific domains, and does not use manually assigned descriptors for documents nor domain-specific thesauri. Given the ranked list of documents provided by a search, the system dynamically decides on the most appropriate attributes for the given set of documents, and generates a concept lattice on the fly. It selects attributes automatically: first, a number of measures is proposed to evaluate the quality of a concept lattice for the task and, then, the proposed measures are used to compare a number of strategies to automatically select attributes. The results show that concept lattices can be very useful to group relevant information in free-text search tasks.

The Information Flow Framework (IFF) [32] is designed to provide semantic interoperability between various object-level ontologies with concept lattices. Semantic integration of object-level ontologies in IFF is represented with its fusion construction. Assume any IFF language determines a lattice of theories ($fibre(L)$), and any diagram of theories (T) has a base diagram of languages ($base()$) of the same shape, the base diagram of languages $L = base(T)$ is computed first, then the fusion language $L = \sum L$ with the given language fusion cocone $\lambda : L \to L$ is formed, and after that, the diagram of theories T is moved from the lattice of theory diagrams $fibre(L)$ to the lattice of theories using the direct image function to obtain the homogeneous diagram of theories $dir(\lambda(T))$. With these, the meet (union) of the diagram $dir(\lambda(T))$ within the lattice $fibre(L)$ is calculated to obtain the fusion theory. Finally, the language fusion cocone can be obtained.

A method is introduced in [50] to guide the knowledge discovery process between people, documents, and topics combining concept lattices and ontologies. A domain ontology is built to function as a filter to clean the attribute set, and then, the reduced attribute set can be used in concept lattices. By descending in the ontology from the top concept and increasing the granularity, a family of lattices can be built to help organise knowledge that can be extracted from the data.

The system OntoRefiner [45] tries to help users to navigate in numerous retrieved documents after a search querying a semantic portal which integrates a large number of documents. The filtering process is based on a dynamic clustering technique, a concept lattice, and a domain ontology. The Galois lattice structure provides a sound basis for the query refinement process. However, its construction as a whole is a very costly process. Therefore, this system was introduced to avoid the construction of the whole Galois lattice by applying a domain ontology. Navigating through a Galois lattice is similar to a navigating from one query to another. Experimentally, the system was shown to be efficient.

The method Topia proposed by [44] consists in building a helpful and informative hierarchical document structure from the returns of an RDF search, since the hierarchical structure helps readers to see relationships between different pieces of information that would otherwise remain unnoticed. As stated in [44], the RDF structure is a node-edge graph, while the document structure is a sequenced hierarchy. The concept lattice clustering algorithm is one potential means of transforming semantic structure to document structure by treating RDF property value assignments as characteristics of their subjects.

The authors of [2] have implemented the identification of knowledge flows in a tool called BlogTrace. It uses semantic web technology to represent weblogs and linkage structures. A weblog can be viewed as a collection of postings. Each posting has a precise hyperlink on the web, a title, and a publication date. Meaningful terms are extracted from the weblog terms. Co-occurrence matrices, a statistical version of formal concept lattices, are then built by the "occurs in" relation between terms as intent and postings as extent. The mutual concept-dependent linkages between individual bloggers can, therefore, be identified.

The need to support classification and semantic annotation of services constitutes an important challenge for service-centric software engineering [8]. The approach in [8], starting from textual service descriptions in the form of documents formulated in the Web Service Description Language (WSDL), performs an automatic classification and, then, applies formal concept analysis to build a service concept lattice for each specific domain.

8.5 Using Semantic Web Techniques for FCA Applications

"The challenge for research in formal concept analysis is to bring together more complex kinds of knowledge representation with the charm of the natural appearance of a concept lattice out of the Galois connection induced by a binary relation" [10].

Some ontology representation languages are based on Description Logic (DL), a subset of first-order logic [3]. In DL, a knowledge base consists of a Terminological Box (T-Box) and an Assertional Box (A-Box). The T-Box defines concepts and relationships between concepts, whereas the A-Box consists of assertions describing a situation [3]. A multi-context in a knowledge base can be transformed to a binary-relation context through conceptual scaling. Thus, a concept lattice can be derived from a knowledge base with the help of description logic.

The research presented in [16] is dedicated to ontology-guided information retrieval in a corporate memory. The ontology is constituted by documents semantically described by RDF annotations. A particular approach of concept formation, where each concept is defined in extension by a subset of resources, and in intension by a set of descriptions shared by these resources, is introduced. Given the RDF graph gathering all annotations, a concept lattice can be built from the power-context family. In order to deal with the intrinsic exponential complexity of such a task, the concept hierarchy is incrementally built by increasing at each step the maximum size of the RDF resource descriptions.

An important problem in FCA applications is lattice size, which grows, in the worst case, exponentially with the size of a formal context. To reduce the size of lattices, a pre-processing procedure is performed to deal with original large datasets. The vector spaces of the data are firstly extended with concepts of a suitable ontology, such as WordNet [41], then the documents are clustered using a kind of k-Means algorithm. Finally, formal concept analysis is applied directly on the documents. Clustering the objects before applying FCA is, however, an abstraction that might involve loss of information.

8.6 Conclusion

An overview on the mutual utilisations of formal concept analysis theory and semantic web mining was given. In particular, we have described how to employ formal concept analysis theory in semantic web techniques, how to apply both concept lattices and semantic web techniques to information retrieval, and how to generate concept lattices with semantic web techniques. In summary, it is evident that there is still a high potential for studying the relations holding between concept lattice theory and semantic web mining techniques.

References

1. Alberink M, Rutledge L, Veenstra M (2003) Clustering Semantics for Discourse Generation. In: Proc 1st Int Workshop on Hypermedia and the Semantic Web http://homepages.cwi.nl/~media/publications/htsw2003topia.pdf
2. Anjewierden A, de Hoog R, Brussee R, Efimova L (2005) Detecting Knowledge Flows in Weblogs. In: Dau F, Mugnier M-L, Stumme G (eds) Proc 13th Int Conf on Conceptual Structures, Kassel University Press, pp 1–12
3. Baader F, Horrocks I, Sattler U (2005) Description Logics as Ontology Languages for the Semantic Web. In: Hutter D, Stephan W (eds) LNAI 2605, Springer, pp 228–248
4. Barwise J, Seligman J (1997) Information Flow: The Logic of Distributed Systems. Cambridge Tracts in Theoretical Computer Science, Vol 44, Cambridge University Press
5. Bashir FI, Mansoor H (2003) Ontology Construction for Structured Textual Data. http://multimedia.ece.uic.edu/~faisal/downloads/research/cs580-proj6.pdf
6. Berners-Lee T, Fischetti M (1999) Weaving the Web: The Original Design and Ultimate Destiny of the World Wide Web. Harper
7. Berners-Lee T, Hendler J, Lassila O (2001) The Semantic Web. Scientific American 284 (5): 35–43
8. Bruno M, Canfora G, Di Penta M, Scognamiglio R (2005) An Approach to Support Web Service Classification and Annotation. In: Proc IEEE Int Conf on e-Technology, e-Commerce and e-Service, pp 138–143
9. Cigarrán JM, Gonzalo J, Peñas A, Verdejo F (2004): Browsing Search Results via Formal Concept Analysis: Automatic Selection of Attributes. In: Proc 2nd Int Conf on Formal Concept Analysis, pp 74–87
10. Cimiano P, Hotho A, Stumme G, Tane J (2004) Conceptual Knowledge Processing with Formal Concept Analysis and Ontologies. In: Eklund P (ed) LNAI 2961, Springer, pp 189–207
11. Cole R, Eklund P, Stumme G (2000) CEM — a Conceptual Email Manager. In: Ganter B, Mineau GW (eds) LNAI 1867, Springer, pp 438–452
12. Compton P, Jansen R (1990) A Philosophical Basis for Knowledge Acquisition. Knowledge Acquisition 2 (3): 242–257
13. Concept Explorer (2003) Free software. http://www.sourceforge.net/projects/conexp
14. Corbett D (2003) Comparing and Merging Ontologies: a Concept Type Hierarchy Approach. In: Zhong N et al (eds) LNAI 2871, Springer, pp 75–82
15. Dean M, Connolly D et al (eds) (2002) OWL Web Ontology Language 1.0 Reference. W3C Working Draft 29. http://www.w3.org/TR/owl-ref/

16. Delteil A, Faron C, Dieng R (2002) Building Concept Lattices by Learning Concepts from RDF Graphs Annotating Web Documents. In: Priss U, Corbett D, Angelova G (eds) LNAI 2393, Springer, pp 191–204
17. De Souza K, Davis J (2004) Aligning Ontologies and Evaluating Concept Similarities. In: Meersman R, Tari Z (eds) LNCS 3291, Springer, pp 1012–1029
18. Doan A, Madhavan J, Domingos P, Halevy A (2002) Learning to Map Between Ontologies on the Semantic Web. In: Proc 11th Int WWW Conf. `http://citeseer.ist.psu.edu/doan02learning.html`
19. Farquhar A, Fikes R, Pratt W, Rice J (1996) Collaborative Ontology Construction for Information Integration. Stanford University WWW Archive. `http://www.isrl.uiuc.edu/~amag/langev/paper/farquhar95collaborativeOntology.html`
20. Fensel D (2000) The Semantic Web and its Languages. IEEE Intelligent Systems 15 (6): 67–73
21. Fonseca F (2001) Ontology-driven Geographic Information Systems. PhD thesis, University of Maine
22. Ganter B (1996) Attribute Exploration with Background Knowledge. In: Proc Conf on Order and Decision Making. `http://citeseer.ifi.unizh.ch/ganter96attribute.html`
23. Ganter B, Stumme G (2003) Creation and Merging of Ontology Top-levels. In: de Moor A, Lex W, Ganter B (eds) LNAI 2746, Springer, pp 131–145
24. Gauch S, Chaffee J, Pretschner A (2004) Ontology-Based User Profiles for Search and Browsing. Web Intelligence and Agent Systems 1 (3/4): 219–234
25. Gauch S, Pretschner A (1999) Ontology Based Personalized Search. In: Proc 11th IEEE Int Conf on Tools with Artificial Intelligence, Chicago, pp 391–398
26. El Gohary N, El Diraby T (2005) Achieving Ontology Interoperability using Formal Concept Analysis: an Approach to Inter-organizational Collaboration. In: Proc 22nd CIB-W78 Conf on Information Technology in Construction, Dresden. `http://itc.scix.net/data/works/att/w78-2005-D5-2-El-Gohary.pdf`
27. Gruber TR (1993) A Translation Approach to Portable Ontology Specification. Knowledge Acquisition 5 (2): 199–220
28. Gu T (2003) Using Formal Concept Analysis for Ontology Structuring and Building. `http://www.ntu.edu.sg/home5/pg04247176/Using_FCA_for_ontology_building_and_structuring.pdf`
29. Haav HM (2003) An Application of Inductive Concept Analysis to Construction of Domain-specific Ontologies. In: Proc 29th Int Conf on Very Large Databases. `http://dbis-conference.informatik.tu-cottbus.de/vldb2003/h_haav.pdf`
30. Heflin J (2004) Web Ontology Language (OWL) Use Cases and Requirements. `http://www.w3.org/TR/2004/REC-webont-req-20040210/`
31. Heflin J, Hendler J (2000) Searching the Web with SHOE. In: Artificial Intelligence for Web Search, AAAI Press, pp 35–40
32. IEEE Standard Upper Ontology Working Group (2006) `http://suo.ieee.org/IFF/`
33. Kalfoglou Y, Schorlemmer M (2003) IF-Map: An Ontology-Mapping Method based on Information-Flow Theory. Journal of Data Semantics 1 (1): 98–127
34. Kent RE (2003) Semantic Integration in the IFF. `http://www.ontologos.org/Papers/SI2003/SI2003.pdf`

35. Kim M, Compton P (2001) Formal Concept Analysis for Domain-Specific Document Retrieval Systems. In: Stumptner M, Corbett D, Brooks M (eds) Proc 14th Int Joint Conf on Artificial Intelligence, pp 237–248
36. Le Grand B, Soto M, Dodds D (2001) XML Topic Maps and Semantic Web Mining. In: Proc Workshop Semantic Web Mining at ECML/PKDD-2001. http://semwebmine2001.aifb.uni-karlsruhe.de/online/semwebmine08.pdf
37. Madrid JM, Gauch S (2002) Incorporating Conceptual Matching in Search. In: Proc 11th Conf on Information and Knowledge Management. http://hagrid.icesi.edu.co/~jmadrid/docs/KeyConcept-CIKM02.pdf
38. Neumann G, Backofen R, Baur J, Becker M, Braun C (1997) An Information Extraction Core System for Real World German Text Processing. In: Proc 5th Conf on Applied Natural Language Processing, Washington, DC, pp 209–216
39. Oosthuizen GD (1993) Rough Sets and Concept Lattices. In: Proc Int Workshop on Rough Sets and Knowledge Discovery, Banff, pp 24–31
40. Papazoglou MP, Krämer BJ, Yang J (2003) Leveraging Web-Services and Peer-to-Peer Networks. In: Eder J, Missikoff M (eds) LNCS 2681, Springer, pp 485–501
41. Princeton University Cognitive Science Laboratory (1998) http://wordnet.princeton.edu/
42. Priss U (2006) Formal Concept Analysis in Information Science. In: Cronin B (ed) Annual Review of Information Science and Technology 40, pp 521–543
43. Quan TT, Hui SC, Cao TH (2004) FOGA: a Fuzzy Ontology Generation Framework for Scholarly Semantic Web. In: Proc Knowledge Discovery and Ontologies Workshop. http://olp.dfki.de/pkdd04/quan-final.pdf
44. Rutledge L, van Ossenbruggen J, Hardman L (2005) Making RDF Presentable: Integrated Global and Local Semantic Web Browsing. In: Proc WWW2005, Chiba pp 199–206
45. Safar B, Kefi H, Reynaud C (2004) OntoRefiner: a User Query Refinement Interface Usable for Semantic Web Portals. In: Proc European Conf on Artificial Intelligence Workshop on Application of Semantic Web Technologies to Web Communities. http://www.lri.fr/~kefi/articles/ECAI'04.pdf
46. Sowa JF (1984) Conceptual Information Processing in Mind and Machine. Addison-Wesley
47. Stanford Medical Informatics (2005) Protege Ontology Editor. http://protege.stanford.edu/
48. Stumme G, Maedche A (2001) FCA-Merge: Bottom-up Merging of Ontologies. In: Proc 17th Int Joint Conf on Artificial Intelligence, Seattle, WA, pp 225–230
49. Stumme G, Taouil R, Bastide Y, Pasqier N, Lakhal L (2002) Computing Iceberg Concept Lattices with Titanic. J on Knowledge and Data Engineering 42 (2): 189–222
50. Szathmaty L, Napoli A (2004) Knowledge Organisation and Information Retrieval using Galois Lattices. In: Proc Modeling, Computation and Optimization in Information Systems and Management Sciences, Hermes Science Publishing, pp 611–618
51. Tane J, Schmitz C, Stumme G (2004) Semantic Resource Management for the Web: an E-Learning Application. In: Proc 13th World Wide Web Conference, New York. http://www2004.org/proceedings/docs/2p1.pdf
52. Tepfenhart WM (1998) Ontologies and Conceptual Structures. In: LNCS 1453, Springer, pp 334–350

53. TopicMaps.Org XTM Authoring Group (2001) XTM: XML Topic Maps 1.0: TopicMaps.Org Specification
54. Wille R (1982) Restructuring Lattice Theory: an Approach based on Hierarchies of Concepts. In: Rival I (ed) Ordered Sets, pp 445–470
55. World Wide Web Consortium (1999) RDF. http://www.w3.org/TR/REC-rdf-syntax/
56. World Wide Web Consortium (2000) RDFS. http://www.w3.org/TR/2000/CR-rdf-schema-20000327/
57. World Wide Web Consortium (2004) Web Ontology Language (OWL). http://www.w3.org/2004/OWL
58. Web Service Description Language (2001). http://www.w3.org/TR/wsdl
59. Zhao Y, Halang WA, Wang X (2006) Rough Ontology Mapping in E-Business Integration. In: Lu J, Ruan D, Zhang GQ (eds) E-Service Intelligence — Methodologies, Technologies and Applications. Springer book series on Computational Intelligence
60. Zhou BY, Hui SC, Fong CM (2005) Web Usage Mining for Semantic Web Personalization. In: Proc Workshop on Personalization on the Semantic Web, pp 66–72. http://www.win.tue.nl/persweb/Camera-ready/9-Zhou-full.pdf
61. Zhu X, Gauch S, Gerhard L, Kral N, Pretschner A (1999) Ontology-based Web Site Mapping for Information Exploration. In: Proc 8th Int Conf on Information Knowledge Management, pp 188–194

9

A Fair Off-line Electronic Payment System

Weidong Qiu

School of Information Security Engineering
Shanghai Jiaotong University, 200030 Shanghai, P.R. China
qiuwd@sjtu.edu.cn

Summary. Some privacy-protecting electronic payment systems have been proposed to realise revocable anonymity among users. Building on these results, a fair off-line electronic cash system (e-cash, for short) is presented featuring that all users are organised into a single group. This novel e-cash system is more efficient and more secure. It is able of coin tracing and owner tracing. The anonymity of the system can be revoked under certain conditions by an off-line trusted third authority. The system is untraceable (under normal cases), unlinkable, and resistant to double spending. Furthermore, compared with the previous e-cash systems, this payment protocol is computationally more efficient.

Key words: Electronic payment, electronic cash, privacy protection, revocable anonymity, fair off-line payment.

9.1 Introduction

Ubiquitous computing is one of the most promising technologies arising in recent years. It pervades the users' enviroment by taking advantage of multiple homogeneous or heterogeneous information services including stationary and mobile ones, seamlessly interconnected through wired or wireless networks, to provide a class of computing which can take place anytime and anywhere. Ubiquitous computing facilitates electronic banking and commerce and, thus, lets them develop more rapidly. Financial transactions are a core element in this business, as most of us check their account balances or carry out payments using on-line Internet services. Accepting electronic payments is essential to electronic commerce. For low-cost airlines, for instance, accepting electronic payments for tickets through credit card or direct debit authorisation is an integral component of their business model.

Various secure network payment schemes have been developed by universities, research institutes, and commercial organisations [6, 17, 18, 22]. Some of these solutions underwent small-scale testing, and some of them

W. Qiu: *A Fair Off-line Electronic Payment System*, Studies in Computational Intelligence (SCI) **42**, 177–195 (2007)
www.springerlink.com © Springer-Verlag Berlin Heidelberg 2007

proved to fail, because unconditional privacy protection systems could be misused by criminals for blackmailing or money laundering [31]). Therefore, new technologies, including new security tools, new cryptographic algorithms, and new protocols need to be developed to protect privacy during transaction, and to make payment systems more secure, more efficient, and more acceptable to organisations and individuals.

In this chapter, the core characteristics of electronic payment systems are described, the current electronic payment systems are classified, and novel results on a secure and completely anonymous electronic cash (e-cash) payment system are presented, which allows to revoke anonymity under certain suspicious cases and, thus, to prevent the system from being misused by criminals.

9.2 Characteristics of Electronic Payment Systems

Several attempts can be found in the literature to describe the core characteristics of electronic payment systems [1]. Most of them are presented from a technology point of view. In this section, we summarise the main characteristics of a general electronic payment system:

Applicability For a payment system it is decisive that it can be applied in real-life situations for payments over open networks. Whether an electronic payment system is applicable depends on the extent to which customers and merchants accept it. Applicability of an electronic payment system may vary from country to country, depending on the cryptographic tools used in the system and its overall efficiency.

Usability Paying with an electronic payment system should not be a complex task. Therefore, the usability of electronic payment systems is a key to acceptance. In other words, a usable electronic payment system will be easy for consumers to operate.

Security Since Internet services are provided today over open networks, the infrastructure supporting electronic commerce and payment systems must be particularly resistant to security attacks on data integrity, confidentiality, authenticity, and undeniability.

Reliability Electronic payment systems should be stable and run smoothly; otherwise nobody would trust and deploy them.

Anonymity A consumer's identity and personal information should be kept secret. It should not be possible to discover a customer's identity, or to trace an individual's purchasing activities. For electronic payment systems, anonymity is a necessity to protect a consumer's privacy. Some cryptographic tools, such as group signature [10] and blind signature [14], are utilised for this purpose.

Revokable Anonymity Although anonymity is a very important issue in electronic payment systems, under certain conditions economic criminals may misuse this property for blackmailing or money laundering

[31]. So, it is strongly recommended to develop anonymity-revocable systems, i.e., under some suspicious circumstances the anonymity of a person involved in an electronic payment transaction can be removed to discover his or her identity. Such systems are called fair electronic payment systems.

Undeniability For electronic payment systems, any transaction carried out should not be deniable.

Scalability As the commercial use of the Internet grows, the demands placed on payment infrastructure will increase, too. The payment infrastructure as a whole should scale to be able to handle additional users and merchants, while avoiding performance degradation.

Interoperability A payment system is interoperable if it does not depend on a single organisation, is open, and allows as many interested parties to join as necessary. This can be achieved by means of open standards for the technology used. Examples of initiatives fostering interoperation are the CAFE project [4] and the European Union's ACTS Project SEMPER [29].

In this chapter, we concentrate on electronic cash systems, which implement one important form of electronic payment systems. Most electronic cash systems proposed so far satisfy most of the requirements listed above except revocable anonymity. We shall address this problem by making use of properties of group signatures so that we can recover the identities of malicious consumers.

9.3 Classification of State-of-the-Art Payment Mechanisms

Before going further into the details of our electronic payment system, we present a classification of such applications. This will be helpful to understand various electronic payment systems including the electronic cash system proposed here.

There are several ways to distinguish between different electronic payment systems. Based on the way in which money transfer is organised, or based on the type of information that is exchanged [35], existing electronic payment systems can be classified into four groups (cp. Fig. 9.1): credit card-based ones [3], electronic cheque [30], electronic cash [23], and micropayment systems [21]. Furthermore, electronic cash systems may be subdivided into on-line and off-line ones, according to whether banks are involved or not in payment transactions. On-line systems are not generally acceptable, since the banks will be bottlenecks in the systems. Therefore, most researchers focus on off-line systems.

Electronic cash resembles conventional cash, where parties exchange electronic tokens that represent values, just as banknotes do. In credit card-based systems, data on bank accounts of customers are transferred over open

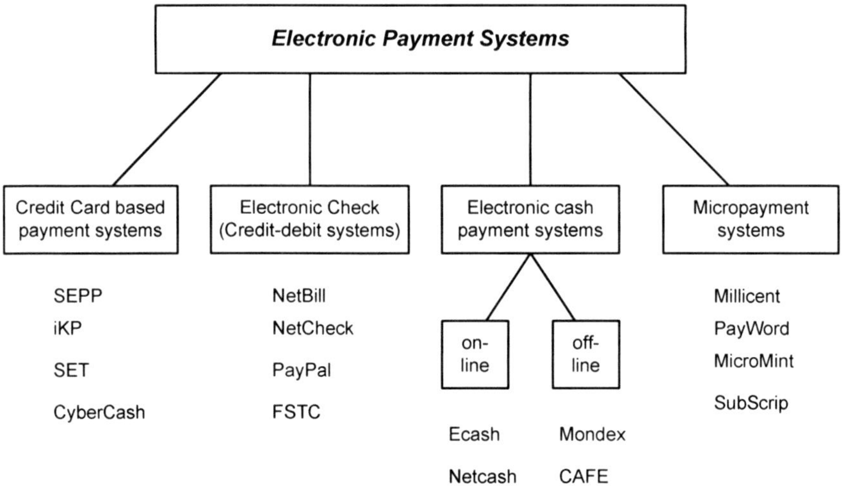

Fig. 9.1: Classification of electronic payment systems

networks, and money is represented by numbers in the accounts. Micropayment systems are a special group in which the monetary value of transactions is small and, therefore, lower security requirements need to be satisfied.

9.4 A Privacy Protecting Fair Electronic Cash System

We focus our work on electronic cash systems for its huge potential market in the future. Electronic cash systems are similar to the national currencies in circulation. Thus, they are relatively easily accepted by the general public. But different from these currencies, there is the danger that electronic coins could be spent more than once, i.e., that a malicious user could make more than one purchase with the same electronic coin. Therefore, resistance to double spending is a necessary requirement for electronic cash systems. A simplified diagram showing the model of electronic cash systems is shown in Fig. 9.2.

An electronic cash system consists of a set of protocols between three or four interacting parties, viz., a bank, a customer (the payer), a merchant (the payee), and in some cases a trusted third party. The customer and the merchant have both an account with the bank. The goal of the system is to transfer money securely from the customer's account to the merchant's account to make a deal. In doing so, five different phases can be identified:

1. a registration phase involving the third party, the bank, and the customer,
2. a withdrawal phase involving the bank and the customer to gain coins,

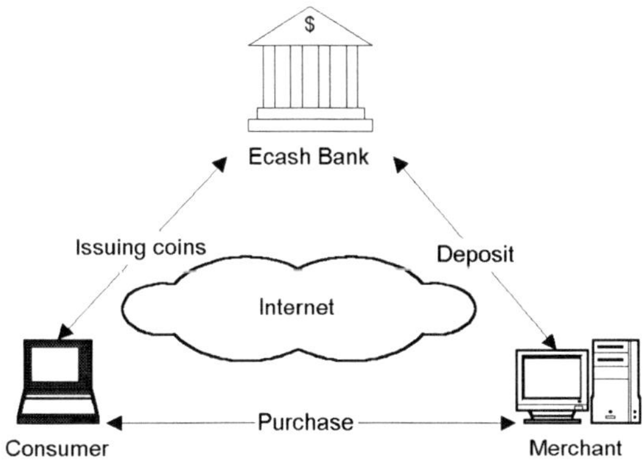

Fig. 9.2: Electronic cash architecture

3. a payment phase involving the customer and the merchant,
4. a deposit phase involving the merchant and the bank, and
5. a tracing phase involving the merchant, the bank, and the third party.

In an off-line system, each phase occurs in a separate transaction, whereas in an on-line system, payment and deposit take place in a single transaction involving all three parties bank, merchant, and customer. These three have different security requirements. The bank wants to make sure that for any account credited, another account has been debited. The merchant, receiving a payment, wants to be assured that the bank will accept to credit his or her account with the received amount. Finally, the customer wants to be sure that coins he or she has withdrawn will be accepted for payment. Furthermore, the customer may require that his or her privacy to be protected.

Many electronic payment systems offering unconditional anonymity have been proposed in recent years (cp., e.g., [5, 7]). These systems protecting privacy unconditionally may, however, be misused by criminals for blackmailing or money laundering [31]. Therefore, it has been proposed that anonymity revokable systems should be designed to prevent misuse. They involve trusted third parties such as governments or authorities of law enforcement.

The concept of fair e-cash systems was put forward independently by Brickell [6] and Stadler [32] in 1995. It offers a compromise between the need to protect the privacy of users and to effectively prevent the misuse by criminals. On one hand, neither banks nor merchants can obtain the identities of users on their own. But on the other hand, under some special circumstances, e.g., when criminal activities are suspected, with the help of a bank a trusted third party can revoke the anonymity of the transaction and reveal the user's identity.

Based on the system of Brands [5], Brickell proposed a fair e-cash system [6], in which a trustee must be involved in the transactions. Camenisch extended his anonymous payment system [7] to be a fair one [8]. In Camenisch's system, users have two accounts in their bank: a personal and an anonymous one. In addition, it is an on-line system, which means that all three parties bank, merchant, and user are involved in a single transaction. Camenisch improved the system's efficiency in [9]. After initially publishing a public key, the trustee needs neither be involved in the opening of an account nor in any withdrawal or deposit transaction.

In 1996, Frankel, Tsiounis, and Yung proposed a fair off-line e-cash system (FOLC) [18]. Davida et al. improved the efficiency of the FOLC system [17]. Using interactive knowledge proofs, these systems need more communication among bank, users, and merchants. Furthermore, both systems of Camenisch and Davida [9, 18] have to search in a database at the bank to be able of owner tracing. This, however, contradicts practice, where the trusted third party should be responsible for revoking anonymity rather than the bank. Yang and Wang proposed a fair electronic payment system with a hardware-based electronic wallet [36], on which there is an anonymous account where users can store their money withdrawn from their accounts in a bank. The hardware-based wallet is composed of a PC and a smart-card rendering it somewhat inconvenient and expensive for the users. Sander, Pfitzmann, and Kügler proposed a new way to offer conditional anonymity without any trustees [22, 26, 28]. They abandoned the idea of a trusted third party (or trustees) involved. These systems are not based on blind signatures, and their security relies on the ability of the bank to maintain the integrity of a public database [28]. Furthermore, these systems lack the ability of tracing, e.g., owner tracing and coin tracing. Kügler's system [22] solves the tracing problems, but is an on-line one, which has been considered to be inefficient and vulnerable as well having a bottleneck.

Since the group signature scheme was first introduced by Chaum and van Heyst in 1991 [15], many different group signatures have been proposed [10, 12, 11, 16]. A group signature scheme allows the members of a group to sign messages anonymously on behalf of the group. The group has a common open public key, which can be used to verify the signature without revealing any information of the signer's identity. Furthermore, the group manager has the ability to remove the anonymity of signees.

In 1999, Traoré presented a new group signature scheme [33], and constructed a privacy-protecting fair off-line e-cash system. In this system, the properties of group signatures are used to realise the revokable anonymity of electronic cash. But the bank can trace issued coins as, during the execution of the payment protocol, the bank can obtain all information needed to re-compute users' electronic coins. In addition, the efficiency of the group signature scheme used is low.

Therefore, in this chapter, utilising Camenisch's highly efficient group signature scheme [12], we devise a new fair off-line e-cash system being much more efficient than Traoré's, in particular with respect to the payment protocol. It requires less communication by using non-interactive proofs of knowledge. The anonymity of users can be revoked in this system resistant to double-spending, and it has the ability to trace the owner of an electronic coin (owner tracing).

9.4.1 Notation and Building Blocks

In this subsection, we describe some notations and building blocks for proofs of knowledge which will be used throughout this chapter:

- $x =?y$ is a predicate determining whether x and y are equal or not.
- $N = \{n \mid n = pq,\ p < q,\ p = 2p' + 1, q = 2q' + 1\}$ is a set where p, q, p', q' are all prime numbers represented with equal bit length.
- $[a, b]$ denotes the set $\{a, a + 1, ..., b - 1, b\}$.
- m denotes a plain message that is related to the signature of proof of knowledge, and may be the empty string ζ.
- $c[i]$ denotes the i-th bit position of the string c.
- $x \in_R E$ means that x is chosen uniformly at random from the set E.
- $G =< g >=< g_1 >=< g_2 >$ is a multiplicative group with the binary representation of its order having the bit length l_g; g, g_1, g_2 are the group's generators.
- $\epsilon,\ l_g,\ l_1,\ l_2,\ k,\ \tilde{l}$ are security parameters, where k is a fixed output length of a secure one-way hash function $H(\cdot) : \{0,1\}^* \rightarrow \{0,1\}^k$; l_g is the bit length of the group's order; $\epsilon > 1$, $k, l_1, l_2 < l_g$, $\tilde{l} = \epsilon(l_2 + k) + 1$.
- $M(G, z) = \{(u, e) | z = u^e, u \in G, e \in \{2^{l_1} - 2^{l_2}, \cdots, 2^{l_1} + 2^{l_2}\}, e \in primes\}$.

According to the literature [12], signatures on proof of knowledge (SPK for short) are signature schemes derived from statistical zero-knowledge proofs of knowledge[1]. Under some assumptions (e.g., modified strong RSA assumption and Diffie-Hellman decision assumption [12]), Fujisaki and Camenisch proved that the schemes considered in this chapter are valid honest-verifier statistical zero-knowledge proofs of knowledge [19].

In the sequel, we introduce three building blocks, borrowing some notations from [12]. The first two show the knowledge of a discrete logarithm that lies in a certain interval. The third one, group-signature block, shows how a group member can generate the group's signatures.

Definition 1. *A proof of knowledge of the discrete logarithm of y with respect to g, which also proves that $log_g y$ is in $[2^{l_1}\ , 2^{l_1} + 2^{l_2}]$, is a pair $(c, s)\ \in\ \{0,1\}^k \times [-2^{l_2+k},\ 2^{\epsilon(l_2+k)}]$ satisfying $c\ =\ H(g||y||g^{s-c2^{l_1}}y^c||m)$. This proof of knowledge is denoted by*

$$SPK\{(\alpha) : y = g^\alpha \wedge (2^{l_1} - 2^{\epsilon(l_2+k)+1} < \alpha < 2^{l_1} + 2^{\epsilon(l_2+k)+1})\}(m).$$

[1] The approach to security analysis on these schemes, defined as random oracle model, was formalised in [27]. As shown by Camenisch [12], although it was proven that this method does not work on general protocols, the schemes considered here are still believed to be valid.

If the discrete logarithm $x \in [2^{l_1}, 2^{l_1} + 2^{l_2}]$ is known such that $y = g^x$ holds, we can compute that proof as follows:

- choose $r \in_{\mathbb{R}} \{0,1\}^{\epsilon(l_2+k)}$ and compute $t := g^r$,
- compute the pair $c := H(g||y||t||m)$ and $s := r - c(x - 2^{l_1})(in\mathbb{Z})$.

A verifier can check whether the equation $c =?H(g||y||g^{s-c2^{l_1}}y^c||m)$ and whether $s \in [-2^{l_2+k}, 2^{\epsilon(l_2+k)}]$ hold to decide on accepting the proof.

By adding a discrete logarithm knowledge proof, the following definition is easily obtained from Definition 1.

Definition 2. *A proof of knowledge of the discrete logarithm of y_1 with respect to g, and of the discrete logarithm of y_2 with respect to h, which also proves that $log_g y_1$ is in $[2^{l_1}, 2^{l_1} + 2^{l_2}]$, is a triple $(c, s_1, s_2) \in \{0,1\}^k \times [-2^{l_2+k}, 2^{\epsilon(l_2+k)}] \times \mathbb{Z}$. We denote this proof of knowledge by*

$$SPK\{(\alpha,\beta) : y_1 = g^{\alpha} \wedge y_2 = h^{\beta} \wedge (2^{l_1} - 2^{\epsilon(l_2+k)+1} < \alpha < 2^{l_1} + 2^{\epsilon(l_2+k)+1})\}(m).$$

Definition 3. *A septuple $(c, s_1, s_2, s_3, a, b, d) \in \{0,1\}^k \times [-2^{l_2+k}, 2^{\epsilon(l_2+k)}] \times [-2^{l_g+l_1+k}, 2^{\epsilon(l_g+l_1+k)}] \times [-2^{l_g+k}, 2^{\epsilon(l_g+k)}] \times G^3$ satisfying*
$c = H(g||h||y||z||a||d||z^c b^{s_1-c2^{l_1}} / y^{s_2}||a^{s_1-c2^{l_1}}/g^{s_2}||a^c g^{s_3}||d^c g^{s_1-c2^{l_1}} h^{s_3}|| m)$
is a group signature of a message $m \in \{0,1\}^*$ with respect to the knowledge $u^e = z$ (see [12]), and can be denoted by:

$$\begin{aligned} SPK\{(\eta,\vartheta,\xi) \; : \; z &= b^{\eta}/y^{\vartheta} \wedge a^{\eta}/g^{\vartheta} \\ &= 1 \wedge a \\ &= g^{\xi} \wedge d \\ &= g^{\eta}h^{\xi} \wedge (2^{l_1} - 2^{\epsilon(l_2+k)+1} < \eta < 2^{l_1} + 2^{\epsilon(l_2+k)+1})\}(m). \end{aligned}$$

As a member of the group, one can sign a message $m \in \{0,1\}^*$ on the group's behalf as follows:

1. choose an integer $w \in \{0,1\}^{l_g}$, and compute $a := g^w, b := uy^w, d := g^e h^w$,
2. choose $r_1 \in_{\mathbb{R}} \{0,1\}^{\epsilon(l_2+k)}, r_2 \in_{\mathbb{R}} \{0,1\}^{\epsilon(l_g+l_1+k)}, r_3 \in_{\mathbb{R}} \{0,1\}^{\epsilon(l_g+k)}$,
3. compute $t_1 := b^{r_1}(1/y)^{r_2}, t_2 := a^{r_1}(1/g)^{r_2}, t_3 := g^{r_3}, t_4 := g^{r_1}h^{r_3}$,
4. compute $c := H(g||h||y||z||a||b||d||t_1||t_2||t_3||t_4||m)$, and
5. compute $s_1 := r_1 - c(e - 2^{l_1}), s_2 := r_2 - cew, s_3 := r_3 - cw, s_1, s_2, s_3 \in \mathbb{Z}$.

The septuple $(c, s_1, s_2, s_3, a, b, d)$ is the result of group signature $SPK\{(e, we, w)\}(m)$. It is easy to show that this result satisfies the equation in Definition 3. Whether the equation $c = H(g||h||y||z||a||d||z^c b^{s_1-c2^{l_1}}/y^{s_2}||a^{s_1-c2^{l_1}} /g^{s_2}||a^c g^{s_3}||d^c g^{s_1-c2^{l_1}} h^{s_3}|| m)$ holds will indicate to accept the signature or not. In addition, the interval of s_1, s_2, s_3 should also be checked when verifying the group signature.

Before we turn to our electronic cash system, we introduce some hard problems and two assumptions on which the security of our system is based. These hard problems are believed to be infeasible to solve.

Discrete Logarithm Problem Given an element g in a finite group G and another element $h \in G$, find an integer x such that $g^x = h$.

Modified Strong RSA Problem Given $G,\ z \in G,\ M \in M(G, z)$, find a pair (u, e) such that $u^e = z,\ e \in \{2^{l_1} - 2^{l_2}, \cdots, 2^{l_1} + 2^{l_2}\}$ and $(u, e) \notin M$.

Modified Strong RSA Assumption Any probabilistic polynomial-time algorithm solves the *Modified Strong RSA Problem* only with negligible probability.

Diffie-Hellman Decision Problem Given a group G, a generator g of G, and three elements $a, b, c \in G$, decide whether there exist integers x, y such that $a = g^x,\ b = g^y$, and $c = g^{xy}$.

Diffie-Hellman Decision Assumption The two distributions

- G, g, g^x, g^y, g^{xy}
- G, g, g^x, g^y, g^z

are computationally indistinguishable. Where g is a generator of group G and x, y, z are chosen at random from $\{1, \cdots, |G|\}$.

9.4.2 Fair and Off-line E-Cash System

Basically, an e-cash system is composed of a set of three protocols in which the three kinds of participants user, merchant, and bank play rôles: the *withdrawal protocol* involves user and bank, the *payment protocol* user and merchant, and the *deposit protocol* merchant and bank. In an off-line system, these three protocols proceed separately, whereas in the on-line case, the bank participates in a single transaction in which payment and deposit protocols take place simultaneously. In the fair system presented here, one more party, viz., the trusted third party, and two more protocols, for registration at the trusted third party, and the tracing protocol for revoking anonymity, are added.

In analysing the security of an e-cash system, the following aspects need to be considered:

- anonymity — or revokable anonymity for fair e-cash systems,
- unforgeability,
- undeniability,
- unlinkability,
- double-spending resistance.

Unforgeability ensures that nobody can produce coins in the name of others. Unlinkability is important in the sense that nobody can judge whether two coins were produced by the same user.

Just as mentioned before, a fair e-cash system should keep users anonymous. On the other hand, under certain conditions the system should be able to revoke the anonymity of users to help tracing suspects. This allows the trusted third party, e.g., a law enforcement authority, to identify those users who are misusing electronic coins for criminal activities, while protecting the privacy of the honest ones. For this reason, our system needs to comprise an additional tracing protocol. Generally, two mechanisms exist to revoke the anonymity of users:

Coin tracing given the bank's view of the withdrawal protocol, the trusted third party can compute the electronic coin and trace it, allowing to trace blackmailed electronic coins;

Owner tracing to prevent money-laundering in privacy-protecting e-cash systems, with the bank's view of the payment protocol, the trusted third party can reveal the owner's identity of the suspicious electronic coins.

In the sequel, we propose a fair off-line e-cash system using the building blocks given in the previous section. As in [33], all users are organised into a group and the trusted third party acts as group manager responsible for removing the anonymity of users.

System Parameters

To establish the system's parameters, the **Trusted Third Party (TTP)** performs the activities

1. Choose a cyclic group $G = < g >$. Let l_g be the bit length of the group's order, and let it be large enough. A possible group can be a cyclic subgroup of $\mathbb{Z}^*_{n'}$, where $n' = pq,\ \ p, q \neq 1(mod\ 8),\ \ p \neq q\ (mod\ 8),\ \ p = 2p' + 1,$ $q = 2q' + 1$ and p, q are large random primes with order nearly $2^{l_g/2}$. The method to prove n' to be a product of two large primes can be found in [13].
2. Choose two integers z and h at random from the group G, satisfying the *Modified Strong RSA Assumption* and *Diffie-Hellman Decision Assumption*[2].
3. Select randomly an integer $x \in \{0, \ldots, 2^{l_g} - 1\}$ as secret key, and then compute the public key $y := g^x$.
4. Select a secure one-way hash function $H(\cdot)$.
5. Choose security parameters: $\hat{l},\ l_1,\ l_2,\ \epsilon$. An example set of parameters is[3]:
$$\epsilon = \frac{9}{8},\ l_g = \hat{l} = 1200,\ l_1 = 860,\ l_2 = 600,\ k = 160.$$
6. Publish $z,\ g,\ h,\ G,\ l_g,\ y$, and prove that z, h were chosen randomly.

and the **Bank (B)** the following ones:

1. Choose an RSA modus n, a public exponent e_b, and the corresponding private key d_b, where $e_b,\ d_b$ satisfy $e_b \cdot d_b \equiv 1\ mod\ (\varphi(n))$.
2. Publish n, e_b.

[2] Further information on these assumptions and on security analysis of the building blocks and group signature scheme can be found in [12].

[3] More choices can be found in [24, 25].

Registration Protocol at the Trusted Third Party

First of all, users should register at the TTP to obtain a membership license of the group. The protocol is shown in Fig. 9.3. The license will be used in the withdrawal protocol, in the payment protocol, and in the process of registration protocol at the bank. We derived the registration protocol from the member registration protocol of Camenisch's group signature scheme. We do not need blind signatures in our registration protocol. Users still need to prove that $\tilde{e}$ is the product of two primes (the methods can be derived from [34]). Furthermore, $\hat{l}, l_1, l_2$ must be chosen so that $\tilde{e}$ cannot be factored. The range constraint on e is to prevent dishonest users from generating the licenses themselves without the TTP being involved [12]. The registration protocol in the TTP also ensures that the third party cannot learn the secret e and cannot counterfeit the honest users. In addition, without knowing the factoring of n', dishonest users cannot obtain legal licenses by themselves.

User | **TTP**

User chooses
- $\hat{e} \in_R \{2^{\hat{l}-1}, \cdots, 2^{\hat{l}} - 1\}$
- $e \in_R \{2^{l_1}, \cdots, 2^{l_1} + 2^{l_2} - 1\}$
- $\hat{e}, e \neq 1 \ (mod\ 8),\ \hat{e} \neq e \ (mod\ 8)$

computes:
- $\tilde{e} := \hat{e}e \ (mod\ n')$
- $\tilde{z} := z^{\hat{e}} \ (mod\ n')$
- $WW = (c, s_1, s_2) := SPK\{(\alpha, \beta)\}(\tilde{z})$

User → TTP: $\tilde{e},\ \tilde{z},\ WW$

TTP:
- verifies the proofs of knowledge WW and makes sure that the user chooses the $\tilde{e}, \tilde{z}$ appropriately
- computes $u := \tilde{z}^{1/\tilde{e}}(\ mod\ n')$
- stores $(u, \tilde{e}, \tilde{z})$ and user's identity

TTP → User: u

- checks the equation $\tilde{z} \equiv u^{\tilde{e}} (mod\ n')$ to make sure that $z \equiv u^e (mod\ n')$ holds
- (u, e) is the license.

Fig. 9.3: Registration protocol at the trusted third party

Registration Protocol at the Bank

Now, users should establish an account at the bank where they deposit and withdraw money. To do so, users and the bank carry out the protocol as shown in Fig. 9.4. In this, YY is verified to make sure that a user has been registered at the trusted third party, and that his or her identity has been authenticated by the TTP. The user's account ID_U is anonymous.

User	**Bank**
- chooses randomly $x_U \in_R \{0, \cdots, l_g\}$ - computes $ID_U := g^{x_U}$ - computes:	
$ZZ = (c, s) := SPK\{(x_U) : ID_U = g^{x_U}\}(\zeta)$ $YY = (c, s_1, s_2, s_3) := SPK\{(e, we, w)\}(ID_U)$	
$\xrightarrow{ZZ,\ YY,\ ID_U}$	
	- verifies YY - verifies the proofs of knowledge, stores ID_U and amount B of money in the user's database

Fig. 9.4: Registration protocol at the bank

Withdrawal Protocol

A user should show his or her account ID_U to the bank before withdrawing an amount V from the account. If there are enough funds left on that account, the bank and the user can carry out the withdrawal protocol as described in Fig. 9.5. In the protocol, $T_1, T_2, T_3, P_1, P_2, P_3, P_4$ are used as non-interactive proofs of knowledge. Furthermore, the user's license is contained in our electronic coin $U_\$$.

Payment Protocol

To make purchases, users and merchants perform the payment protocol as described in Fig. 9.6, which consists of two parts. First, the merchant verifies that the user's electronic coin $U_\$$ holds the signature from the bank. Then, the user convinces the merchant that he or she owns a valid license from the TTP, and that the license information has been embedded into the electronic coin $U_\$$ by the non-interactive proofs of knowledge block defined in Section 9.4.1. Let $m = ID_M || date/time$ where $date/time$ is the date and time of the payment. The merchant confirms that the user owns a valid license and that the license was embedded in the electronic coin $U_\$$ by verifying YY (checking the equation defined in Definition 3 of Section 9.4.1). The merchant verifies the equation $T = (T_1||T_2||T_3||P_1||P_2|| P_3||P_4)$ to ensure the integrity of user's

User		**Bank**
- chooses $w \in_R \{0,1\}^{l_g}$ $r \in_R \mathbb{Z}_n^*$ $r_1 \in_R \{0,1\}^{\epsilon(l_2+k)}$ $r_2 \in_R \{0,1\}^{\epsilon(l_g+l_1+k)}$ $r_3 \in_R \{0,1\}^{\epsilon(l_g+k)}$ - computes $T_1, T_2, T_3, P_1, P_2, P_3, P_4$ as described in Definition 3 - computes $T := T_1 \| T_2 \| T_3 \| P_1 \| P_2 \| P_3 \| P_4$ $S := r^{e_b} H(T)$ - Let TT denote the septuple $(T_1, T_2, T_3, P_1, P_2, P_3, P_4)$		
	S →	
		- computes $s' := S^{d_b} (mod\, n)$ - stores TT in withdrawal database - B := B -V
	← s'	
- obtains the blind signature: $s := s'(r)^{-1} (mod\, n)$ - the electronic coin is $U_\$ = (T, s)$		

Fig. 9.5: The withdrawal protocol

electronic coin $U_\$$. Finally, the merchant sends the electronic coin $U_\$$, TT, YY, $H(T)$ to the bank to proceed with the deposit protocol described in the following section.

User		**Merchant**
- computes: $YY = (c, s_1, s_2, s_3) := SPK\{(e, we, w)\}(m)$		
	$U_\$, YY, TT, ID_U$ →	
		- checks the equations: $s^{e_b} \equiv H(T) (mod\, n)$ $T =?(T_1 \| T_2 \| T_3 \| P_1 \| P_2 \| P_3 \| P_4)$ - verifies YY

Fig. 9.6: The payment protocol

Deposit Protocol

After the user spent his or her electronic coin $U_\$$, the bank should credit the merchant the corresponding value by the deposit protocol. As mentioned in the last section, the merchant sends a transcript of the payment protocol's execution to the bank. Just as the merchant did, the bank verifies the signature of the electronic coin $U_\$$ to make sure that this electronic coin was issued by itself. If the signature is valid, the bank searches the pair $(U_\$, s_3)$ in the deposit database. This results in two possibilities:

1. No items exist in the database
   The bank verifies the embedded license using the equation
   $$c = H(g||h||y||z||T_1||T_2||T_3||z^c T_2^{s_1 - c2^{l_1}}/y^{s_2}$$
   $$||T_1^{s_1 - c2^{l_1}}/g^{s_2}||T_1^c g^{s_3}|T_3^c g^{s_1 - c2^{l_1}} h^{s_3}||m).$$
   If the equation holds, the bank stores the quintuple $(T_1, T_2, c, s_3, date/time)$ in the deposit database and transfers the amount V to the merchant's account.
2. There is an item in the database
   This shows that there must be dishonesty between user and merchant. If ID_M and $date/time$ are the same then, without doubt, the merchant tries to be credited twice. Otherwise, the user is double-spending his or her electronic coin. There should be another (c', s_3') in the database. The bank can compute the randomly chosen integer w as $w := \frac{s_3 - s_3'}{c' - c}$ and the license's information $u := T_2/y^w$, which it sends to the TTP. The latter, in turn, will obtain the user's identity corresponding to the computed u from the database.

Tracing Protocol

In anonymity-revokable e-cash systems there are two kinds of tracing mechanisms to revoke anonymity (see [17]). In our system, owner tracing works as follows. In some dubious cases, the trusted third party can expose the identity of a specific electronic coin's owner with the information obtained from the bank during the execution of the deposit protocol. To this end, the bank gives the view of T_1, T_2 to TTP during protocol execution. The trusted third party computes the license information $u := T_2/T_1^x$, searches the database for the user's identity corresponding to the specific license information u and, then, returns the identity to the bank. Since the identity of the specific electronic coin's owner can be exposed, the owner tracing protocol can prevent criminal activities such as illegal purchases and money laundering described in [31].

9.4.3 Security Analysis

Some important aspects of a fair off-line electronic payment system's security have been introduced in the previous section. Here, we analyse the security of the proposed scheme in some detail.

Revokable anonymity During the execution of the withdrawal and payment protocols, the license's content, u, e, is either encrypted using the El Gamal algorithm [20] or appears with the representation of g^e, g^u. There is no other efficient way to obtain the value (u, e) except to break the El Gamal encryption algorithm or to solve the discrete logarithm problem. Also, (c, s_1, s_2, s_3) do not reveal any information about the user's license. Learning whether a specific electronic coin belongs to the user owning the license (u, e) is equivalent to judge whether the equation $log_g T_1 = log_y(T_2/u')$ holds or not. This is difficult to decide according to the Diffie-Hellman decision assumption (see [19, 12]).
The bank and the merchant can only obtain three equations from the payment and deposit protocols, viz.,
$T_1 = g^w, T_2 = uy^w, T_3 = g^e g^w,$
$P_1 = T_2^{r_1}(1/y)^{r_2}, P_2 = T_1^{r_1}(1/g)^{r_2}, P_3 = g^{r_3}, P_4 = g^{r_1} g^{r_3},$
$s_1 = r_1 - c(e - 2^{l_1}), s_2 = r_2 - cew, s_3 = r_3 - cw.$
Obviously, it is impossible to obtain the values of five unknown elements (e, r_1, r_2, r_3, w)[4] from the only three equations listed above. So it is difficult to obtain the license's part u ($u = T_2/y^w$) without knowing w.
Since no useful information can be derived from the withdrawal, payment, and deposit protocols, we conclude that our scheme can protect the user's privacy and keep the scheme anonymous. However, as we have shown in the last section, the anonymity of users can be revoked in co-operation of bank and trusted third party in certain special cases.

Unforgeability The choice of the system's secure parameters makes it difficult to factor the integer $\tilde{e}$, which means that even the trusted third party cannot know the user's secret e during the execution of the registration protocol in TTP. Therefore, the trusted third party is unable to disguise itself as a registered user. The bank needs the user's secret value g^u to compute the electronic coin $U_\$$. As we have pointed out, the bank cannot obtain any information about a user's license (u, e) during the execution of the withdrawal and deposit protocols. So, it is impossible for the bank to generate the electronic coin $U_\$$ for itself and spend it. Merchants and dishonest users cannot do better than the bank, since they cannot obtain more useful information than the bank during a transaction. Under the *Modified Strong RSA Assumption*, nobody can forge a valid license (u, e) if the factors of the system parameter n are unknown. In our scheme, only the trusted third party knows the factors of n'.

---

[4] e is an important part of the user's license(u, e).

Undeniability According to [12], T_1, T_2 and T_3, computed in the withdrawal protocol, prove unconditionally the knowledge of the license (u, e). The user owning a license cannot deny the transaction, since no other person can generate such T_1, T_2 and T_3 without knowing the license (u, e). This ensures that the system is undeniable.

Unlinkability Various random integers, $w \in_{\mathbb{R}} \{0,1\}^{l_g}$, $r \in_{\mathbb{R}} \mathbb{Z}_n^*$, $r_1 \in_{\mathbb{R}} \{0,1\}^{\epsilon(l_2+k)}$, $r_2 \in_{\mathbb{R}} \{0,1\}^{\epsilon(l_g+l_1+k)}$, $r_3 \in_{\mathbb{R}} \{0,1\}^{\epsilon(l_g+k)}$, are selected as the same user withdraws two different electronic coins from the bank. The useful information we can obtain in the withdrawal and payment protocols on any two different electronic coins of the same user, $T_1, T_2, T_3, P_1, P_2, P_3, P_4$ and $T_1', T_2', T_3', P_1', P_2', P_3', P_4'$, is different. To know whether these two different electronic coins belong to the same user is equivalent to decide whether the equation $log_g(T_1/T_1)' = log_y(T_2/T_2') = log_h(T_3/T_3')$ holds. This is also difficult under the *Diffie-Hellman Decision Assumption*.

Double-spending Resistance is guaranteed by the deposit protocol presented.

Traceability is guaranteed by the tracing protocol presented.

9.5 Conclusion

A fair off-line electronic cash system based on non-interactive proofs of knowledge has been presented. Combined with a group signature tool, the system provides conditional user anonymity. A user's anonymity can be revoked by processing owner tracing in co-operation between bank and trusted third party. This fair off-line electronic cash system is secure as analysed above, simple and straightforward, and more efficient than other ones.

Both Traoré's and this one are electronic cash systems comprising cryptographic primitives of group signature. But our system is more efficient in its payment protocol. In Traoré's system, three non-interactive proofs of knowledge blocks are used, while only one is needed in the payment protocol presented here. This is advantageous as the payment protocol is the one most frequently used in e-cash systems. Furthermore, the proof of knowledge of roots of representations used in Traoré's system involves more computation. In 2001, Maitland and Boyd [23] introduced an electronic cash system using a new group signature scheme [2]. However, the efficiency of the group signature used in the scheme is low, and the scheme is only able of owner tracing.

Finally, we pointed out some weaknesses of electronic cash systems based on group signatures, which leave room for further research. All electronic cash systems utilising group signatures are not able of coin tracing. The reason is that, during execution of the withdrawal protocol, the e-cash bank signs completely blindly the electronic coins. Therefore, no tracing information can be integrated into the electronic coins. A possible way to achieve coin tracing is to use a partially blind signature scheme.

Furthermore, in all these systems various signing keys of the bank have to be deployed for different electronic coin denominations. This is because the bank cannot gain any information including the value in issuing coins. Another fact is that in all these systems users and merchants have to withdraw and deposit electronic coins at the same bank. In other words, they do not support a multiple-banks model. Currently, there are many researchers working on this topic.

Acknowledgement

To a large extent, this work was performed during a post-doctoral research stay at the Faculty of Electrical and Computer Engineering of FernUniversität in Hagen, and was supported by the German Academic Exchange Service (DAAD) through a scholarship. The author is particularly grateful to Prof. F. Kaderali for his support, and to Ms. B. Cubaleska for helpful discussions.

References

1. Asokan N, Janson PA, Steiner M, Waidner M (1997) The State of the Art in Electronic Payment Systems. J of IEEE Computer 30: 28–35
2. Ateniese G, Camenisch J, Joye M, Tsudik G (2000) A Practical and Provably Secure Coalition-resistant Group Signature Scheme. In: Proc CRYPTO'00, LNCS 1880, pp 255–270, Springer
3. Bellare M, Garay J, Hauser R, Herzberg A, Krawczyk H, Steiner M, Tsudik G, Waidner M (1995) iKP — A Family of Secure Electronic Payment Protocols. In: Proc 1st Workshop on Electronic Commerce, USENIX Association, pp 89–106
4. Boly J-P et al. (1994) The ESPRIT project CAFE. In: Proc ESORICS'94, LNCS 875, Springer, pp 217–230
5. Brands S (1993) Untraceable Off-line Cash in Wallet with Observers. In: Proc CRYPTO'93, LNCS 773, Springer, 302–318
6. Brickell E, Gemmell P, Kravitz D (1995) Trustee-based Tracing extensions to Anonymous Cash and the Making of Anonymous Change. In: Proc 6th Annual ACM-SIAM Symp on Discrete Algorithms, ACM, pp 457–466
7. Camenisch J, Piveteau J, Stadler M (1994) An Efficient Payment System Protecting Privacy. In: Proc ESORICS'94, LNCS 875, Springer, pp 207–215
8. Camenisch J, Piveteau J, Stadler M (1996) An Efficient Fair Payment System. In: Proc 3rd ACM Conf on Computer and Communications Security, ACM Press, pp 88–94
9. Camenisch J, Maurer UM, Stadler M (1996) Digital Payment Systems with Passive Anonymity-revoking Trustees. In: Proc ESORICS'96, LNCS 1146, Springer, pp 33–43
10. Camenisch J (1997) Efficient and Generalized Group Signatures. In: Proc EUROCRYPTO'97, LNCS 1233, Springer, pp 465–479

11. Camenisch J, Stadler M (1997) Efficient Group Signatures for Large Groups. In: Proc CRYPTO'97, LNCS 1296, Springer, pp 410–424
12. Camenisch J, Stadler M (1998) A Group Signature Scheme with Improved Efficiency. In: Proc ASIACRYPTO'98, LNCS 1514, Springer, pp 160–174
13. Camenisch J, Michels M (1999) Proving in Zero-knowledge that a Number is the Product of Two Safe Primes. In: Proc EUROCRYPTO'99, LNCS 1592, Springer, pp 107–122
14. Chaum D (1983) Blind Signature for Untraceable Payments. In: Proc CRYPTO'82, Plenum, NY, pp 199–203
15. Chaum D, van Heyst E (1991) Group Signatures. In: Proc EUROCRYPTO'91, LNCS 547, Springer, pp 257–265
16. Chen L, Pedersen TP (1988) New Group Signature Schemes. In: Proc EUROCRYPTO'94, LNCS 950, Springer, pp 171–181
17. Davida G, Frankel Y, Tsiounis Y, Yung M (1997) Anonymity Control in E-cash Systems. In: Proc FC'97, LNCS 1318, Springer, pp 1–16
18. Frankel Y, Tsiounis Y, Yung M (1996) Indirect Discourse Proofs: Achieving Efficient Fair Off-Line E-cash. In: Proc ASIACRYPTO'96, LNCS 1163, Springer, pp 286–300
19. Fujisaki E, Okamoto T (1997) Statistical Zero-knowledge Protocols to Prove Modular Polynomial Relations. In: Proc CRYPTO'97, LNCS 1294, Springer, pp 16–30
20. El Gamal T (1985) A Public Key Cryptosystem and a Signature Scheme based on Discrete Logarithms. In: Advances in Cryptology: CRYPTO '84, LNCS 196, Springer, pp 469–472
21. Glassman S, Manasse MS, Abadi M, Gauthier P, Sobalvarro P (1995) The Millicent Protocol for Inexpensive Electronic Commerce. In: Proc 4th Int World Wide Web Conf, O'Reilly, pp 603–618
22. Kügler D, Vogt H (2002) Fair Tracing without Trustees. In: Proc FC'01, LNCS 2339, Springer, pp 136–148
23. Maitland G, Boyd C (2001) Fair Electronic Cash Based on a Group Signature Scheme. In: Proc ICICS'01, LNCS 2229, Springer, pp 461–465
24. Misarsky JF (1997) A Multiplicative Attack using LLL Algorithm on RSA Signatures with Redundancy. In: Proc CRYPTO'97, LNCS 1294, Springer, pp 231–234
25. Misarsky JF (1998) How to Design RSA Signatures Schemes. In: Proc PKC'98, LNCS 1431, Springer, pp 14–28
26. Pfitzmann B, Sadeghi A (2000) Netbill: Self-escrowed Cash Against User Nlackmailing. In: Proc FC'00, LNCS 1962, Springer, pp 42–45
27. Pointcheval D, Stern J (1996) Security Proofs for Signature Schemes. In: Proc EUROCRYPTO'96, LNCS 1070, Springer, pp 387–398
28. Sander T, Tashma A (1999) Auditable, Anonymous Electronic Cash. In: Proc CRYPTO'99, LNCS 1666, Springer, pp 555–572
29. SEMPER (2006) `http://www.semper.org`
30. Sirbu M, Tygar JD (1995) Netbill: An Electronic Commerce System Optimized for Network Delivered Services. In: Proc IEEE COMPCON'95, pp 20–25
31. von Solms S, Naccache D (1992) On Blind Signatures and Perfect Crime. J of Computer and Security 11: 581–583
32. Stadler M, Piveteau J, Camenisch J (1995) Fair Blind Signatures. In: Proc EUROCRYPTO'95, LNCS 921, Springer, pp 209–219
33. Traoré J (1999) Group Signature and Their Relevance to Privacy-protecting Offline Electronic Cash Systems. In: Proc ACISP'99, LNCS 1587, Springer, pp 228–243

34. Van de Graaf J, Peralta R(1988) A Simple and Secure Way to Show the Validity of Your Public Key. In: Proc CRYPTO'87, LNCS 293, Springer, pp 128–134
35. Wayner P (1997) Digital Cash: Commerce on the Net. AP Professional, London.
36. Yang B, Wang Y (1999) Fair Payment System with Electronic Wallet. J of Computer 22: 792–796

10

A UML Profile to Model Safety-Critical Embedded Real-Time Control Systems

Shourong Lu, Wolfgang A. Halang

Faculty of Electrical and Computer Engineering
FernUniversität in Hagen, 58084 Hagen, Germany
{shourong.lu|wolfgang.halang}@fernuni-hagen.de

Summary. To enable the description of safety-related software, a UML profile based on standards is presented. An inherently safe structure associated with well proven constructs, e.g., subsets of the real-time programming language PEARL according to DIN 66253-2 and of Function Block Diagrams according to IEC 61131-3 and IEC 61499, resp., as well as a safety-oriented execution architecture based on concepts taken from PEARL are established. These constructs are ordered in nested sets to fulfill the respective requirements of the four Safety Integrity Levels of IEC 61508. The genuine extension mechanisms of UML are employed in newly defining several stereotypes to incorporate domain-specific elements into the framework of UML, and then collecting them in a profile, which makes available in UML the elegant and suggestive concepts of PEARL and Function Blocks for describing the behaviour of distributed embedded real-time systems and for capturing their essential semantic concepts.

Key words: Embedded systems, architectural modeling, safety mechanisms, UML, PEARL.

10.1 Introduction

Nowadays, embedded systems are commonplace, and often safety-critical with respect to functional and non-functional requirements. In developing such computer controlled systems, the assurance of safety — during the entire life-cycle — becomes more important. Safety-related requirements need to be specified, hazard and risk analyses need to be performed, and the systems as well as the safety structures must be certified against safety standards before being put into operation.

Safety must be considered at an early stage right from the conceptual phase, i.e., safety analysis ought be on top of the agenda, as it is an integral part of a system's design. Safety analysis teams and engineers produce safety programmes and perform various types of hazard analyses using such techniques as Hazard and Operability Analysis (HazOp), Fault Tree Analysis (FTA), Event Tree Analysis (ETA), or Failure Mode and Effect Analysis

S. Lu and W.A. Halang: *A UML Profile to Model Safety-Critical Embedded Real-Time Control Systems*, Studies in Computational Intelligence (SCI) **42**, 197–218 (2007)
www.springerlink.com © Springer-Verlag Berlin Heidelberg 2007

(FMEA) [17]. Architecture is also a crucial element in the life-cycle of safety-critical systems, as indicated by the possibility to provide well-known safety mechanisms such as forms of design diversity (e.g., N-version programming) [27].

Safety standards define safety management schemes for the development of safety-critical systems. Over the years, several standards for safety-related applications appeared, such as RTCA DO-178B [23] for civil aviation, or the British Defence Standards 00-55 and 00-56 [20]. These standards specify how software and systems should be developed in order to adhere to the requirements of specific industrial areas. The standard most often referred to for computerised systems is IEC 61508 [14]. It is concerned with the safety of software-intensive systems. Here, Safety Integrity Levels (SIL) play an important rôle, to give a certain form to safety requirements and their implication on the development of software. A SIL is defined as a discrete level to specify the safety integrity requirements of safety functions, and can be derived from a risk assessment.

To cope with the complexity and safety-related requirements of embedded real-time systems, it may be beneficial to combine in their development process well-proven subsets of programming language structures fulfilling the respective requirements of the four Safety Integrity Levels of IEC 61508 with modern techniques of object-orientation, which could enable designers to both design abstract solutions based on the systems requirements, and implementation solutions based on appropriate subsets.

To this end, here we shall extend UML by constructs oriented at the well-designed and industry-proven facilities available in the "Process and Experiment Automation Realtime Language" (PEARL) [3, 4]. PEARL has been enhanced towards distributed systems and object-oriented design of applications having to meet severe safety requirements [10]. Our extensions will also be oriented at the high-level graphical language Function Block Diagram (FBD) for programmable logic controllers as defined in the international standard IEC 61131-3 [13, 18]. We aim to exploit the ideas incorporated in these languages and the safety elements of the cited standards as well as software-based structural patterns fostering dependability for the model-based and object-oriented development of safety-related embedded real-time control systems. The elements in safety-related programming are modeled by UML components and classes, and safety-related conditions and guards are assigned to components as tagged values and constraints.

This chapter is organised as follows. In Section 10.2 a short overview of the fundamentals of UML and IEC 61508 for safety applications is given. In Section 10.3 SIL-related programming language constructs, and correspondingly defined nested subsets of PEARL for safe system development are addressed. In Section 10.4 a UML profile for subsets of programming languages oriented at PEARL and Function Block Diagrams according to IEC 61131-3 and IEC 61499 is defined, with which safe architectures can be constructed.

After this, in Section 10.5 a novel approach to specify safe architectures by UML models is explored and, finally, conclusions are drawn.

10.2 Fundamentals of UML and IEC 61508 for Safety-Related Applications

The Unified Modeling language (UML) has become the industrial standard for object-oriented modeling and design [21]. It provides constructs to deal with varying levels of modeling abstraction addressing both the static and dynamic aspects of systems. A system specification in UML is represented by diagrams (Package-, Component-, Class-, Object-, Use-case-, Collaboration-, Sequence-, State-, Statechart- and Activity diagrams) [5, 6, 22]. Its built-in extensibility is a powerful feature of UML, providing mechanisms like stereotypes, tagged values, and constraints, with which the semantics of model elements can be customised and extended [1]. As such, quite naturally UML is the first choice when it comes to select a language for describing designs of complex systems.

The mechanism *stereotype* provides a way to define virtual subclasses of UML metaclasses with additional semantics. It represents a new modeling element that defines additional values based on tag definitions. Derived from a pre-existing base class, it is specific to a problem and can be applied to other model elements. That is, any model element may be attached by one or more particular stereotypes, in which case it receives the tag values of the stereotypes. The construct *tag definition* is used to specify new kinds of properties that may be attached to model elements. The actual properties of individual model elements are specified using *tagged values*. These may either be simple data type values or references to other model elements. Finally, *constraints* are semantic restrictions represented as text expressions usually formulated in the Object Constraint Language (OCL) [21]. Constraints are attached to one or more UML model elements to extend their semantics, allowing to add new rules or modify existing ones. A *profile* is a stereotyped package that contains model elements customised for a specific domain or purpose by extending the metamodel using stereotypes, tag definitions, and constraints.

On the other hand, in UML there is no adequate support for architecture descriptions, for modeling real-time and safety aspects, and for meeting the needs of heterogenous application domains such as mechatronics. This shortcoming has widely been recognised, and given rise to prominent extensions such as UML-RT [26] or ROOM [25]. Neither of them, unfortunately, really fulfills the requirements of automation engineering. Striving to make UML a tool useful to design distributed architectures, to model real-time and safety-related operation, and to express domain-specific specifications, in this chapter UML will be extended by constructs oriented at the well-designed and industry-proven facilities available in the real-time programming language

PEARL. The reason for the orientation at PEARL is that this language is characterised by the clearest concepts, and that it provides the richest and most comprehensive set of real-time features.

The international standard IEC 61508 [14], "Functional Safety of Electrical/Electronic/Programmable Electronic Safety-related Systems", is a framework for developing safety-critical systems and domain-specific software. Its aim is to ensure that electrical or electronic equipment carrying out safety functions provides the safety integrity required. For instance, equipment must be designed, installed, operated and maintained in such a way as to provide the required safety functions and safety integrity, taking into account the possibility of failure due to random hardware faults and systematic faults in both hardware and software. To achieve this, first it is necessary to determine, on the basis of an analysis of the hazards and risks associated with the application, what the required safety integrity is. This results in a specification of the safety functions in terms of both functionality and safety integrity. In IEC 61508, functional safety is defined as the ability of a safety-related system to carry out the actions necessary to achieve a safe state for the equipment under control, or to maintain a safe state for this equipment.

Most safety standards employ a notion of categorising the levels of safety assurance required for functions, systems, subsystems or components according to their criticality to safety. Accordingly, in IEC 61508 four distinct levels of safe performance, called Safety Integrity Levels (SIL), are introduced to specify the safety requirements of safety-related systems. Safety integrity is defined as the likelihood of a safety-critical system to satisfactorily perform the required safety functions under all stated conditions within a standard period of time [14]. Different degrees of safety integrity depend on what kind of risk a safety system is tackling, how willing we are to accept that risk, how many measures we have, as well as how effective they are.

The standard follows an integrated approach to achieve and assess the functional safety of systems and all their components. It provides a unified approach for all activities in the safety lifecycle that is rational and consistent, which covers all phases of the lifecycle for all system components, from initial concept, via hazard analysis and risk assessment, development of safety requirements, specification, design and implementation, operation and maintenance, and modification, to final decommissioning. The safety lifecycle aims to deal in a systematic manner with all activities needed to qualify a system for a certain Safety Integrity Level. The part IEC 61508-3 recommends to use the V-shaped lifecycle model approach for software design and development of safety-related systems [14]. Fig. 10.1 shows the arrangement of development steps in the V-model, and Fig. 10.2 shows the V-model in relation to safety-related activities. The V-model is the German development standard for data processing systems [12].

In Fig. 10.1, the design and development process starts with specifying the software safety requirements as indicated on the left branch of the V-model. It describes the implementation path, starting out from high-level

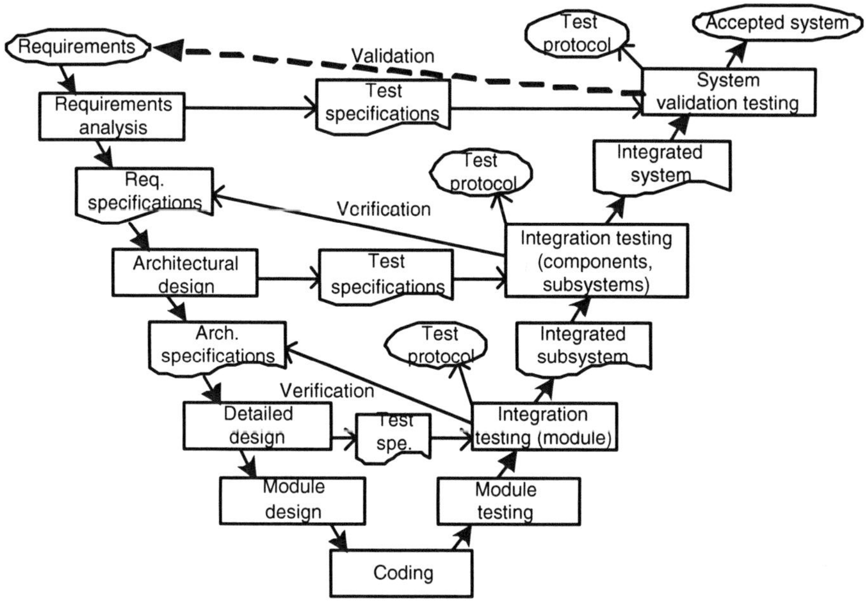

Fig. 10.1: V-model of software development process

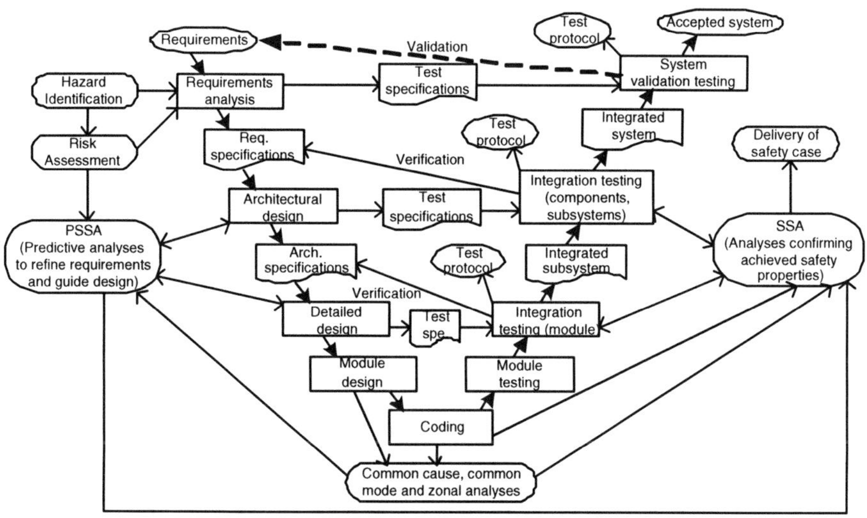

Fig. 10.2: Safety-related activities in the V-model

requirements, and becoming more detailed with each step towards the creation of actual production code. With each phase in the left branch, a verification step is associated in the right branch. The verification steps ensure that a system and its subsystems meet the design specification after they are designed and built. The testing activities in the right branch validate the software against the design specifications given in the left branch of the V-lifecycle.

In Fig. 10.2, the round-cornered boxes around the V-model show the safety tasks which are associated with the following development activities.

Hazard identification is the first step in the safety process, and sets the context for all subsequent activities. Its rôle is to identify the potential hazards in the proposed system so that they can be managed and controlled.

Risk assessment (Hazard analysis) examines each of the identified hazards to determine how much of a threat they pose, i.e., the likelihood of potential accidents.

Preliminary system safety assessment (PSSA) is used for the safety activities which accompany the design activities in the development process.

System safety analysis (SSA) is the main confirmatory safety activity, producing the evidence demonstrating that the safety requirements have been met.

Common cause or Common mode analysis is a range of techniques to identify non-independence of failures. They are applied throughout the development process in support of both PSSA and SSA. The final safety activity in a development process is the delivery of a

Safety case which is a comprehensive and defensible argument that the system is acceptably safe to use in a given context, supported by the necessary evidence.

The V-model identifies the major elements of the development process, and emphasises a top-down approach to design, and a bottom-up approach to testing. In descriptions of the V-model it is explained that such a model is just an approximation to the development process. In practice, the various stages are not always performed in such a strictly sequential manner. There are usually some iterations, with a series of operations performed repeatedly until a satisfactory result is obtained. Combining the V-model and UML can provide support for the implementation and testing of safety software for embedded systems.

10.3 SIL-related Programming Language Constructs

The safety-related evaluations of programming language constructs reported in [9, 11] are summarised in Table 10.1. It lists language constructs and typical programming methods as well as some code structures, which are suitable

Table 10.1: Typical programming methods

SIL	Typical programming method	Code structures and goals
SIL 4	Cause-effect table	No sequential program No transformation Specification = machine code **Goal:** Verification by consensus
SIL 3	Function block diagrams based on verified libraries	No declarations of variables No loops Use of library procedures only **Goal:** Match FBD programming
SIL 2	Language subsets enabling formal verification	No usage of pointers No multitasking Pre/post-conditions for any procedure Invariants for any procedure **Goal:** Allow formal verification
SIL 1	Languages with safe constructs	No time-unbounded constructs No unstructured constructs Structured synchronisation Structured handling of all exceptions **Goal:** Guarantee schedulability

for safety-related control applications having to meet the requirements of the different SILs according to IEC 61508.

The general principle to adhere to in designing any safety-critical system is to keep it as simple as possible. The more safety-critical a systems is, the more simple the related control software needs to be. Therefore, the utilisation of those programming concepts and architectural features appears to be mandatory which support the process of verification. A well structured software is easier to analyse and test, consequently less likely to contain errors. Most modern programming languages encourage programmers to use block structures, modular programming and language subsets, as well as selected features such as strong typing, runtime constraint and parameter checking, which can be employed to increase dependability.

For safety-related programming, language subsets ought to be used which comprise only the constructs being inherently safe and easy to understand, and leave all other features out. The advantage of this subset approach is that the bounds of a subset can be flexible, to allow for the use of some features in a limited and controlled way. Such uses must be carefully designed, controlled and justified, to ensure that the integrity of software is not adversely affected.

Cause-effect tables are ideal to program safety-related systems having to meet the highest safety requirements. The method aims to parameterise multipurpose controllers thus configured to perform specific functions. The rows of cause-effect tables are associated with events, the occurrence of which are controlled by Boolean pre-conditions. By marking fields that belong to certain columns in such tables, and which are associated with specific actions, respectively, users select pre-conditions and specify that these actions shall be performed when all corresponding pre-conditions become true in the sense of logical conjunction. Since specifications are formulated in a commonly understandable, but nevertheless formal manner, which can directly be interpreted and executed by machines without requiring complicated transformations and corresponding verification of correct implementation, this paradigm lets the need for generally difficult software verification simply disappear. Therefore, in automation technology, one strives to employ cause-effect tables as widely as possible.

Function Block Diagrams (FBD) are oriented at signal-flow and design-objects, and defined within the standard IEC 61131-3 [13]. The basic language elements in FBD are instances of functions and function blocks which, for the purpose of program composition, are interconnected by lines between their inputs and outputs. Functions and function blocks represent high-level application-oriented and re-usable language elements. Before being released, all functions and function blocks contained in a library are verified employing appropriate, usually formal methods. Safety licensing needs to be carried out only once per function block for a certain application area after a suitable set of function blocks has been identified. In general, rather few library elements are sufficient to formulate all programs in a particular area of process automation. For any given application program, only the correct implementation of the corresponding software architecture, i.e., the interconnection pattern of invoked functions and function block instances, needs to be verified based on the verified properties of the architecture's constituent elements. For this purpose, the commonly understandable method of diverse back translation [16] can be employed.

As they enable easy, trustworthy, and economic verification of automation software, the programming methods "cause-effect tables" and "function block diagrams" on the basis of verified libraries are suitable to meet the requirements of the two upper Safety Integrity Levels SIL 4 and SIL 3, respectively.

To provide a common framework for textual programming, increasingly restrictive, nested subsets of PEARL were proposed in [10] for each of the four Safety Integrity Levels, viz., HI-PEARL, Safe PEARL, Verifiable PEARL, and Table PEARL. In the latter subset, cause-effect tables controlling systems with SIL 4 requirements are formulated, which can be verified by visual inspection instead of employing formal proofs. Control software for SIL 3 systems is constructed graphically based on already proven function blocks, and the interconnection patterns are verified by diverse back translation. In the subset Verifiable PEARL, function block diagrams can be expressed textually. The subset Safe PEARL for SIL 2 enables formal program verification based only the provision of pre- and post-conditions as well as invariants. Finally, programs formulated in the subset HI-PEARL for the lowest SIL 1 can easily be analysed for schedulability, which is assured by eliminating some unsafe language constructs such as jumps and unbounded loops. Although not as easily as their SIL 2 counterparts, HI-PEARL programs are verifiable as well.

10.4 UML Profiles for Safety Mechanisms

In this chapter, we are aiming to build a set of UML extensions that support safe architectures meeting relevant safety criteria. The leading idea is to design systems in such a way that component faults do not result in system failures. In the following, we shall address SIL-related programming language constructs, define a UML profile for programming language subsets oriented at PEARL and Function Block Diagrams, and build a safe architecture based on PEARL concepts.

10.4.1 Safe Subsets Oriented at PEARL

PEARL is one of the very few genuine high-level real-time languages. With industrial real-time applications in mind, its development began in the late 1960s by a group of industrial companies and research institutes. Incorporating decades of experience, its later version PEARL90 was standardised in 1998 [4], and its extension for distributed systems in 1989 [3]. PEARL has been extended towards safe object-oriented and distributed applications (Table PEARL, Verifiable PEARL, Safe PEARL, HI-PEARL) [10]. Owing to its clear concepts, combining features from classical programming languages and special ones for real-time systems, PEARL is particularly suitable for industrial process control and embedded applications.

PEARL supports both periodic (time-driven) and aperiodic (event-driven) tasks, semaphores for synchronisation, device-level input/output, and modules. Further, there is explicit support for temporal variables with the data types *clock* and *duration* and associated operations. These data types describe time instants and periods, and also appear in time schedules. The schedule option in PEARL is associated with six multitasking commands for the control of task state transitions, viz., *activation*, *termination*, *suspension*, *continuation*, and *resumption* as well as *annihilation* of scheduled activations (called prevention). A more detailed description can be found in [7].

PEARL for distributed systems includes elements to describe hardware and software configurations. As shown in Fig. 10.3, an architecture description consists of *station division*, *configuration division*, *net division*, and *system division*, which describe different associated layers of a system design. In a *station division*, a system's processing nodes are introduced. A station is treated as black box with connections for information exchange. There may be more stations in a system, each one of them being uniquely identified, and associated with state information for re-configuration purposes. The basic components of a station are its processing elements, such as Proctype, Workstore, and Devices. In a *net division*, the physical connections between stations are given by listing the point-to-point connections between their interfaces. A *system division* encapsulates the assignment of symbolic names to hardware devices. The items from the station and net divisions are listed here. A *configuration division* deals with software architecture. The largest executable program component that is associated with a station and its state is a collection consisting of modules. The latter consist of tasks, which communicate through ports and connection lines. Ports are further described as import and export ports, in which it is stated, which data structures and task references are shared with other modules. A task is described by its scheduling parameters. The connections between ports are described by their directions and line attributes.

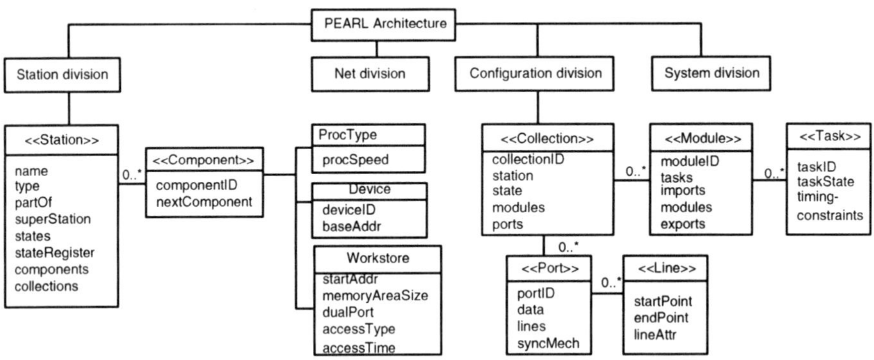

Fig. 10.3: Basic architecture description in PEARL

After this short introduction to PEARL given to provide a basic understanding, we now return to the four safety-related subsets which are shown in Fig. 10.4.

Table PEARL provides just a single executable statement, which is used to formulate rules constituting cause-effect tables.

Verifiable PEARL matches FBD, but is a textual language whose constructs allow for parameter passing and procedure invocation, only.

Safe PEARL is an inherently safe language subset restricted to the executable statements procedure call, assignment, conditional selection, and loop with a compile-time bounded number of iterations.

HI-PEARL was derived from PEARL90 by eliminating all unsafe language constructs which may lead to non-determinisitic behaviour.

The main element dealing with safety issues in the above-mentioned subsets of PEARL is the module. Used to describe software architectures,

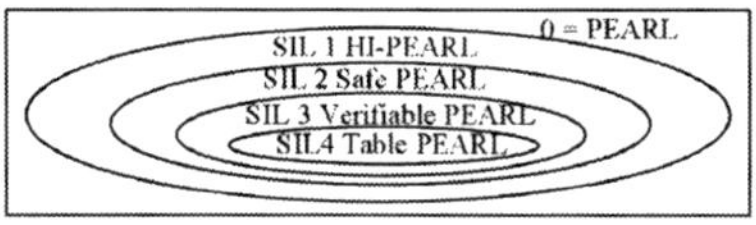

Table PEARL

```
<module> ::= Module [<var-decl>;]* [<rule>;]*  MODEND.
<var-decl> ::= DCL <name> : <type>;
<rule> ::= IF <bool-expression>
           THEN
           <variable> := <expression>
           FIN
```

Verifiable PEARL

```
<module>::=MODULE[(<name>)]
           [(<problem>)] MODEND

<problem>::= PROBLEM
[<proce-spec>]* [<var-decl>]*[<task-decl >];
<proc-spec>::= SPC<name>: PROC<lst-of-par> <return-type>:
               [GLOBAL];
<var-decl>::= DCL <name>: <type>;
<task-decl>::= <name>: Task[GLOBAL];<proc-stmt>+ END
<proc-stmt>::=<proc-name> <lst-of-par>
<list-of-par>::= (<par> [,<par>]*)
<par>::= <const>|<var-name>
```

Safe PEARL

```
<module>::=MODULE[(<name>)]
           [<mod-extension>]; [<interface-spec>]*
           [<system-part>] [<problem-part>] MODEND;

<mod-extension>::= EXTEND (<name>) [, <name>]+)
<interface-spec>::=INTERFACE (<name>)  <interface-decl>*
<interface-decl>::= <interface-var-decl> | <interface-proc-del>
<interface-var-decl>::= SPC <name>: <type> READ;
<interface-proc-decl>::= SPC <name>: PROC <lst-of-par> <return-type>

<system-part>::= SYSTEM;  <name> : <name>;

<problem-part>::= PROBLM; [<declarations>]+
<declarations>::= <var-decl> | <proc-decl>
<var-decl>::= DCL <name> : <type>;
<proc-decl>::= <name>: PROC <lst-of-par> <return-type> <body> END
<lst-of-par>::= (<par-decl>) [,<par-decl>]*)
<proc-decl>::= <name>: type
<body> ::= [<var-decl>]+[<statement-seq>]
<statement-seq>::= [<statement>;]*
<statement>::=(<statement-seq>) <assignment-statemet>
              |<conditional-statement> |<while-statement>
<assignment-statement>::=<variable>:=<expression>
<conditional-statement>::= IF<bool-expression>
                           THEN <statement-seq>
                           [ELSE <statement-seq>]FIN
<while-satement>::=WHILE <bool expression>
                   REPEAT<statement-seq> END
```

Context **PearlInterface** inv:

```
self.feature -> forAll (f |
  (f.parameter -> size >=1) and
  (f.operation = Operation) )
```

Context **PearlPort** inv:

```
self.owned. Port () impies
    (self.required -> size () =1  and
    (self.provided ->size()=1)
```

PEARL Collection:
Context **PearlCollection** inv:

```
self.baseClass = Component implies
self.ownedElement.contents -> forAll (m|
  m.OclIsKindOf( PearModule) and
  m.Port.isKindOf (PearlPort))
```

Hi-PEARL Module for SIL1
Context **PearlModule** inv:

```
self.baseClass = Classifier and
self.ownedElement -> notExit (unSafeBehaviors)
```

Table PEARL Module for SIL 4:
Context **TablePearlModule** inv:

```
self.baseClass = Classifier and
self.ownedElement.contents  -> forAll ( E |
E.OclIsKindOf(BollenExpression)
```

Verifiable PEARL Module for SIL 3:
Context **VerifiablePearlModule** inv:

```
self.baseClass = Classifier and
self.ownedElement -> forAll( E | E = TaskStatement and
  (exist (p: Parameter |
   p.type = PrimitiveType) ->size >= 1) and
   self. Interfaces() = 0 )
```

Safe PEARL Module for SIL 2:
Context **SafePearlModule** inv:

```
self.baseClass = Classifier and
self.ownedElement -> forAll( E | E = TaskStatement and
 (exist (p: Parameter |
   p.type = PrimitiveType) ->size >= 1) and
   self.Interfaces.isKindOf (PearlInterface))
```

Fig. 10.4: PEARL subsets matching the four SILs and corresponding UML stereotypes

modules are, according to the different requirements of the SILs, constructed with procedures, interfaces, various variables, and conditional statements. A PEARL program is structured by grouping modules into collections. The latter are either statically distributed or dynamically allocated to system nodes (stations); they form the elements for dynamic re-configuration. Collections communicate by point-to-point message exchange on the basis of the port concept, only. Messages are sent to ports, or received from ports, which are known only locally in their own collections. Thus, ports form the input and output interfaces of collections.

The collection of PEARL is closely matched by the component concept [24]. A component is defined in a problem domain. It can be understood as an encapsulated software entity with an explicit interface to its environment.

At the specification level, a component provides services to other components and possibly requires services from other components in order to deliver those promised. From the component developer's point of view, a component has a unique identifier and set of properties. It encapsulates a well-defined piece of the overall application functionality. An application is assembled from collaborating components accessing each other through well defined component interfaces. The external view on a component is a set of provided and required interfaces, which may be exposed via ports. A component may also have an internal view in the form of a realisation, which is a set of class instances or subcomponents that co-operate to implement the services exposed by the component's provided interfaces while relying on the services of its required interfaces. These concept can be used to specify both logic and physical components [2, 24].

To make the concepts of the four SIL-related subsets of PEARL available for modeling with UML, we create a stereotype named PearlModule, using Classifier of UML 2.0 as its base class. Constraints allow us to specify claims about architectures, their components, and their expected execution behaviour. Formulated as expressions in the Object Control Language (OCL), these constraints can be pre- or post-conditions, but also invariants. Thus, we can also describe Table PEARL programs by OCL expressions. The parameter passing of Verifiable PEARL is mapped to the UML feature tagged values. To match Safe PEARL's and HI-PEARL's interface services, we employ both the offered and the required interfaces as known in UML 2.0. The defined stereotypes are also shown in Fig. 10.4.

10.4.2 Safe Subsets Oriented at Function Blocks

Function blocks are defined within the IEC 61131-3 [13, 15] and IEC 61499 [18] standards' software model as shown in Fig. 10.5. In IEC 61131-3, an overall application is composed of nested functional components called *Program Organisation Units (POUs)*, that are *Programs* (in higher levels of the hierarchy), *Function Blocks (FBs)*, i.e., re-usable software blocks defined by a data set (input, output parameters and local variables) and an internal algorithm, and *Functions* that differ from FBs only by the lack of permanent internal variables and in having a single output parameter. Unlike the FBs of IEC 61131-3, a function block according to IEC 61499 generally consists of two parts, namely, Execution control that creates a processing of events with control inputs and outputs (control flow), and Algorithm with data inputs and outputs and internal data (data flow and processing). FBs of IEC 61499 require distinct data interfaces and event interfaces.

In IEC 61131-3, a program can contain several instances of the same FB, each one independent of the others as their internal data are allocated in separate memory areas. Moreover, the same FB can be used in several different programs, reducing effort for code re-writing. An instance of a function block is specified by input, output, in-out, and internal variables, and by an

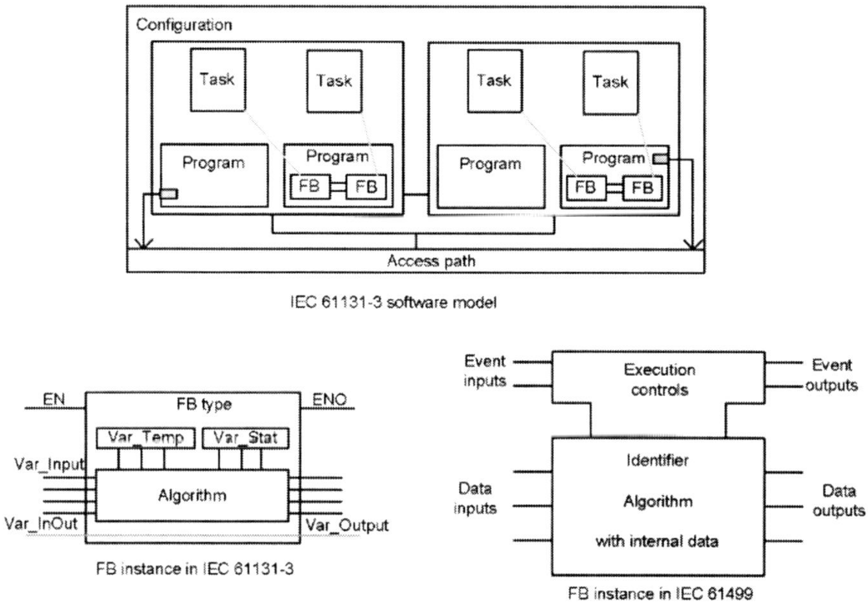

Fig. 10.5: FB instances in IEC 61131-3 and IEC 61499

internal behaviour. The input variables can only be written from the outside of the FB. From the inside they can be read, only. Output variables can be read and written from the FB's inside, and only be read from the outside. In-out variables are special shared variables. If their data types match, output variables can be connected to input variables by connectors. Assignment of data values to interface variables is the only means to communicate with other FBs. Function blocks have internal state information that persists the execution of FB instances. A characteristic feature of FBs is the separation of external interfaces and internal implementation. In IEC 61499, the function block concept is more related to execution control and scheduling, because IEC 61499 addresses distributed control systems, in which it is very important that software modules residing on different devices follow a well predictable execution sequence. Periodic or event-driven real-time tasks can be associated with function blocks.

To make the concept of function blocks according to IEC 61131-3 available in UML, we use the component architecture of UML 2.0 to define FB stereotypes. Hereby, we can model the internal structure of FBs with Parts and Connectors, and model the variable-interfaces of FBs with Ports and Interfaces. The input and output variables can be specified by connecting in-ports and out-ports, respectively. Out-ports contain the results of computations based on the inputs and current states of function blocks. The in-out variables are modeled by in-out ports with *contracts* added to specify the data exchange protocols applying. The correspondingly defined UML component stereotypes formulated in OCL are depicted in Fig. 10.6. The ports in the UML components are direct counterparts of the FB interface variables. The stereotypes FBInPort and FBOutPort are inheriting from FBPort. FBOutPort requires one FBInterface and FBInPort provides one FBInterface. The stereotype FBInterface is needed to model input and output variables of FBs; it may contain operations.

1)

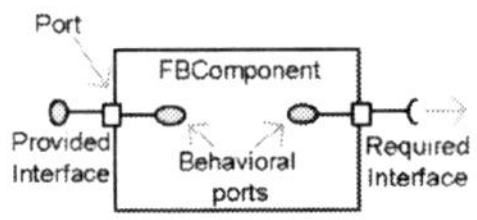


```
context FBComponent inv:
    self.feature->forAll( f |
     (f.classifier = FBPort) )
```

2)

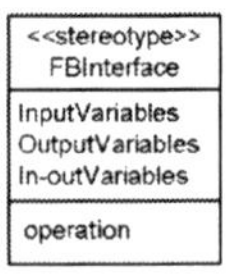


```
context FBInterface inv:
 (self.feature->size = 1) and
 (self.feature->forAll( f |
          (f.name = 'OperationName') and
          (f.classifier = Operation) and
          (f.parameter->size = 1)) )
```

3)

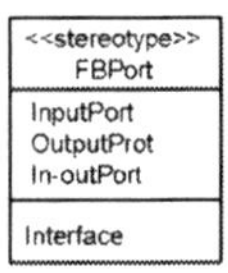


```
context FBInPort inv:
(self.required->size = 0) and
     (self.provided->size = 1) and
     (self.provided->forAll(p |
       (p.classifier = FBInterface)))
```

```
context FBOutPort inv:
   (self.provided->size = 0) and
   (self.required->size = 1) and
   (self.required->forAll( r |
        (r.classifier = FBInterface)))
```

Fig. 10.6: Stereotypes providing the function block paradigm

10.5 A Safety Architecture based on PEARL Concepts

In the PEARL methodology, hardware and software architectures are described separately. A hardware architecture is composed of processing nodes, called stations, whose descriptions also mention their components with their properties. A software architecture is organised in the form of collections of program modules, combining program tasks, functions, and procedures of the application software. The collection is the unit of software to be mapped onto stations, i.e., at any time there is exactly one collection assigned to run on a station. Thus, the collection is also the unit of dynamic re-configuration. To manage collections, a Configuration Manager (CM) [8] was defined, which is located between the real-time operating system, if any, and applications. A UML profile for software architectures oriented at the concepts of the real-time language PEARL was also defined in [19]; the corresponding stereotypes are shown in Fig. 10.7.

PEARL Element	UML Element	Stereotype	Icon
Station	Node	<<PStation>>	
Component	Class	<<PComponent>>	Workstore, Device, Proctype
Connection	Class	<<PConnection>>	
Collection	Component	<<PCollection>>	
Port	Class	<<PPort>>	
Module	Class	<<PModule>>	
Task	Class	<<PTask>>	
Line	Class	<<PLine>>	

Fig. 10.7: UML stereotypes oriented at PEARL

10.5.1 An Architecture for Real-Time Applications

An application architecture with a configuration manager is depicted in Fig. 10.8. It consists of a set of stations, collections, and a set of static or dynamic links that may be established during application execution. At runtime, the CM allocates collections to stations according to the application architecture. The latter is defined as a class *ArchitectureDefinition*, which stores the relevant information about the system architecture and forms (a part of) the application. This information is represented in UML models by the classes *StationInfo* and *CollectionInfo*, respectively. The operations *getComponentInfo()* and *getConnectorInfo()* are used to inquire about stations and collections accessed in order to carry out the actual deliveries. The configuration manager's Application Programming Interface (API) is independent on a station's type, but its message routing is affected by the rôle the station plays in a system, which is determined by its type. Given this information, the configuration manager serves as a system observer or monitor, too. It monitors relevant signals, as defined in the transitions leading out of the current state. When the logical conditions for a state transition are satisfied, the CM initiates the structural transition process.

In the application architecture, the important element is the configuration manager which deals with all necessary components as well as their actions. The CM can be specified by its functions, architectural configuration, and behavioural characteristics employing UML notations to capture all information necessary to synthesise a system.

Functional specification Execution at each processing node (station) starts with initiating the CM. First, it loads the collections of task objects, and activates the initial collection by triggering the latter's initialisation-task objects. In stations without a real-time operating system, the main task

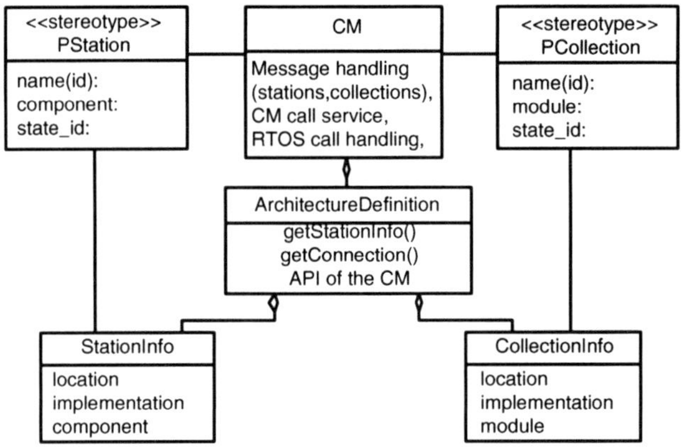

Fig. 10.8: An application architecture

of the collection is started and control delegated by the CM, whereas otherwise the CM acts as a front-end to the operating system functions. It also uses appropriate system calls and system ports to transfer system requests to or from nodes enabled by real-time operating systems to schedule the collection's tasks.

Besides local execution, the CM is also responsible for communication with other stations, and for co-operation among the tasks of the same collection. Hence, it must establish port-to-port connections through the interfaces of the station, and appropriate local port connections for inter-task communication. Synchronisation and system service requests are handled on the same station, in case the station is configured to run a real-time operating system. Otherwise, these requests are forwarded to an appropriate station through a proprietary port.

In addition, the CM is responsible to maintain hardware and software consistency and dependable behaviour of the system. This feature is closely associated with runtime re-configuration, which depends on different operation scenarios of the target system. To perform runtime re-configurations, the CM contains tables (architecture data) and scenarios dependent on operation state for each processing node, captured by the behavioural state machine that defines the transitions at which re-configurations are to occur. Tied to these state-based descriptions is the information necessary to appropriately configure the system's hardware and software components.

Functional configuration involves time-sharing of resources between computational tasks. The primary consideration here is to preserve the state and the intermediate results when a computation is swapped out, which is the responsibility of the real-time operating system for tasks, whereas the collection contexts are the responsibility of the CM. The internal state and the intermediate results must be restored when the computation is swapped back in. To address these issues, the representation of internal state and intermediate results needs to be considered, and the runtime execution environment needs to provide appropriate API functions and variables for observation, change and restoration of state and data.

Architectural configuration specification The objective of architectural configuration is to modify hardware and computation by re-allocating resources to computations. The need for this type of re-configuration arises in situations where some resources become unavailable, either due to faults or due to re-allocation to higher-priority jobs, or due to shutdowns in order to minimise power usage. For a system to keep functioning in spite of faults, the hardware needs to be modified and the computational tasks need to be re-assigned.

An architecture re-configuration specifies which instances of which components are needed, how their resource requirements will be satisfied, as well as the configuration of container services, and how they apply to components and their instances. A device specification describes the available features of the target device (and operating system), on which the resulting application is to be deployed. This ultimately determines which container features are available in the target system, as well as how these features are implemented.

Architectural re-configuration involves relocating computations. Safe transitioning of a computation's internal state is the most important consideration here. The state transitioning could be further complicated due to relocation of computation from hardware to software. ¿From the structural point of view, the CM represents a component. All application components are instantiated by it, since their functionality is required to be addressed by the CM. Therefore, the structure of the CM can be expressed as a set of components communicating through ports, connectors and contracts. As shown in Fig. 10.9, component architectures are structured by subcomponents, connectors, ports, and interfaces. A component owns a unique identifier and a set of properties, and defines a set of communication ports, which provide interfaces. Components can exchange data with each other through ports and connectors, only. A component can be a composite one containing other component(s).

The composition of a system can be expressed in terms of components, with a component providing services to other components. A component within a system can be either a simple component, or itself be composed of a group of other components. The components provide and obtain services through ports. The interaction between ports can take many forms, e.g., buffered message passing, one-to-many event dissemination, or synchronous request-reply communication. Dynamic configuration management usually means changes to component inter-connections or replacement of one or more software components performed at runtime.

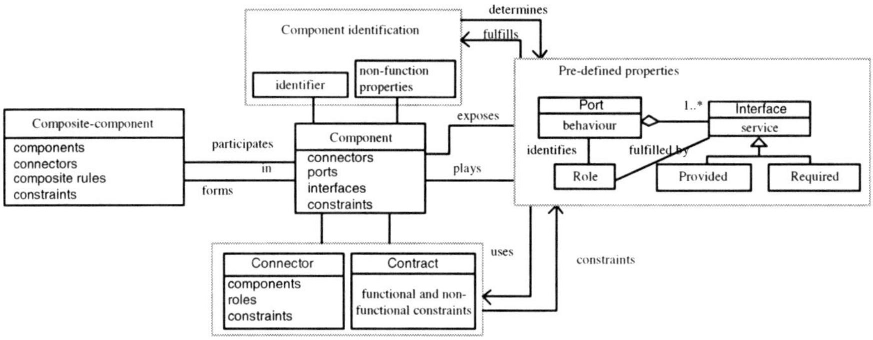

Fig. 10.9: Component structure diagram

Behavioural specification The behavioural aspect can be expressed by employing a statechart diagram, which provides an overview on the possible states of the processing node, and also presents the conditions, under which the processing node should change scenarios. Behavioural modeling defines the modes in which a system will operate, and specifies the conditions under which mode changes occur. The modes and state transitions can be specified from the behavioural aspect by a state transition graph.

A configurable system operates in discrete modes, with specific transitions between them. Therefore, a statechart representation can be chosen to model the adaptive operational behaviour. The modeling objects provided are states, events, and transitions. States represent operational modes, events represent the causes of mode-shifts, and transitions and transition rules define the pre-conditions and the consequences of mode-changes.

10.5.2 An Architecture for Safety-Related Applications

In the PEARL methodology as addressed above, an executable program is a collection of modules, each of which is composed of a set of tasks responding to events. Tasks represent the (computing) processes of a running system. A safety-oriented architecture oriented at PEARL concepts is defined in Fig. 10.10, which illustrates the structure of an application architecture enhanced to design dependable systems. The element *EventDetection* is defined to provide fail-safe behaviour. The various architectural elements have the following semantics.

Configuration Manager controls the execution path of a system. Since the main flow of control is based on events, it receives events from *EventArray* and dispatches them to *Events*. The CM processes events, decides

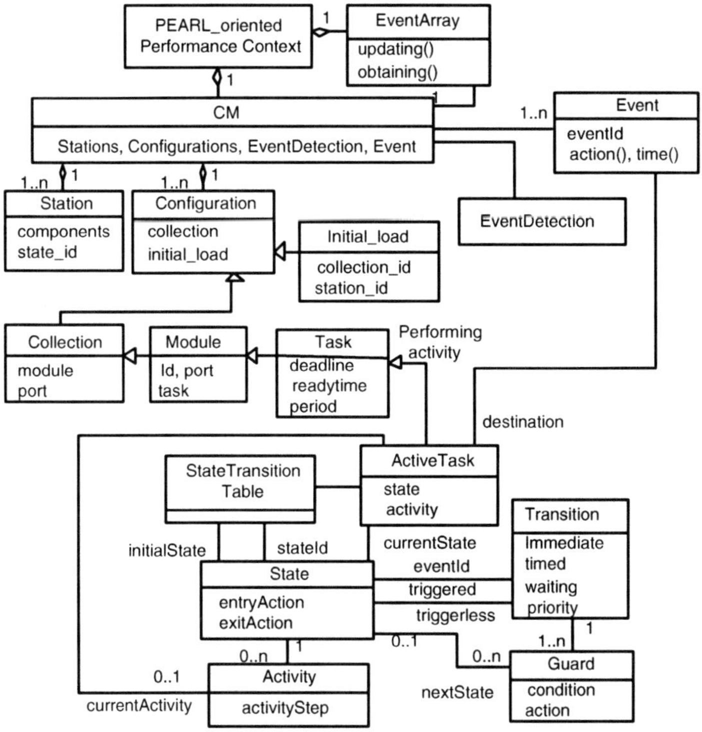

Fig. 10.10: Safety architecture diagram

which tasks to activate, and keeps a list of active tasks currently executing activities. Switching between dispatching events and executing activities allows other active tasks in the system to process, and also allows activities to be pre-empted by incoming events.

Class EventArray represents an ordered array of events. Its operations are related to updating and fetching events from *performanceContext*.

Event is capable of triggering a mode-change and is defined by transition rules. The latter are Boolean equations composed of event variables. When such an expression is satisfied, the transition from one mode to another is enabled, and system re-configuration is to be performed. The class *Event* defines the functions that dispatch an event to its destination task. Each event carries an identifier and points to a task. Events trigger transitions from one state to another.

Configuration classes are larger parts of simulation models, which are described in terms of modules and of input and output ports. A program is structured by grouping modules into collections. Collections communicate by message exchange via their ports. The ports can be mapped internally to variable values. Collections are either statically distributed or dynamically allocated to system nodes.

Module consists of a collection of tasks, that together provide a desired functionality. The interaction and dependencies between tasks are described by communication and synchronisation constructs.

Task is the execution of a procedure within a given time-frame, i.e., it has a trigger condition and a response time.

Transition is a directed relationship between a source and a target state. It has tagged values for real-time performance, and can be guarded by Boolean expressions, which are evaluated when an event instance is dispatched by a state machine. If a guard is true at that time, the corresponding transition is enabled, or otherwise disabled.

Class EventDetection is used to detect fault occurrences. With the help of the class CM it checks for such fault events as depicted in the sequence diagram of Fig. 10.11. In the diagram, horizontal lines represent event notifications between objects, and vertical ones represent their temporal sequence. When the CM accepts an event from *EventArrary*, *EventDetection* uses its detection method to check the event, before returning a result message to the CM. If the event is true, then the CM dispatches it to the class *Event*.

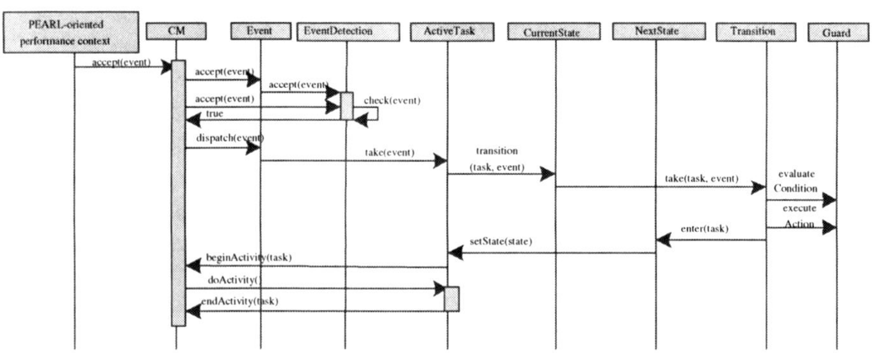

Fig. 10.11: Sequence chart of activity scenario

10.6 Conclusion

With the help of a profile, UML can be adapted to application areas for which standard UML is not specific enough. In this chapter, a set of UML stereotypes was introduced providing the functionality of safe programming language subsets needed to design safety-related systems. Using UML's syntax, these extensions can be integrated into the standard UML framework, since they are based on the UML metamodel. The new safety-related stereotypes can be assigned to UML elements. Oriented at the safety facilities found in the prevailing safety standards, the stereotypes can be incorporated into models of safety-related embedded real-time control systems. Their advantage is to allow for high-level descriptions during early phases of the development process.

The UML profile for safety mechanisms can provide facilities to capture safety requirements of safety-critical applications, as well as dependability structures which deal with safe elements and desired measures. The benefit of using UML in modeling safety mechanisms is that UML provides numerous diagrammatic techniques to comfortably describe process and system models of any kind. Thus, for each aspect to be modeled the most expressive technique can be selected by the user.

Acknowledgement

This work was financially supported by a Matching Funds Scholarship of the German Academic Exchange Service (DAAD) and Institut für Automation und Kommunikation e.V. Magdeburg.

References

1. Booch G, Rumbaugh J, Jacobson I (1999) The Unified Modeling Language User Guide. Addison-Wesley
2. Cheesman J, Daniels J (2001) UML Components — A Simple Process for Specifying Component-based Software. Addison-Wesley
3. DIN (1989) Norm 66253 Teil 3: Mehrrechner-PEARL. Beuth, Berlin Köln
4. DIN (1998) Norm 66253-2: Programmiersprache PEARL90. Beuth, Berlin Köln
5. Douglass BP (1999) Doing Hard Time: Developing Real-Time Systems with UML, Objects, Frameworks, and Patterns. Addison-Wesley
6. Douglass BP (1999) Real-time UML: Developing Efficient Objects for Embedded Systems, 2nd edn. Addison-Wesley
7. Gumzej R (1999) Embedded System Architecture Co-Design and its Validation. Doctoral thesis, University of Maribor
8. Gumzej R, Halang WA (2003) A Configuration Manager for Embedded Real-Time Applications. In: Proc Co-design for Embedded Real-time Systems, pp 24–29
9. Halang WA, Frigeri AH (1998) Methods and Languages for Safety Related Real Time Programming. In: Ehrenberger WD (ed) Computer Safety, Reliablity and Secuity, LNCS 1516, Springer, Berlin Heidelberg New York, pp 196–208
10. Halang WA, Pereira CE, Frigeri AH (2001) Safe Object Oriented Programming of Distributed Real Time Systems in PEARL. In: Proc 4th IEEE Int Symp on Object-Oriented Real-Time Distributed Computing, IEEE Computer Society Press, Los Alamitos, pp 87–94

11. Halang WA (2003) Software Dependability Considered as the Main Problem of Embedded Systems. In: Proc IEEE Int Conf on Industrial Technology, IEEE, Piscataway, pp 14–21
12. IABG (1997) Entwicklungsstandard für IT-Systeme des Bundes. `http://www.v-modell.iabg.de/`
13. International Electrotechnical Commission (1992) Standard IEC 61131-3: Programmable Controllers, Part 3: Programming Languages. Geneva
14. International Electrotechnical Commission (1998) Standard IEC 61508: Functional Safety of Electrical/Electronic/Programmable Electronic Safety-related Systems. Geneva
15. John KH, Tiegelkamp M (2001) IEC 61131-3: Programming Industrial Automation Systems. Springer, Berlin-Heidelberg-New York
16. Krebs H, Haspel U (1984) Regelungstechnische Praxis 26: 28–33
17. Leveson NG (1995) Safeware: System safety and computers. Addison-Wesley
18. Lewis R (2001) Modelling control systems using IEC 61499, Applying function blocks to distributed systems. IEE Control Engineering Series, No 59
19. Lu S (2003) An PEARL orientierte Spezifikation verteilter eingebetteter Systeme mit UML-Stereotypen. In: Holleczek P, Vogel-Heuser B (eds) Verteilte Echtzeitsysteme. Reihe Informatik aktuell, Springer, Berlin Heidelberg, pp 13–22
20. Ministry of Defence (1997) Defence Standard 00-55 Issue 2: Requirements for Safety-Related Software in Defence Equipment
21. Object Management Group (2001) Unified Modeling Language specification V1.4
22. Object Management Group (2003) Unified Modeling Language: Superstructure. OMG document ptc/2003-08-02
23. RTCA (1992) Standard DO-178B: Software Considerations in Airborne Systems and Equipment Certification, Washington, DC
24. Szyperski C, Gruntz D, Murer S (2002) Component Software — Beyond Object-Oriented Programming. 2nd edn. Addison-Wesley
25. Selić B, Gullekson G, Ward P (1994) Real-Time Object Oriented Modeling. Wiley
26. Selić B, Rumbaugh J (1998) Using UML for Modeling Complex Real-time Systems. `http://www.objectime.com/uml`
27. Torres-Pomales W (2000) Software Fault Tolerance: A Tutorial. In: NASA/TM-2000-210616

Part III

Distributed Systems Management, Effective Communication and Security

11

Probabilistic Management of Distributed Systems

Jianguo Ding

Software Engineering Institute
East China Normal University, Shanghai 200062, P.R. China
jianguo.ding@ieee.org

Summary. With the growth in size, heterogeneity, pervasiveness, and complexity of applications and network services, the effective management of distributed systems has become more important and more difficult. Due to the inherent complexity of large-scale distributed systems, it is not possible to build a precise model of all hardware and software components and their mutual dependencies. Therefore, an efficient distributed system needs to work effectively even in face of incomplete management information, uncertain error situations, and dynamic changes in open distributed systems. We use Bayesian networks to model certain properties and dependencies between managed objects, and propose to perform probabilistic inference in the Bayesian model. In case of faults or failures, the presented approach can track the strongest causes and trace the strongest routes between particular effects and their causes. Core properties of this approach are empirically analysed in a simulation environment.

Key words: Distributed systems management, management under incomplete knowledge, Bayesian networks, probabilistic backward inference.

11.1 Introduction

With the growth in size, heterogeneity, pervasiveness, and complexity of applications and network services, the effective management of distributed systems has become more important and more difficult. Individual hardware defects, software errors, or combinations of such defects and errors in different system components may cause the degradation of services provided by other (remote) components in a distributed system or even their complete failure due to functional dependencies between managed objects. Hence, an effective distributed fault detection mechanism is needed to support quick decision-making by systems managers, and to allow for a partial automation of fault management.

Although the Open Systems Interconnection (OSI) management standard provides a framework to manage faults in heterogeneous open systems, it

J. Ding: *Probabilistic Management of Distributed Systems*, Studies in Computational Intelligence (SCI) **42**, 221–248 (2007)
www.springerlink.com © Springer-Verlag Berlin Heidelberg 2007

does not address methodological issues to detect and diagnose faults. To fill this gap, much research effort in the past decade has been focused on improving a management system in fault detection and diagnosis. Rule-based expert systems have so far been the major approach to the alarm correlation problem in fault detection [29, 18, 25]. This approach suits well-defined problems in cases where the environment is not very dynamic, but it does not adapt well to evolving distributed system environments [17]. Case-based reasoning [40] and coding-based methods [61, 41] offer potential solutions for fault identification and isolation, but cannot deal with uncertain or unstable situations in distributed systems. Finite State Machines (FSMs) were used to model fault propagation behaviours and duration [5, 50, 8, 58]. But this approach exhibits difficulties in scaling up to large and dynamic distributed systems. Kätker and Geihs provide model traversing techniques for fault isolation in distributed systems [30], however, their approach suffers from a lack of flexibility, especially when fault propagation is complex and not well structured. Most of these solutions are very sensitive to "noise" (such as loss of management information, delay in information collection and response, misinterpretation of alarms). As a consequence, they are unable to effectively deal with incomplete and imprecise management information in uncertain and dynamic environments. Probabilistic reasoning is another effective approach employed for fault detection in [11, 26, 56, 12].

Most of the current commercial management software, such as IBM Tivoli, HP OpenView, SunNet Manager, Cabletron Spectrum, and Cisco Works network management software, supports the integration of different management domains, collects information, performs remote monitoring, generates fault alarms, and provides statistics on management information. However, it lacks facilities for exact fault localisation and automatic execution of appropriate fault recovery actions. Experience in distributed systems' management shows that a typical metric for on-line fault identification is 95% fault location accuracy, while 5% of faults cannot be located and recovered in due time. Hence, for large distributed systems with thousands of managed components, it may be rather time-consuming and difficult to resolve the remaining five percent of problems in due time by an exhaustive search for the root causes of failure. In addition, exhaustive detection procedures may interrupt important system services.

Due to the complexity of distributed systems, it is mostly impossible to build precise models for fault management. During severe failures a number of entities might be unreachable causing unavailability, loss, or delay of network fault management messages. Also a transient entity failure may result in a number of unreliable alarms. A well-designed strategy for fault management of distributed systems should, therefore, operate efficiently in the case of redundant, incomplete and unreliable information. Because of losses or delays in data collection, it is also difficult to obtain complete and precise management information. In addition, due to complex dependency relationships between managed objects and inscrutable cause-effect relationships among faults and alarms, it is impossible to obtain a full and exact understanding of a managed system's behaviour from the viewpoint of systems management.

In daily management, specialist or expert knowledge is very important, but quantitatively expert knowledge is often expressed in imprecise ways, such as "very high", "normal", or "sometimes". When expert knowledge is to be incorporated into a management system, probabilistic approaches are needed for the quantitative expression of this kind of expert knowledge.

In this chapter, strategies to improve fault management in distributed systems in the face of uncertainty and imprecise management information are investigated, and inference facilities coping with fault localisation in distributed systems are presented. Section 11.2 introduces basic concepts and tasks of fault management, while Section 11.3 describes a novel approach to build management strategies on top of a Bayesian network model and related probabilistic reasoning techniques. Section 11.4 presents a probabilistic inference scheme for fault management, and Section 11.5 finally reports on measurements gained in a simulation environment.

11.2 Fault Management in Distributed Systems

Fault management consists of a set of functions that enable the detection, isolation, and correction of abnormal behaviour in a monitored system [2]. It includes fault management, performance management, security management, accounting management, and configuration management.

With the growth of the systems managed, it is estimated that the management centre of a medium size regional operator will be receiving tens of thousands of alarm notifications per day, which will render "manual" processing of all of them practically unfeasible [46]. Traditional manual management techniques are equally apt to harm systems rather than enhance them [57]. In addition, many of the notifications received do not contain original information, but are caused by components that are just affected by a fault. In fact, the occurrence of a single fault in a distributed system sometimes results in an avalanche of multiple alarm notifications sent from different components. Also a single notification may be generated by multiple faults. The dependency relationship between managed objects contributes to this situation. The fault of a given component may affect several other components, causing the fault's propagation. Hence, in a distributed system, it is difficult to identify the root cause from notifications received directly [12]. In fact, the problem of fault localisation is NP-complete [32]. Besides this, the following aspects must be taken into account as well.

1. Fault evidence may be ambiguous, inconsistent, redundant, noisy, and incomplete [16, 25, 32, 12].

Ambiguity in a set of alarms messages stems from the fact that the same alarm may be generated as a result of many different faults. Inconsistency results from a disagreement among devices with regard to the facts related to distributed system operation; one device may perceive a component to operate correctly, while another may consider the component faulty [16]. Noise can be made up of meaningless or redundant information, by streaming alarms, occasional spikes, frequent oscillations, and repeated occurrences. Incompleteness is a consequence of alarm loss or delays [25]. It is essential that a fault management system is able to create a consistent view of network operations even in the presence of ambiguous, inconsistent, or incomplete information [16].

2. A fault management system should provide means to represent and interpret uncertain data within the system knowledge and fault evidence [11, 16, 25, 35, 12].

A set of alarms generated by a fault may depend on many factors such as dependencies among distributed system devices, current configurations, services in use since fault occurrence, presence of other faults, or values of other distributed system parameters. Due to these factors, the system knowledge may be subject to inaccuracy and inconsistency. Fault evidence may also be inaccurate because of spurious alarms, which are generated by transient problems or as a result of overly sensitive fault detection mechanisms. When spurious symptoms occur, the management system may not be sure which alarms observed should be taken into account in the fault localisation process.

3. Fault management should be able to deal with hidden and complex dependencies [13].

In many approaches to distributed systems' management, the model of dependencies adopted very often presumes that, when a supported feature fails, all the components that depend on this feature will fail, too. Very often the strategy adopted in event correlation demands the construction of a complete model of the distributed system managed. The simplifications adopted in such models may render some components of the managed system "invisible" to the correlation process. This allows a fault occurred in an "invisible" system component to simulate the occurrence of a fault in another system component.

4. Fault management in large distributed systems should be performed in a distributed fashion [3, 31, 61].

Distributed systems are becoming more and more complex in terms of size, speed, and level of heterogeneity. It would be computationally prohibitive to process large volumes of information necessary to perform the fault localisation in such systems. It is also impractical to assume that the fault localisation process has access to the information of the entire system. Many researchers [3, 31, 61] have concluded that the fault localisation process in large distributed systems should be performed in a distributed fashion by a group of event management nodes, which divide data and processing efforts among them. Each of the managers governs a subset of system hardware or software components within boundaries marked by protocol layers or network domains. Errors propagate horizontally between peer devices within the same layer or vertically from the upper layers down to lower layers and vice versa between related services [59]. They may cross boundaries of management domains. As a result, the fault management system may be provided with indications of faults that did not happen in its management domain, or it may be unable to detect all symptoms of faults existing in its management domain [31, 59].

5. Fault management has to take into account temporal factors and dynamic changes in distributed systems [14].

In distributed systems, dynamic changes are unavoidable because of the continuous evolution of hardware, software, and distributed applications. Static management strategies can sometimes not work efficiently in dynamic environments. Distributed dynamic systems are apt to be unmanageable if the management systems cannot identify and deal with dynamic changes in due time. In dynamic environments, changes are pertinent to a time factor. Fault management should, therefore, concern the temporal factor of dynamic changes to catch the trend of changes occurring over time.

Each of the tasks listed above presents challenges to fault management. Solving these problems efficiently remains an open research area. These challenges also make it hard for a human system administrator to manage and understand all the tasks of smoothly operating a distributed system. The skills learned from any one distributed system may prove insufficient in managing a different one, thus rendering it difficult to generalise the knowledge gained from any given distributed system.

11.3 Bayesian Networks

As more and more knowledge-based systems are being developed for a large variety of problems, it becomes apparent that the knowledge required to solve these problems is often not precisely defined; it is of an imprecise nature instead. In fact, many real-life problem domains are fraught with uncertainty. Probabilistic models based on directed acyclic graphs (DAGs) have a long and rich tradition, which began with the geneticist Sewall Wright (1921).

Variants have appeared in many fields; in cognitive science and artificial intelligence, such models are known as Bayesian networks. Their initial development in the late 1970s was motivated by the need to model the top-down (semantic) and bottom-up (perceptual) combination of evidence in reading. The capability for bidirectional inference combined with a rigorous probabilistic foundation led to the rapid emergence of Bayesian networks as the method of choice for uncertain reasoning in artificial intelligence and expert systems, replacing earlier, ad-hoc rule-based intelligence and expert systems, replacing earlier, ad-hoc rule-based schemes [48, 54, 23, 28].

11.3.1 Applications

Bayesian Networks (BNs) are effective means to model probabilistic knowledge by representing cause-effect relationships cause-effect relationship between key components of a system managed. BNs can be used to generate useful predictions about future faults, and support decisions even in the presence of uncertain or incomplete information. BNs have been applied in various areas: [45, 60, 53] use them in medical diagnosis, [34] describes the application of BNs in fault diagnosis in Diesel engines, [9] presents methods in distributed web mining from multiple data streams based on the model of BNs, [22] describes an application of BNs for information retrieval taking into account a users' areas of interest, [7] presents a system that utilises a BN to debug very complex computer programs, [6] utilises BNs in encoding, representing, and discovering knowledge through processes that seek new knowledge on a given domain based on inferences on new data or on the knowledge already available [36], [4] uses BNs for map learning, and [55] applies the model of BNs to image sensor fusion.

11.3.2 Fundamental Concepts

A Bayesian network can be defined as a triplet (V, L, P), where V is a set of variables (nodes of the DAG) which represent propositional variables of interest, L is a set of causal links between the variables (the directed arcs between nodes of the DAG), which represent informational or causal dependencies among the variables, and P is a set of probability distributions defined by: $P = \{p(v \mid \pi(v)) \mid v \in V\}$; $\pi(v)$ denotes the parents $v_1, \ldots, v_n \in V$ of node v with $(v_1, v), \ldots, (v_n, v) \in L$. The dependencies are quantified by conditional probabilities for each node given its parents in the network. The network supports the computation of the probabilities of any subset of variables given evidence about another subset. In BNs, the information included in one node depends on the information of its predecessor nodes. Any direct predecessor v' of v with $(v', v) \in V$ denotes an effect node, while v represents its causes. Note that an effect node can also act as a causal node of other nodes. Causal relationships also have a quantitative side, namely their strengths. This is expressed by attaching numbers (probabilities) to the links.

A BN provides a complete description and very compact representation of a domain. It encodes joint probability distributions (JPD) in a compact manner. An important advantage of BNs is the avoidance of building huge JPD tables that include permutations of all network nodes. Rather, for an effect node, only the states of its immediate predecessor need to be examined. A complete joint probability distribution over n binary-valued attributes requires $2^n - 1$ independent parameters to be specified. In contrast, a BN over n binary-valued attributes, in which each node has at most k parents, requires at most $2^k n$ independent parameters. To make this concrete, suppose we have 20 nodes ($n = 20$) and each node has at most 5 parent nodes ($k = 5$). Then, the BN requires only 640 numbers, but the full joint probability distribution requires a million. Clearly, such a network can encode only a very small fraction of the possible distributions over these attributes, since it has relatively few parameters. The fact that the structure of a BN eliminates the vast majority of distributions from consideration indicates that the network structure itself encodes information about the domain. This information takes the form of conditional independence relationships that hold between attributes in the network.

11.4 Probabilistic Inference in Fault Management

As discussed before, uncertainty is unavoidable in distributed systems. The cause-effect relationship between symptoms and possible causes is inherently non-deterministic. Probabilistic models are considered to provide an accurate representation of such phenomena and, hence, BNs are appropriate models for probabilistic fault management. From the management perspective, in case of a fault, a typical action is to trace the root causes from detected symptoms suggested by alarm events. In this context, backward inference in a probabilistic model, which infers from effects to causes, is a proper basis for distributed fault management.

11.4.1 Bayesian Networks for Fault Management

The development of a diagnostic BN requires a deep understanding of the cause-effect relationships in a domain. One advantage of BNs is that the knowledge modeled provides a clear insight in the relationship between effects and causes. Other than in neural networks, this knowledge is not just represented as a black box. In addition, compared with other logic models, BNs also provide a more fine-grained quantitative evaluation of probabilistic phenomena.

In distributed systems, some faults may be directly observable, i.e., they are problems and symptoms at the same time. However, many types of faults are unobservable due to their intrinsically unobservable nature, or the lack of management functionality necessary to provide indications of a fault's existence. Some faults may be partially observable. A management system normally indicates an occurrence of a fault but the information provided is often not sufficient to precisely locate the fault.

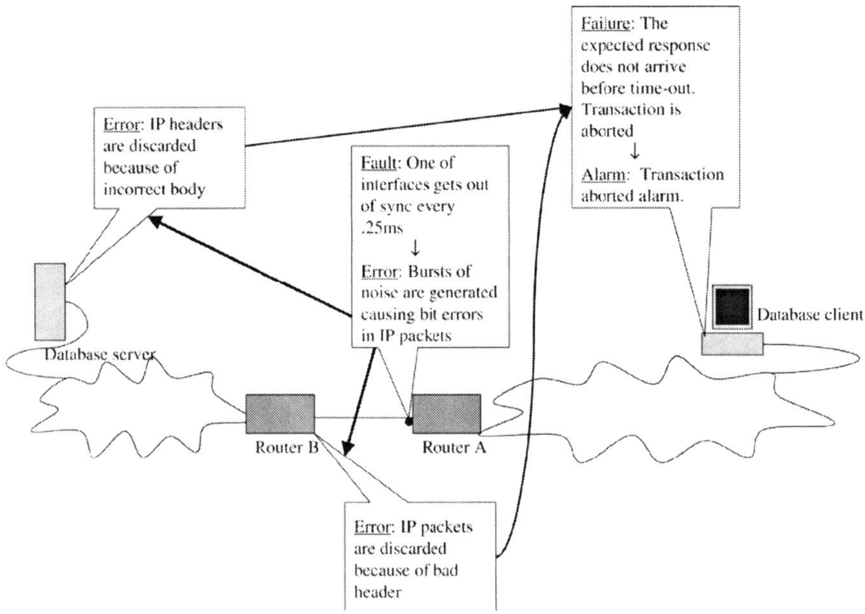

Fig. 11.1: Fault propagation in a distributed system

In Fig. 11.1, a simple distributed system is presented in a situation in which a client accesses a remote database server. An interface of one of the routers between the client and the server turns intermittently out of synchronisation causing bursts of bit errors in transmitted IP datagrams. As a result, many IP datagrams passing through the router are rejected by the next router because of header errors, or by the server itself because of a corrupted datagram body. The client does not receive any response to its query and times out. This example illustrates how a seemingly invisible fault manifests itself through a failure at a location distant from the location of the fault. Since most faults are not directly observable, the management system has to infer their existence from information provided by the received alarm messages. The information carried within reported alarms may include the following: the identity of the object that generated the alarm, the type of failure condition, a time stamp, an alarm identifier, a measure of severity of the failure condition, a textual description of the failure and other relevant information [27, 52].

Hence, in a distributed system, a single fault may cause a number of alarms that are delivered to the network management centre. Multiple alarms may be a result of (1) fault re-occurrence, (2) multiple invocations of a service provided by a faulty component, (3) generating multiple alarms by a device for a single fault, (4) detection of and issuing a notification about the same network fault by many devices simultaneously, and (5) error propagation to other devices causing them to fail and, as a result, generate additional alarms [27]. It may be argued that typical distributed systems provide plenty of information necessary to infer the existence of faults [61].

Due to the dense knowledge representation of BNs, they can represent large amounts of interconnected and causally linked data as they occur in distributed systems. Besides the arguments already brought up in favour of the BN model for fault management, the following benefits are worth mentioning:

- Due to their grounding in probability theory, BNs have the capability to handle noisy, transient, and ambiguous data, which is unavoidable in complex distributed systems.
- BNs support inference techniques based on the combination of:
  - statistical data empirically surveyed during network operation;
  - subjective probabilities supplied by specialists, and
  - information (like evidences or alarms) received from the distributed system.
- BNs provide a compact and well-defined problem space, because they use an exact solution method for any combination of evidences or set of faults. By the evaluation of a BN, it is possible to obtain approximate answers, even when the existing information is incomplete or imprecise; as new information becomes available, BNs allow a corresponding improvement in the precision of correlation results.
- BNs are abstract mathematical models. In distributed systems' management, BNs can be used at different levels for different management intentions, or they can be used for particular applications or services. When a connection service is considered, for instance, the physical topology would be the basis to construct a BN, while, when a distributed service is taken into account, logical dependencies between service and between managed objects will provide the foundation for constructing a BN model.

Designing Bayesian Networks for Distributed Fault Management

We represent uncertainty about the dependencies between the components of a distributed system by assigning probabilities to the links in the dependency or causality graph [32, 35]. Some commonly accepted assumptions in this context are: (a) given a fault a, the occurrences of faults b and c that may be caused by a are independent; (b) given the occurrences of faults a

and b that may cause event c, whether a actually causes c is independent of whether b causes c (the "or" relationship between alternative causes of the same event); and (c) root faults are independent of one another.

When a distributed system is modeled as a BN, two important processes need to be resolved:

1. Ascertain the dependency relationship between managed entities.

A distributed system consists of a number of managed objects. An object is a component of the distributed system that has a separate and distinct existence. At the physical level, an object can be a network, a node, a switch, a layer in a protocol stack, a virtual link, a physical device like an optical fibre, a piece of cable, a hardware component, and others. At the logic level, an object can be a software component or a service, such as a process, a piece of code, a URL, a servlet, or a service request. Objects in a distributed system consist of other objects down to the level of the smallest objects that are considered indivisible. An indivisible object is defined as a terminal object that cannot be decomposed further. The concept of division and appropriate level of division are system- and application-dependent.

Objects in a distributed system may depend on each other in rather complex ways. These dependencies are very important for the alarm correlation and fault identification process. In most cases, a failure in one object has side effects on other objects that depend on it. For example, a link failure has an effect on other resources in the network, such as that connections on the various layers using the link will experience timeouts. The knowledge of these dependencies provides valuable information for the purpose of alarm correlation and fault localisation.

Dependency analysis explores causal dependencies among objects and data items with the goal to trace the fault symptoms back to the cause. This is an often-used trouble-shooting technique applicable to any system that is based on collaboration of independent entities. For instance, deadlocks in databases may be diagnosed by following transactions that are blocked waiting for other transactions to occur. In computing, there exist many different kinds of dependencies. However, not all references and interactions actually represent causal dependencies that are relevant for diagnosis. Hence, the dependencies, which are pertinent to the purpose of management, need to be taken into account. The dependencies among distributed entities can be modeled by assigning probabilities to the links in the dependency graph [32, 35]. This dependency graph can be transformed into a BN with certain special properties [24].

Two models are useful to capture the dependency between co-operating entities in distributed systems:

Functional model (from the viewpoint of users). A functional model defines generic service dependencies and establishes the principal constraints to which the other models are bound. A functional dependency is an association between two components, typically captured first at design time, which says that one component requires some services from another. The functional dependence between logical objects is determined by the implementation and functional support relationships and originates in a graph, from which it is possible to correlate a set of state changes (which may be considered as the "signature" of a problem) to the original cause of the problem. The functional model is utilised by a "network state estimator" to correlate the changes in the network state. The state changes are reported by the received alarms, to which information exogenous to the network (such as those related to climatic situations) is added.

Structural model (from the viewpoint of system implementers). The structural model contains detailed descriptions of software and hardware components that realise services. A structural dependency contains detailed information, and is typically captured first at deployment or installation time.

2. *Obtaining dependency measurements.*

The faults and anomalies in distributed systems can be identified based on the statistical behaviour of variables maintained in the management information base (MIB) and the recordings in log files. BNs represent causes and effects between observable symptoms and an unobserved problems, so that, when a set of evidences is observed, the most likely causes can be determined by inference mechanisms.

Single-cause (fault) and multi-cause (fault) are two kinds of general assumptions when considering the dependencies between managed entities in distributed systems' management. In BNs, a non-root node may have one or several parents (causal nodes). Single-cause means that any of the causes must lead to the effect indicated. So the dependencies between causes and the effect in a single-cause fault are denoted by:

$$P(\overline{e} \mid c_1, \ldots, c_n) = \begin{cases} 100\% \text{ if } \exists i, i \in [1, n], \text{such that } c_i = F(False); \\ 0\% \quad \text{otherwise.} \end{cases} \quad (11.1)$$

where e denotes the effect node, and $c_1, \ldots, c_n$ denote the causes of e.

In contrast, a multiple-cause fault can be observed when several causes need to to occur simultaneously to generate some effect. So, dependency measurement includes various possibilities based on the particular problem domain. In the above description, the states of the objects are identified as T(*True*) or F(*False*). In complex systems, it is possible that managed objects hold more than two states. The measurement of dependencies between managed objects can be obtained by the following methods:

- management information statistics,
- the empirical knowledge of experts, and
- targeted experiment.

In [21], Hasselmeyer argues that the dependencies among distributed co-operating components should be maintained and published by specific services, and proposes a schema that allows these dependencies to be identified. Some researchers have performed useful work to discover dependencies from the application point of view on distributed systems [20, 33, 19].

Despite all the methods cited in this section, one should notice that obtaining dependency information in an automatic fashion is still an open research problem.

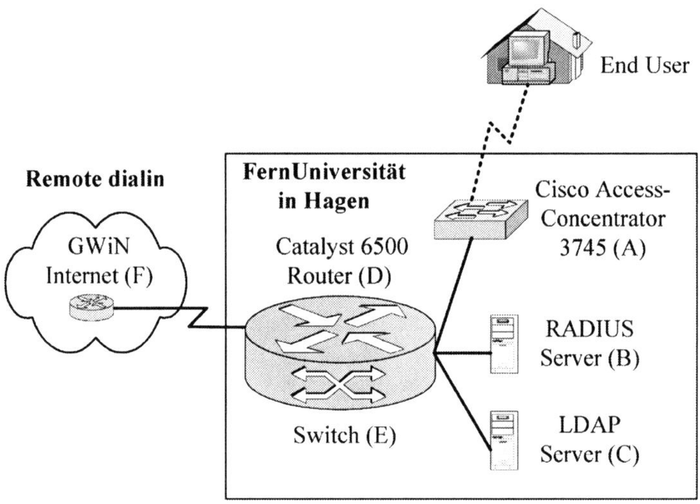

Fig. 11.2: Example of a campus network

Fig. 11.2 shows part of the campus network of FernUniversität in Hagen. We shall use this example to illustrate some of the techniques mentioned before. When examining the connection service for end users in the example presented in Fig. 11.2, the corresponding BN is shown in Fig. 11.3. Here the dependency probability for the connection service is derived from the record of the connection failure between objects, thus adaptations to load-balancing mechanisms within a router do not interfere with this model and can be ignored. The arrows in the BN denote the dependency from causes to effects. The weights of the links denote the probability of the dependency between the objects. When one node has several parents (causes), the dependency between the parents and their child can be denoted by a joint probability distribution . In this example, the components F and E are the causes for component D. The annotation $P(\overline{D}|\overline{E}F) = 100\%$ denotes the probability of the non-availability of component D, which is 100% when component F is in order but component E is not. Other annotations can be read similarly. In this example, some evidences, such as the status of component D, are easily detected through regular monitoring. But the causes of a failure of component D are not always obvious. One important task in management is to infer hidden factors from the available evidence.

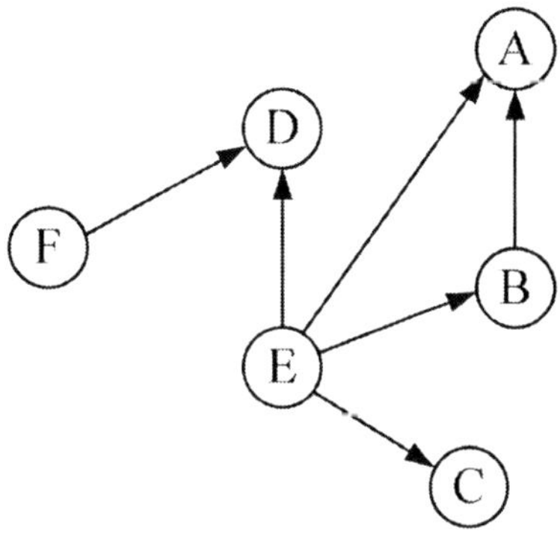

Component A: Access-Server

Component B: RADIUS-Server

Component C: LDAP-Server

Component D: Router-Connection to Internet-Provider

Component E: Switch for LAN Access

Component F: GWiN and Internet Access

$$
\begin{aligned}
p(\bar{A}) &= 0.0092\% \\
p(\bar{B}) &= 0.0097\% \\
p(\bar{C}) &= 0.0097\% \\
p(\bar{D}) &= 0.376\% \\
p(\bar{E}) &= 0.063\% \\
p(\bar{F}) &= 0.035\% \\
p(\bar{A}|BE) &= 0\% \\
p(\bar{A}|\bar{B}E) &= 100\% \\
p(\bar{A}|B\bar{E}) &= 100\% \\
p(\bar{A}|\bar{B}\bar{E}) &= 100\% \\
p(\bar{D}|EF) &= 0\% \\
p(\bar{D}|\bar{E}F) &= 100\% \\
p(\bar{D}|E\bar{F}) &= 20\% \\
p(\bar{D}|\bar{E}\bar{F}) &= 100\% \\
p(\bar{B}|\bar{E}) &= 100\% \\
p(\bar{B}|E) &= 0\% \\
p(\bar{C}|\bar{E}) &= 100\% \\
p(\bar{C}|E) &= 0\%
\end{aligned}
$$

Fig. 11.3: Bayesian network for the system depicted in Fig. 11.2

In terms of precision, the behaviour of a BN reflects the quality and the level of detail of its structure, which stems from the object system model. Another factor that affects the BN model is the precision of the value of the conditional probabilities.

11.4.2 Probabilistic Inference in Bayesian Networks for Distributed Fault Management

The semantics of a BN determines the conditional probability of any event given any other event. When computing such a conditional probability, the conditioning event is called the evidence, while the event for which we want to determine its conditional probability given the evidence is called the query. The general capability of a BN to compute conditional probabilities allows it to exhibit many particular patterns of reasoning (by inference). In case of fault management, we only consider backward inference (evidential reasoning), which is the basic operation of fault diagnosis.

Basic Model of Backward Inference in Bayesian Networks

The fault localisation algorithm based on the fault propagation model (FPM), which we shall introduce now, should return a number of fault hypotheses that best explain a set of symptoms observed.

The most common approach towards reasoning with uncertain information about dependencies in distributed systems is probabilistic inference, which traces causes from effects. The task of backward inference amounts to find the most probable instances of some hypothesis variables, given the observed evidence. Let E be the set of effects (evidence) we can observe, and C be the set of causes. Then, the inference from effects to causes is denoted by $P(c|E')$, $E' \subseteq E$, $c \in C$.

Before discussing the backward inference in BNs, a simplification model will be examined. In BNs, one node has one or several parents (if it is not a root node). We denote the dependency between parents and their child by a joint probability distribution (JPD).

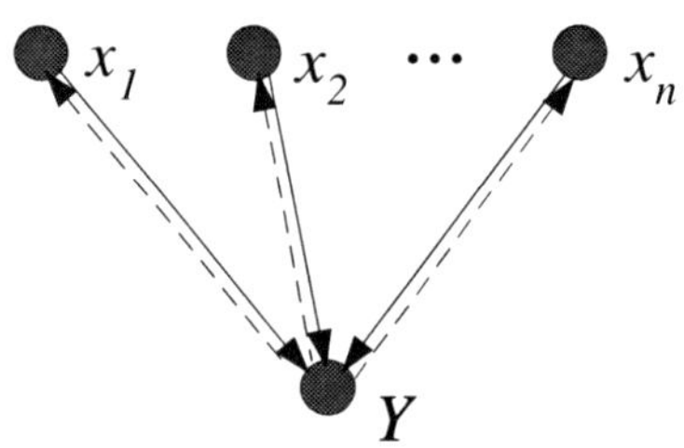

Fig. 11.4: Basic model for backward inference in Bayesian networks

Fig. 11.4 shows the basic model for backward inference in BNs. Let $X = (x_1, x_2, \ldots, x_n)$ be a set of causes and Y be the effect of X. A solid arrow denotes the causal relationship from a cause to an effect. A dashed arrow denotes the inverse relationship from an effect to a cause. According to the definition of BNs, the following variables are known: $P(x_1), P(x_2), \ldots, P(x_n)$, $P(Y|x_1, x_2, \ldots, x_n) = P(Y|X)$. Here, $x_1, x_2, \ldots, x_n$ are mutually independent, so that the following equations hold:

$$P(X) = P(x_1, x_2, \ldots, x_n) = \prod_{i=1}^{n} P(x_i) \tag{11.2}$$

$$P(Y) = \sum_X [P(Y|X)P(X)] = \sum_X [P(Y|X) \prod_{i=1}^{n} P(x_i)] \tag{11.3}$$

By Bayes' theorem we conclude:

$$P(X|Y) = \frac{P(Y|X)P(X)}{P(Y)} = \frac{P(Y|X)\prod_{i=1}^{n} P(x_i)}{\sum_{X}[P(Y|X)\prod_{i=1}^{n} P(x_i)]} \tag{11.4}$$

which computes to:

$$P(x_i|Y) = \sum_{X-\{x_i\}} P(X|Y) \tag{11.5}$$

According to Eqs. (11.2) – (11.5), the individual conditional probability $p(x_i|Y)$ can be achieved from the JPD $p(Y|X)$, $X = (x_1, x_2, \ldots, x_n)$.

In Fig. 11.3, E, F are the parents of D. The calculation of backward inference from D to E and F is shown in Table 11.1. The backward dependency from effects to causes can be obtained from Eq. (11.5). The dashed arrowed lines in Fig. 11.4 denote the backward inference $P(x_i|Y)$ from effect Y to an individual cause x_i, $i \in [1, 2, \ldots, n]$.

Table 11.1: Calculation of backward inference in Bayesian networks

$P(E)$ $P(F)$	$P(\overline{E})[0.063\%]$ $P(\overline{F})[0.035\%]$	$P(\overline{E})[0.063\%]$ $P(F)[99.965\%]$	$P(E)[99.937\%]$ $P(\overline{F})[0.035\%]$	$P(E)[99.937\%]$ $P(F)[99.965\%]$	
$P(E,F)$	0.000%	0.063%	0.035%	99.900%	1
$P(\overline{D}\|E,F)$	100.000%	100.000%	20.000%	0.000%	
$P(\overline{D},E,F)$	0.000%	0.063%	0.007%	0.000%	0.0007
$P(E,F\|\overline{D})$	0.000%	90.000%	10.000%	0.000%	1

When a fault is detected in component D of the system modeled in Fig. 11.3, then based on Eqs. (11.7) and (11.8) below we obtain $P(\overline{F}|\overline{D}) = 90.000\%$ and $p(\overline{E}|\overline{D}) = 10.000\%$. This can be interpreted as follows: when component D is not available, the probability of a fault in component F is 90.00%, and the probability of a fault in component E is 10.00%. In this situation, we can make the hypothesis that F is more likely the cause of the fault in D. (Remember that here the fault related to the connection service is considered, only.)

$$P(E,F|\overline{D}) = \frac{P(\overline{D}|E,F)P(E,F)}{P(\overline{D})} \tag{11.6}$$

$$\begin{aligned} P(\overline{E}|\overline{D}) &= P(\overline{E},F|\overline{D}) + P(\overline{E},\overline{F}|\overline{D}) \\ &= 90.000\% + 0.000\% = 90.000\% \end{aligned} \tag{11.7}$$

$$\begin{aligned} P(\overline{F}|\overline{D}) &= P(\overline{F},E|\overline{D}) + P(\overline{F},\overline{E}|\overline{D}) \\ &= 10.000\% + 0.000\% = 10.000\% \end{aligned} \tag{11.8}$$

Strongest Dependency Route Algorithm for Backward Inference

The objective of any fault management system is to minimise the time to localise and repair a fault. The time to localise a fault is the sum of the time to propose possible fault hypotheses (fault identification) , and the time to perform testing to verify these hypotheses. The time required for testing is affected by the number of managed objects that must be tested. Thus, if the network management system is about to identify the source of a fault, it is desirable that the minimum number of tests be performed. Hence, there are two main aspects, subject to optimisation, of any fault localisation process: accuracy of the hypothesis it provides, and time complexity of the fault identification algorithm it uses. To optimise the time to localise a fault, we should maximise the accuracy of the proposed hypotheses, and minimise the time complexity of the fault identification process. Normally, when some faults are reported in a distributed system, the most urgent tasks are to locate the causes, to bring faulty states back to normal and, possibly, to improve the performance of the system. The key factors that are related to the defect in the system should be identified. The *strongest dependency route* (SDR) algorithm is proposed to resolve these tasks based on probabilistic inference. Before we describe the SDR algorithm, the strongest cause and strongest dependency route are defined as follows with C a set of causes and E a set of effects in a BN:

Strongest Cause For $e_i \in E$, let C_i be the set of causes based on effect e_i. Then, c_k is the strongest cause for effect e_i iff $p(c_k|e_i) = Max[p(c_j|e_i), c_j \in C_i]$.

Strongest Dependency Route Let $R = \{R_1, R_2, \ldots, R_m\}$ be the set of routes from effect $e_i \in E$ to its cause $c_j \in C$, and let M_k be the set of transit nodes, which are located on the path from e_i to c_j in route $R_k \in R$. Iff $p(c_j|M_k, e_i) = Max[p(c_j|M_t, e_i), t = (1, 2, \ldots, m)]$, then R_k is the strongest route between e_i and c_j.

Based on the definitions, the SDR algorithm is described in the sequel.

Pruning of a Bayesian Network

When a concrete problem domain is considered, a common strategy is to omit some variables that are not related to the problem domain. To achieve this, we define a pruning operation as follows:

Pruning Algorithm

Prune $(BN = (V, L, P), E_k \subseteq E, E_k = \{e_1, e_2, \ldots, e_k\})$
    new $BN' = (V', L', P')$;
    $V' = E_k$;   //add E_k to V'
    $L' = \varnothing$;

for $e_i \in E_k$ $(i = 1, \ldots, k)$
  $v_i = e_i$,
  while $v_i \neq NIL$ **do**
    $V' = V' \cup \{\pi(v_i)\}$, //add vertex $\pi(v_i)$ to V'
    $v_i \leftarrow \pi(v_i)$,
    $L' = L' + < \pi(v_i), v_i >$; //add edge $(\pi(v_i), v_i)$ to L'
return BN';

Generally speaking, multiple effects (symptoms) may be observed at a single moment, so $E_k = \{e_1, e_2, \ldots, e_k\}$ is defined as the set of initial effects. In the pruning operation, every step just integrates the current nodes' parents into the sub-BN and omits their brother nodes, because brother nodes are independent from each other. The pruned graph is composed of the effect nodes E_k and all their ancestors, and the end nodes construct the set of causes based on the effect node e_i. The pruning algorithm will cut off the unrelated nodes of the effects considered in a BN. It provides a simplified method to deal with the backward inference in a sub-BN, which is embedded in a large and complex BN.

Strongest Dependency Route (SDR) Algorithm

After the pruning algorithm has been applied to a BN, a simplified sub-BN is obtained. Between every cause and effect, there may be more than one dependency route. The questions now are: which route is the strongest dependency route and, among all causes, which is the strongest cause?

The SDR algorithm is used to trace the strongest dependency route, to locate the causes, and to generate the dependency sequence among the causes based on particular effects in the BN. The algorithm uses product calculation to measure the serial strongest dependencies between effect nodes and causal nodes. In the SDR algorithm, multiple effects can be considered. Suppose $E_k \subseteq E$ with $E_k = \{e_1, e_2, \ldots, e_k\}$. If $k = 1$, the graph will degenerate to a single-effect model. The backward dependency calculation is based on Eqs. (11.2)–(11.5). When several effects (symptoms) E_k are observed at the same time, the effects are instantiated and $P(e_i) = 1$ or $P(\overline{e_i}) = 1$ for $i = (1, \ldots, k)$. Here, only the state $P(e_i) = 1$ is considered, e_i denotes a fault evidence (symptom).

SDR Algorithm:
Input:
$BN = (V, L, P)$;
$E_k = \{e_1, e_2, \ldots, e_k\}$: the set of initial effect nodes in BN, $E_k \subseteq V$

Output:
T: a spanning tree of the BN rooted at $E_k = \{e_1, \ldots, e_k\}$, whose path from e_i to each causal node is a strongest dependency route , and a vertex-labeling giving the dependency probability from e_i to each causal node.

Variables:

$depend[v]$: the strongest probability dependency between v and all its descendants;

$\psi(l)$: the conditional probability of $P(v|u)$, v as the parent of u, sharing the link l; $\psi(l)$ can be calculated from JPD of $P(u|\pi(u))$ based on Eqs. (11.2)–(11.5);

$\varphi(l)$: the temporal variable which records the strongest dependency between nodes.

Initialise the SDR tree T as E_k; // E_k is added as root nodes of T
Initialise the set of frontier edges for tree T as empty;
Write label 1 on e_i //$e_i \in E_k$
While SDR tree T does not yet span the BN
  For each frontier edge l in T
    Let u be the labeled endpoint of edge l;
    Let v be the unlabeled endpoint of edge e (v is one parent of u);
    Let $\psi(l) = P(v|u)$;
    Set $\varphi(l) = depend[u] * \psi(l) = depend[u] * P(v|u)$;
  Let l be a frontier edge for BN that has the maximum φ-value;
  Add edge l (and vertex v) to tree T;
  $depend[v] = \varphi(l)$;
  Write label $depend[v]$ on vertex v;
Return SDR tree T and its vertex labels;

The result of the SDR algorithm is a spanning tree T. Every cause node $c_j \in C$ is labeled with $depend[c_i] = P(c_j|M_k, e_i)$ (with $e_i \in E_x$), M_k is the set of transit nodes between e_i and c_j in route $R_k \in R$.

According to the values of the labels in the cause nodes, a cause sequence can be obtained. This sequence is important for primary fault diagnosis and related maintenance operations. Meanwhile, using Depth-First search on the spanning tree, the strongest route between effect nodes and cause nodes can be achieved, too.

Suppose one reasoning chain of a series of variables (from cause to effect) is: $\delta_0 \rightarrow \delta_1 \rightarrow \ldots \rightarrow \delta_n$, then the JPD $P(\delta_0|\delta_n, \delta_n - 1, \ldots, \delta_1)$ is considered as the backward reasoning value based on the backward serial variables (from effect to cause): $\delta_n \rightarrow \delta_{n-1} \rightarrow \ldots \rightarrow \delta_0$.

Proof of the Strongest Path Property of the SDR Algorithm

Now we prove the core property of the SDR algorithm [15], which is to compute the strongest path among all possible paths from an effect node to a possible cause node. Let the route $< e_i, u_1, u_2, \ldots, u_n, c_j >$ be the strongest dependency route for a given BN, and effect nodes e_i and $< e_i, \delta_1, \delta_2, \ldots, \delta_m, c_j >$ be any route from e_i to c_j. Then

$$P(u_1|e_i)*P(u_2|u_1)*\ldots*P(c_j|u_n) \geq P(\delta_1|e_i)*P(\delta_2|\delta_1)*\ldots*P(c_j|\delta_m) \quad (11.9)$$

The above formula demonstrates that the product of all backward dependencies along the strongest dependence route maintains the maximum of those in all routes. When we define $weight(u, \pi(u)) = -lg(p(\pi(u)|u))$, Eq. (11.9) can be transformed to:

$$\begin{aligned} &weight(e_i, u_1) + weight(u_1, u_2) + \ldots + weight(u_n, c_j) \\ &\leq weight(e_i, \delta_1) + weight(\delta_1, \delta_2) + \ldots + weight(\delta_m, c_j) \end{aligned} \quad (11.10)$$

Lemma 1. *When a vertex u is added to the spanning tree T computed by means of SDR, $d[u] = weight(e_i, u) = -lg(depend[u])$.*

As $0 < depend[\delta_j] \leq 1$, so $d[\delta_j] \geq 0$. Note $depend[\delta_j] \neq 0$, or else there is no dependency relationship between δ_j and its offspring .

Proof: Suppose the contrary, that at some point the SDR algorithm first attempts to add a vertex u to T for which $d[u] \neq weight(e_i, u)$ as shown in Fig. 11.5. Consider the situation just prior to the insertion of u, and the true strongest dependency route from e_i to u. Since $e_i \in T$, and $u \in V \backslash T$, at some point this route must first take a jump out of T. Let (x, y) be the edge taken by the path, where $x \in T$, and $y \in V \backslash T$ (it may be that $x = e_i, y = u$). We now prove that $d[y] = weight(e_i, y)$.

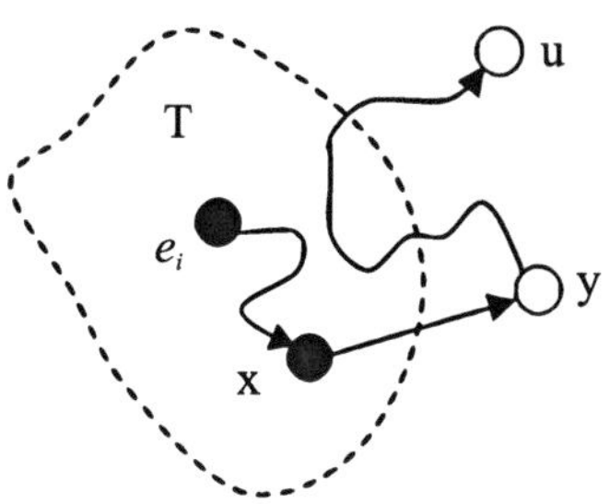

Fig. 11.5: Illustration to the SDR algorithm's proof

We have computed x, so

$$d[y] \leq d[x] + weight(x, y) \quad (11.11)$$

Since x was added to T earlier, by hypothesis we have:

$$d[x] = weight(e_i, x) \quad (11.12)$$

Since $< e_i, \ldots, x, y >$ is a subpath of the strongest dependency route, by Eq. (11.12) we obtain:

$$weight(e_i, y) = weight(e_i, x) + weight(x, y) = d[x] + weight(x, y) \quad (11.13)$$

By Eqs. (11.11) and (11.13) we obtain:

$$d[y] \leq weight(e_i, y) \tag{11.14}$$

Hence,

$$d[y] = weight(e_i, y) \tag{11.15}$$

Now note that y appears midway on the route from e_i to u, and all subsequent edges are positive, we have:

$$weight(e_i, y) < weight(e_i, u) \tag{11.16}$$

and, thus, the following relation holds:

$$d[y] = weight(e_i, y) < weight(e_i, u) \leq d[u] \tag{11.17}$$

Therefore, y would have been added to T before u, which contradicts our assumption that u is the next vertex to be added to T.

Since the calculation is correct for every effect node, it is also true for multiple effect nodes in tracing the strongest dependency route. When the algorithm terminates, all vertices are in T, thus all dependency (weight) estimates are correct.

Complexity Analysis of the SDR Algorithm

To determine the complexity of the SDR algorithm, we observe that every link (edge) in a BN is only calculated once, so the size of the links in a BN corresponds the algorithm's complexity. It is known in a complete directed graph that the number of edges is $\frac{n(n-1)}{2} = \frac{n^2-n}{2}$, where n is the size of the nodes in the pruned spanning tree of a BN. Normally, a BN is an incomplete directed graph. So the number of calculation steps of SDR is less than $\frac{n^2-n}{2}$. The complexity of SDR is $O(n^2)$.

According to the SDR algorithm, multiple effect nodes can be observed. From the spanning tree, the strongest routes between effects and causes can be obtained by a depth-first search. The value of $depend[c_j]$ $(c_j \in C)$ generates a dependency ranking of causes based on E_k. This dependency sequence is a useful reference for fault diagnosis and system recovery.

In terms of precision, the behaviour of a BN reflects the quality and the detailed levels of its structure, which stem from the object network model. Another factor that affects the precision of the Bayesian alarm correlation process are the quality and detailed levels of the alarms to be correlated. These two factors affect the precision in any alarm correlation process, which is independent of the adapted approach. The third factor, i.e., the precision of the conditional probabilities' values, also contributes to the precision of the correlation process.

Related Algorithms for Probabilistic Inference

Various types of inference algorithms exist for BNs. All in all, they can be classified into two types of inferences: exact inference [39, 48, 49], and approximate inference [44]. Each class offers different properties and works better on different classes of problems, respectively. It is very unlikely, however, that a single algorithm can solve all possible problem instances effectively. Any resolution is always based on a particular requirement. It is true that almost all computational problems and probabilistic inferences using general BNs have been shown to be NP-hard by Cooper [10].

In the early 1980s, Kim and Pearl published an efficient message propagation inference algorithm for polytrees [37, 47]. The algorithm is exact and has polynomial complexity in the number of nodes. But it works for singly connected networks, only. Pearl also presented an exact inference algorithm for multiply connected networks, called loop cutset conditioning algorithm [47]. The latter changes the connectivity of a network and renders it singly connected by instantiating a selected subset of nodes referred to as a loop cutset. The resulting singly connected network is solved by the polytree algorithm and, then, the results of each instantiation are weighted by their prior probabilities. The complexity of this algorithm results from the number of different instantiations that must be considered. This implies that the complexity grows exponentially with the size of the loop cutest being $O(d^c)$, where d is the number of values that the random variables can take, and c is the size of the loop cutset. It is thus important to minimise the size of the loop cutset for a multiply connected network. Unfortunately, the loop cutset minimisation problem is NP-hard. A straightforward application of Pearl's algorithm to an acyclic digraph comprising one or more loops invariably leads to insuperable problems [38, 44].

Another popular exact BN inference algorithm is Lauritzen and Spiegelhalter's clique-tree propagation algorithm [39]. It is also called a clustering algorithm. First, it transforms a multiply connected network into a clique-tree by clustering the triangulated moral graph of the underlying undirected graph and, then, performs message propagation over the clique-tree. The clique-tree propagation algorithm works efficiently for sparse networks, but can be extremely slow for dense networks. Its complexity is exponential in the size of the largest clique of the transformed undirected graph. In general, the existing exact BN inference algorithms share the property of run-time exponentiality in the size of the largest clique of the triangulated moral graph, which is also called the induced width of the graph [39]. Furthermore, it is difficult to record the internal nodes and the dependency routes between particular effect nodes and causes. In distributed systems' management, the states of internal nodes and the key routes, which connect effects and causes, are important for management decisions. Moreover, the sequence of localisation for potential faults can be a reference for system managers and, thus, is very useful. It is also important for system performance management to identify the relevant key factors. Few algorithms provide a satisfactory solution for this case.

Compared to other algorithms, the SDR algorithm belongs into the class of exact inferences, and provides an efficient method to trace the strongest dependency routes from effects to causes, and to track the dependency sequences of the causes. It is useful in fault location and beneficial for performance management. Moreover, it can treat multiply connected networks modeled as DAGs.

11.5 Simulation Measurement for Probabilistic Inference

We have developed a simulation program for BNs to verify the effectiveness and efficiency of the proposed approach towards backward inference and prediction properties with respect to fault management. The Twister Generalised Feedback Shift Register Generator (TGFSRG) [42, 43] is used to generate pseudo-random numbers for the simulation. Fig. 11.6 presents the topology of the BN produced by the simulation program. The sample BN has 22 nodes in total, 5 for source nodes, and 3 for effect nodes. The cause set is $C = \{A, B, C, D, E\}$, the effect set is $E = \{T, U, V\}$. In the production of the JPD of the BN, the valid number of tests is 50000.

Suppose the effect nodes $T = 1$, $U = 1$ and $V = 1$ were detected. After the SDR algorithm was applied on the BN in Fig. 11.6, a spanning tree with strongest dependency routes is produced. It is rooted on nodes T, U, V, and holds the entire strongest dependency route as depicted in Fig. 11.7.

The sequence of possible causes of T, U, V and the strongest dependency route can be obtained from the simulation program as follows:

$$\begin{aligned} &P(A = 1|T = 1, U = 1, V = 1) > P(D = 1|T = 1, U = 1, V = 1) \quad (11.18)\\ &> P(C = 1|T = 1, U = 1, V = 1) > P(B = 1|T = 1, U = 1, V = 1)\\ &> P(E = 1|T = 1, U = 1, V = 1) \end{aligned}$$

In the simulation, the comparison of the detection rate between SDR detection and random or exhaustive detection is demonstrated in Table 11.2. From simulation results, we notice that the proposed SDR algorithm not only identifies multiple problems at once, but also holds much lower calculation complexity than related exact inference algorithms.

Theoretically, the simulation program can produce a BN with any number of nodes and any probability distribution. The simulation model is shown to be reasonable close to real-life systems including large distributed systems.

11.6 Conclusions and Future Work

After a detailed study of the challenges and solutions towards fault management in distributed systems, Bayesian networks (BNs) were proposed to rep-

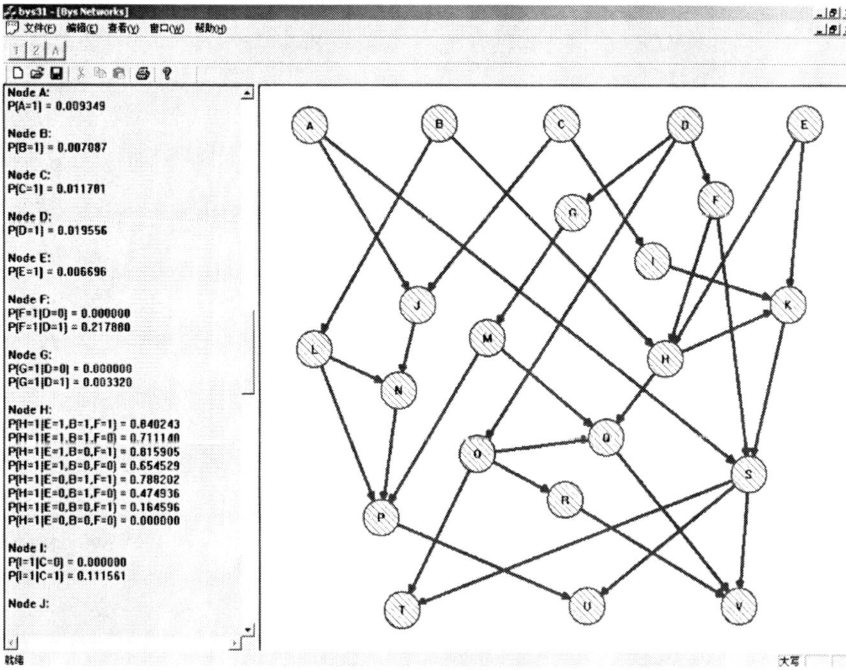

Fig. 11.6: Simulation of a Bayesian network with 22 nodes

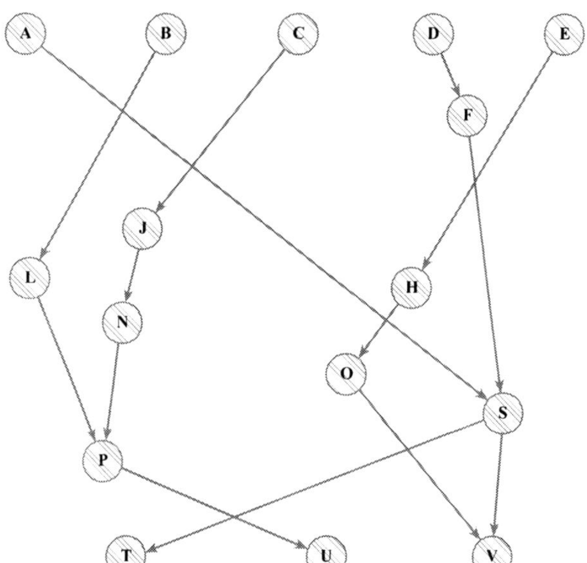

Fig. 11.7: Spanning tree rooted in T, U and V resulting SDR applied to the BN in Fig. 11.6

Table 11.2: Comparison of the detection rates d of SDR and random or exhaustive detection

No. of total detected nodes	1	2	3	4	5
d of SDR detection from T	0.844268	0.978437	0.992216	0.998865	1.000000
d of SDR detection from U	0.582127	0.811281	0.903791	0.995415	1.000000
d of SDR detection from V	0.799663	0.926743	0.960729	0.986949	1.000000
d of SDR detection from T,U,V	0.850160	0.998092	0.999973	1.000000	1.000000
d of random or exhaustive detection	0.200000	0.400000	0.600000	0.800000	1.000000

resent knowledge about its managed objects and their dependencies. Based on this model, probabilistic reasoning was proposed to apply to determine the causes of failures or errors. A novel algorithm, strongest dependency route (SDR), was introduced in this chapter and shown that it provides an efficient method in backward inference to trace causes from effects, which allows for better fault location.

However, the structure, topologies and dependencies between managed objects change dynamically. Therefore, we need to accommodate sustainable changes and maintain a healthy management system based on learning strategies that allow to modify the cause-effect relationship in the corresponding BN. The Bayesian paradigm does not provide, however, direct mechanisms to model temporal dependency in dynamic systems [1, 51]. To cope with temporal effects, we propose to apply dynamic Bayesian networks (DBNs), and further to investigate their prediction capabilities based on a proper variant of the inference techniques for fault management introduced in this chapter. Related research results on these issues can be found in [14].

The inherent uncertainty about dependencies among the components of a large distributed system brings about challenges in all five functional management domains: fault, configuration, accounting, performance, and security management. A priority goal for all domains is, therefore, to design and implement adaptive, flexible, robust, and self-managing strategies that operate under uncertainty. Fault management sets its goals both in fault diagnosis and fault repair to eliminate errors or at least to neutralise their effects. To realise a self-managing system, automatic fault repair mechanisms are on the research agenda. The feedback resulting from diagnosis and repair should provide useful references on preferred modification and improvements. The recurrence of problem hypothesis and repair is helpful to realise a healthy self-improving fault management system.

Acknowledgement

This work was largely performed during a research stay at the Faculty of Electrical and Computer Engineering of FernUniversität in Hagen, and was supported by the German Academic Exchange Service (DAAD) by a personal PhD scholarship.

References

1. Aliferis CF, Cooper GF (1996) A Structurally and Temporally Extended Bayesian Belief Network Model: Definitions, Properties, and Modeling Techniques, In: Proc 12th Conf on Uncertainty in Artificial Intelligence, Morgan Kaufmann, pp 28–39
2. ANSI (1994) T1.215 OAM&P — Fault Management Messages for Interface between Operations Systems and Network Elements
3. Bouloutas AT, Calo SB, Finkel A, Katzela I (1995) Distributed Fault Identification in Telecommunication Networks, J of Network and Systems Management 3 (3): 295–312
4. Basye K, Dean T, Vitter SJ (1997) Coping with Uncertainty in Map Learning. Machine Learning 29 (1): 65–88
5. Bouloutas A, Hart G, Schwartz, M (1992) Simple Finite-State Detectors for Communication Networks. IEEE Trans Communications 40 (3): 477–479
6. Buntine W (1996) Graphical Models for Discovering Knowledge. In: Advances in Knowledge Discovery and Data Mining, ACM, pp 59–82
7. Burnell L, Horvitz E (1995) Structure and Chance: Melding Logic and Probability for Software Debugging. Comm ACM 38 (3): 31–41
8. Chao CS, Yang DL, Liu AC (2001) An Automated Fault Diagnosis System using Hierarchical Reasoning and Alarm Correlation. J of Network and Systems Management 9 (2): 183–202
9. Chen R, Sivakumar K, Kargupta H (2001) Distributed Web Mining Using Bayesian Networks from Multiple Data Streams. In: Proc IEEE Int Conf on Data Mining, pp 75–82
10. Cooper G (1990) Computational Complexity of Probabilistic Inference using Bayesian Belief Networks. Artificial Intelligence 42: 393–405
11. Deng RH, Lazar AA, Wang W (1993) A Probabilistic Approach to Fault Diagnosis in Linear Lightwave Networks, IEEE J on Selected Areas in Communications 11 (9): 1438–1448
12. Ding JG, Krämer BJ, Bai YC, Chen HS (2004) Probabilistic Inference for Network Management. Universal Multiservice Networks, LNCS 3262, Springer, pp 498–507
13. Ding JG, Krämer BJ, Bai YC, Chen HS (2004) An Inference Algorithm for Probabilistic Fault Management in Distributed Systems. In: Proc IFIP Int Conf on Network Control and Engineering for QoS, Security and Mobility, Springer, pp 193–204
14. Ding JG, Krämer BJ, Xu SH, Chen HS, Bai YC (2004) Predictive Fault Management in the Dynamic Environment of IP Networks. In: Proc IEEE Int Workshop on IP Operations & Management, pp 233–239

15. Ding JG, Zhang J, Bai YC, Chen HS (2004) One Backward Inference Algorithm in Bayesian Networks. In: Proc 5th Int Conf Parallel and Distributed Computing: Applications and Technologies, LNCS 3320, Springer, pp 72–75
16. Dupuy A, Schwartz J, Yemini Y, Barzilai G, Cahana A (1989) Network Fault Management: A User's View. In: Integrated Network Management I, North-Holland, Amsterdam, pp 101–107
17. De Franceschi ASM, Kormann LF, Westphall CB (1996) Performance Evaluation for Proactive Network Management. In: Proc IEEE ICC, pp 22–26
18. Frontini M, Griffin J, Towers S (1991) A Knowledge-based System for Fault Localization in Wide Area Networks. In: Integrated Network Management II, North-Holland, Amsterdam, pp 519–530
19. Gao J, Kar G, Kermani P (2004) Approaches to Building Self Healing Systems using Dependency Analysis. In: Proc IEEE/IFIP Network Operations and Management Symp, pp 119–132
20. Gupta M, Neogi A, Agarwal MK, Kar G (2003) Discovering Dynamic Dependencies in Enterprise Environments for Problem Determination. In: Proc 14th IEEE/IFIP Int Workshop on Distributed Systems Operations and Management, LNCS 2867, Springer, pp 221–233
21. Hasselmeyer P (2001) Managing Dynamic Service Dependencies. In: Proc 12th Int Workshop on Distributed Systems: Operations & Management, INRIA Press, pp 141–150
22. Heckerman D, Breese J, Rommelse K (1995) Decision-theoretic Troubleshooting. Comm ACM 38: 49–57
23. Heckerman D, Mamdani A, Wellman M (1995) Real-world Applications of Bayesian Networks. Comm ACM 38 (3): 24–26
24. Heckerman D, Wellman MP (1995) Bayesian Networks, Comm ACM 38 (3): 27–30
25. Hong P, Sen P (1991) Incorporating Non-deterministic Reasoning in Managing Heterogeneous Network Faults. Integrated Network Management II, North-Holland, Amsterdam, pp 481–492
26. Hood C, Ji C (1997) Proactive Network Fault Detection. IEEE Trans Reliability 46 (3): 333–341
27. Houck K, Calo S, Finkel A (1995) Towards a Practical Alarm Correlation System. In: Sethi AS, Faure-Vincent F, Raynaud Y (eds), Integrated Network Management IV, Chapman and Hall, London, pp 226–237
28. Jensen FV (1996) An Introduction to Bayesian Networks. Springer, New York
29. Joseph C, Kindrick J, Muralidhar K, Toth-Fejel T (1989) MAP Fault Management Expert System. In: Meandzija B, Westcott J (eds), Integrated Network Management I, North-Holland, Amsterdam, pp 627–636
30. Kätker S, Geihs K (1997) A Generic Model for Fault Isolation in Integrated Management System. J of Network and Systems Management 5 (2): 109–130
31. Katzela I, Bouloutas AT, Calo SB (1995) Centralized versus Distributed Fault Localization. In: Integrated Network Management IV, Chapman and Hall, London, pp 250–263
32. Katzela I, Schwarz M (1995) Schemes for Fault Identification in Communication Networks. IEEE/ACM Trans Networking 3: 753–764
33. Keller A, Blumenthal U, Kar G (2000) Classification and Computation of Dependencies for Distributed Management. In: Proc 5th IEEE Symp on Computers and Communications, pp 78–83

34. Kirsch H, Kroschel K (1994) Applying Bayesian Networks to Fault Diagnosis. In: Proc 3rd IEEE Conf on Control Applications, IEEE Computer Society Press, pp 895–900
35. Klinger S, Yemini SA, Yemini Y, Oshie, D, Stolfo S (1995) A Coding Approach to Event Correlation. In: Proc 4th IEEE/IFIP Int Symp on Integrated Network Management, IEEE Computer Society Press, pp 266–277
36. Klösgen W (1996) Knowledge Discovery in Database Terminology, Advances in Knowledge Discovery and Data Mining, AAAI/MIT Press, pp 573–592
37. Kim JH, Pearl J (1983) A Computational Model for Combined Causal and Diagnostic Reasoning in Inference Systems. In: Proc 8th Int Joint Conf on Artificial Intelligence, William Kaufmann, pp 190–193
38. Koch FL, Westphall CB (2001) Decentralized Network Management Using Distributed Artificial Intelligence. J of Network and Systems Management 9 (4): 291–313
39. Lauritzen SL, Spiegelhalter DJ (1998) Local Computations with Probabilities on Graphical Structures and Their Application to Expert Systems. J of the Royal Statistical Society, Series B 50: 157–224
40. Lewis L (1993) A Case-based Reasoning Approach to the Resolution of Faults in Communication Networks. In: Integrated Network Management, III, Elsevier Science, Amsterdam, pp 671–682
41. Lo C, Chen SH, Lin B (2000) Coding-based Schemes for Fault Identification in Communication Networks, J of Network and Systems Management 10 (3): 157 –164
42. Matsumoto M, Kurita Y (1992) Twisted GFSR Generators. ACM Trans Modeling and Computer Simulation 2: 179–194
43. Matsumoto M, Kurita Y (1994) Twisted GFSR Generators II. ACM Trans Modeling and Computer Simulation, pp 254–266
44. Neal RM (1993) Probabilistic Inference Using Markov Chain Monte Carlo Methods. Tech Rep CRG-TR93-1, University of Toronto, Department of Computer Science
45. Nikovski D (2000) Constructing Bayesian Networks for Medical Diagnosis from Incomplete and Partially Correct Statistics, IEEE Trans Knowledge and Data Engineering 12 (4): 509–516
46. Nygate YA (1995) Event Correlation Using Rule and Object Based Techniques. In: Proc IFIP/IEEE Int Symp on Integrated Network Management IV, 278–289
47. Pearl J (1986) A Constraint-propagation Approach to Probabilistic Reasoning. In: Uncertainty in Artificial Intelligence. North-Holland, Amsterdam, pp 357–369
48. Pearl J (1988) Probabilistic Reasoning in Intelligent Systems: Networks of Plausible Inference. Morgan Kaufmann
49. Pearl J (2000) Causality: Models, Reasoning, and Inference, Cambridge University Press
50. Rouvellou I, Hart GW (1995) Automatic Alarm Correlation for Fault Identification. In: Proc IEEE INFOCOM95, pp 553–561
51. Santos Jr. ES, Young JD (1999) Probabilistic Temporal Networks: A Unified Framework for Reasoning with Time and Uncertainty, Int J of Approximate Reasoning 20 (3): 263–291
52. Schow PH (1996) The Alarm Information Base: A Repository for Enterprise Management. In: Proc 2nd IEEE Int Workshop on Systems Management, IEEE Computer Society Press, pp 142–147

53. Sebe N, Cohen I, Huang TS, Gevers T (2004) Skin Detection: A Bayesian Network Approach. In: Proc 17th Int Conf. on Pattern Recognition, vol 2, IEEE Computer Society Press, pp 903–906
54. Shafer G, Pearl J (1990) Readings in Uncertain Reasoning. Morgan Kaufmann
55. Singhal A, Jiebo L, Brown C (2000) A Multilevel Bayesian Network Approach to Image Sensor Fusion. In: Proc 3rd IEEE Int Conf on Information Fusion, vol 2, pp 9–16
56. Steinder M, Sethi AS (2001) Non-deterministic Diagnosis of End-to-end Service Failures in a Multi-layer Communication System. In: Proc ICCCN, pp 374–379
57. Studwell TW (2003) Orchestrating Self-managing Systems for Autonomic Computing: The Role of Standards. In: Proc 14th IEEE/IFIP Int Workshop on Distributed Systems Operations and Management, LNCS 2867, Springer, pp 1–2
58. Wang C, Schwartz M (1993) Identification of Faulty Links in Dynamic-routed Networks. IEEE J on Selected Areas in Communications 11 (9): 1449–1460
59. Wang Z (1989) Model of Network Faults. In: Meandzija B, Westcott J (eds), Integrated Network Management I, North-Holland, Amsterdam, pp 345–352
60. Wiegerinck WAJJ, Kappen HJ, Nijman MJ, Ter Braak EWMT, Ter Burg WJP (1999) Approximate Inference for Medical Diagnosis. Pattern Recognition Letters 20: 1231–1239
61. Yemini SA, Klinger S, Mozes E, Yemini Y, Ohsie D (1996) High Speed and Robust Event Correlation. IEEE Communications Magazine 34 (5): 82–90

12

Denial of Service Detection in Mobile Agent Systems with Dependencies in Computations

Biljana Cubaleska[1] and Markus Schneider[2]

[1] Faculty of Electrical and Computer Engineering
FernUniversität in Hagen, 58084 Hagen, Germany
biljana.cubaleska@fernuni-hagen.de

[2] Institute for Secure Telecooperation
Fraunhofer-Gesellschaft, 64295 Darmstadt, Germany
markus.schneider@sit.fraunhofer.de

Summary. In connection with mobile agents, security aspects are extensively being discussed. In this context, denial of service attacks can cause tremendous economic damages and are, therefore, of considerable interest to researchers as well. Here, a special kind of denial of service attack is considered, focusing on malicious hosts that either delete received agents or prevent them from continuing their route. In general, the prevention of such attacks is not possible. A detection method for the *a posteriori* identification of such malicious hosts is presented and discussed. Agent owners can use the output of this procedure when composing future routes of their agents. This method relies on a novel protocol that combines the application of well known cryptographic primitives and a set of behaviour rules. The protocol takes into account given visiting-order constraints for routes, which arise due to dependencies of computation results on an agent's journey. The protocol also works in the case of colluding hosts.

Key words: Internet security, internet agent technology, mobile agent security, malicious host, denial of service attacks, agent deletion.

12.1 Introduction

Mobile and ubiquitous computing refer to an emerging computing paradigm that aims to provide hardware and software means for offering user-friendly information and communication services, anywhere and anytime. Computers will be integrated in objects of everyday life, and data will be distributed over different nodes. The amount of new devices and services makes an efficient use by centralised systems very difficult. The paradigm of mobile agents ideally fits into the framework of ubiquitous computing. A mobile agent thereby constitutes a virtual reflection of the user, and carries personal information based on which he can perform different services. Mobile agents

B. Cubaleska and M. Schneider: *Denial of Service Detection in Mobile Agent Systems with Dependencies in Computations*, Studies in Computational Intelligence (SCI) **42**, 249–268 (2007)
www.springerlink.com © Springer-Verlag Berlin Heidelberg 2007

offer a possibility to encapsulate information of a person and his preferences, likings and habits, and perform location-based services of ubiquitous systems in the name of the user.

Mobile agents are autonomous programs, which migrate through a network of sites to accomplish tasks or to take orders on behalf of their owners. The owner of an agent can instruct it to visit many hosts in a network, and thereby execute some desired tasks for him. After having carried out the instructions, the agent returns to its home and delivers the results it collected during its journey to its owner.

The paradigm of mobile agents causes lots of security threats to all parties involved. Both the visited hosts and the agents are exposed to serious dangers from the opposite side — the mobile agent can attack the host and the host can attack the mobile agent. A visit from an agent is a security risk for the visited host, because a malicious agent can, for example, have some undesired functionality like a Trojan horse, and it can try to get unauthorised access to resources, such as reading private information, or even altering or deleting it. An agent can also disrupt operations of the host platform thereby executing a denial of service attack against the host. There are various possibilities for malicious hosts to attack agents, too, and thereby the agent owners. These attacks can focus on the extraction of private information, stealing digital goods, or modifying agent data.

In this chapter our concern focuses on potential denial of service (DoS) attacks by malicious hosts. In such a denial of service attack, a malicious host can prevent an agent from continuing to migrate to another host or may even delete the agent. As a consequence, all results the agent has collected so far from other hosts are lost. This may happen every time the agent passes through this malicious host, while the agent owner has no possibility to identify the source of the attack. Commonly, the term *denial of service* is used for attacks in which the focus is on exhausting resources with the effect that other entities cannot be served anymore. This differs from the kind of attacks we are considering. But in fact, the goal of these attacks is also denying services to others.

Unfortunately, there is no possibility to generally prevent visiting agents of being *annihilated*. Thus, agent owners need a mechanism which can be used to identify the culprit host. The knowledge obtained by such a mechanism can be used by agent owners for modifying their trust values and building trust policies based on their own experience, e.g., to exclude the corresponding host from future agent migrations. In this chapter, we present a protocol that allows an agent owner to identify malicious hosts executing denial of service attacks as previously described. The method takes into account arbitrary dependencies of computations on an agent's journey to be expressed by visiting-order constraints. The method also enables the agent owner to circumvent the activities of a collusion when malicious parties try to skip a host that should be visited. Improper protocol execution from some host can also be detected. These functionalities are achieved by applying cryptographic primitives such as digital signatures and a fixed set of rules. The contribution of this chapter is an improvement of the method proposed in [4], where the focus was exclusively on independent computations.

This chapter is organised as follows. Some related work in the area of protecting mobile agents against malicious hosts is discussed in the next section. In Section 12.3 we introduce the components of mobile agent systems needed for our approach. A simplified solution to identify hosts that performed a denial of service attack is given in Section 12.4. Then, in Section 12.5 a protocol is presented that solves the problems discussed in previous sections.

12.2 Related Work

Many of the problems concerning the security of mobile agent systems, both protecting hosts from malicious agents and protecting agents from malicious hosts, have been discussed in the literature. While countermeasures directed towards host protection directly evolved from traditional mechanisms employed by hosts, those for agent protection are often totally new concepts. In the following, some of the most important approaches to protect mobile agents are mentioned.

One method for an agent to encapsulate result information in order to achieve privacy and integrity of its computation was proposed by Yee [22]. He uses Partial Result Authentication Codes, which are based on Message Authentication Codes. Karjoth et al. [10] devised a technique to encapsulate partial results which re-formulates and improves the technique of Yee, also offering a forward integrity and confidentiality feature. Execution tracing [19] is a technique to detect unauthorised modifications of an agent by recording the agent's behaviour during its execution on each host. In [16], Sander and Tschudin introduced the concept of computing with encrypted functions and, thus, protecting the integrity and the privacy of an agent's computations. Lee et al. [14] extended the approach proposed by Sander and Tschudin and removed many problems found in the original idea of [16], while preserving most of its benefits.

Hohl introduced in [9] the concept of a blackbox, i.e., an agent that performs the same work as an original agent but has a different structure. This difference allows to assume a certain agent protection time interval, during which it is impossible for an attacker to discover relevant data or to manipulate the execution of the agent. Corradi et al. presented in [3] methods to protect an agent's integrity — both making use of a Trusted Third Party and without one. In [13], Kim et al. suggested an adaptive migration strategy that can be used to avoid mobile agents from blocking or crashing. This is achieved by a route re-ordering algorithm and a backward recovery algorithm. In [20, 21], Westhoff et al. described methods to protect an agent's route against hosts spying out route information. In [7], Ferreira and Dahab introduced a new cryptographic primitive called blinded-key signature that allows to protect private keys carried by agents. They also presented in [8] a protocol that combines some of the above techniques to protect mobile agents. In [5], a novel remote distributed scheme for protecting mobile agents is introduced by Dadon. It is based on fault-tolerant and modest cryptographic techniques, and prevents tampering of any mobile computation that is carried out in an honest-but-curious environment[3]. The concept of self-protected mobile agents was introduced in [1] by Ametller et al.: Agents are enhanced with an independent, fully encapsulated protection mechanism carried by the agents themselves.

One technique to ensure that a mobile agent arrives safely at its destination is through the use of replication and voting. This method was presented by Schneider in [17]. The problem of detecting the *black hole* — a stationary process destroying visiting agents — in an anonymous ring is addressed in [6] by Dobrev et al. However, most of the proposed techniques mentioned above have serious limitations which must be overcome. Some of these problems are addressed in this chapter. In particular, we focus on issues concerning the protection of agents against attacks. Other problems like guaranteeing that a host performs the computations it is supposed to, or ensuring the authenticity of data communicated exceed the scope of this contribution.

12.3 Some Basics on Mobile Agent Systems

In the past years, lots of work has been done in the area of mobile agent systems. Presently, there is a variety of mobile agent systems available, e.g., Aglets (IBM) [12], Ajanta (University of Minnesota) [18, 11], Mole (University of Stuttgart) [2], Concordia (Mitsubischi Electric), Voyager (Objectspace), Odyssey (General Magic), Ara (University of Kaiserslautern), Agent-Tcl (Dartmouth College), Grasshopper (GMD Fokus), or Jumping Beans (AdAstra Engineering).

In the following, we shall not focus on a specific mobile agent system. We shall consider mobile agents in a rather abstract way. This means that exclusively those components of mobile agents will be considered that are of special relevance for the solution presented. This generic approach also implies that the proposed solution can be applied to different agent systems.

12.3.1 Agent Components

In our level of abstraction, a mobile agent a consists of the following components:

$$a^j = (uid, bc, in, out^j, r, vc^{\#(c_j)}, voc). \tag{12.1}$$

[3] This technical term means that the environment, e.g., a server, refrains from active attacks like deletion or modification of data or use of a false identity (honesty), while it performs passive attacks like tapping and storing data (curiosity).

where index j indicates that agent a resides on host c_j after having been executed; uid is a *locally unique identifier* for the agent, and is created by the agent owner in such a way that all his agents have distinct identifiers. We shall use uid later to trace the agent in order to identify the attacker in case the agent does not return to its home. Component bc denotes the *binary code* of the agent to be executed; in represents input data for agent computations, and is given by the agent owner; out^j denotes the *output data* that result from the agent's previous executions, and are produced by the visited hosts contained in the agent after its execution at host c_j. This means that for $j = 1, \ldots, n$ the output data out^{j-1} coming from host c_{j-1} are contained in out^j after the agent's execution at host c_j. Component $r = (c_1, \ldots, c_j, \ldots, c_n)$ describes the agent's *route* as an n-tuple consisting of addresses of hosts that have to be visited. This route is defined by the agent owner, but the hosts do not necessarily have to be visited in the given order. In the following, we assume that agents will exclusively visit hosts that are contained in their route, i.e., dynamic route extensions will not be discussed. Component $vc^{\#(c_j)}$ describes the sequence of already *visited hosts* also including host c_j when the agent resides on c_j. The superscript $\#(c_j)$ denotes the number of hosts that have been visited so far. Thus, $vc^{\#(c_j)}$ is a sequence with $\#(c_j)$ elements:

$$vc^{\#(c_j)} = \underbrace{c_{i_1}, \ldots, c_{i_m}}_{\#(c_j)\text{ elements}} \tag{12.2}$$

where $i_l \in \{1, \ldots, n\}$ and $c_{i_m} = c_j$. Note that in some cases, hosts can be visited multiple times during an agent's journey, i.e., c_{i_k} and c_{i_l} for $k \neq l$ could refer to the same address. The last component of our agent definition, voc, denotes visiting-order constraints that will be discussed in detail in the next subsection.

Before an agent starts its first migration residing at its home h, the sequence of visited hosts $vc^{\#(h)} = vc^0$ is empty. When the first host c_1 is visited, c_1 creates $vc^{\#(c_1)} = vc^1 = (c_1)$ before it releases the agent. If we assume that the next host visited after c_1 is c_3 — because host c_2 as intended in the route r could, e.g., be off-line — then c_3 creates the sequence $vc^{\#(c_3)} = vc^2 = (c_1, c_3)$. In general, for $1 \leq j \leq n$ we have $vc^{\#(c_j)} = vc^{\#(c_j)-1} + c_j$, where $+$ denotes the append operator.

12.3.2 Dependencies of Computations

Concerning the computation of an agent's task at a given host, we can distinguish between *dependent* and *independent* computations. On an agent's journey, there can be tasks for dependent computations and other tasks for independent computations. In the case *independent*, the computation of the agent's results at some host does not require computation results produced elsewhere as an input, and the hosts prescribed in r can be visited in any arbitrary order. In the case *dependent*, a computation is not possible without the result of another computation used as input. For a dependent computation at c_j, the results contained in out^{j-1} calculated from some previous hosts are a necessary input to produce the results to be contained in out^j. In this case, the agent has to visit the hosts in an order ensuring that the inputs needed for the computations at host c_j are already available before the agent visits c_j. So, in case of dependent computations, an agent cannot fulfill its task if a required host is not available or refuses its services. These dependency relations imply some *visiting-order constraints*. Thus, the real order for visiting hosts must match these constraints.

The visiting-order constraints are contained in the agent component voc, which is of special importance when — for some reasons — the hosts will not be visited in the order as originally given in r. We define $voc=(voc_1, \ldots, voc_n)$ as an n-tuple, where n is the number of hosts contained in r. An element voc_i represents the set of hosts whose results are required for the computation at c_i. If $voc = \emptyset$, then there are no such requirements. For instance, if $r = (c_1, c_2, c_3, c_4, c_5)$ and $voc = (\emptyset, \emptyset, \{c_1, c_2\}, \emptyset, \{c_3, c_4\})$, we conclude that the computation at c_1 does not depend on the results of any other hosts. The same holds for c_2. The computation at c_3 depends on the results of c_1 and c_2, the computation at c_4 is independent, and c_5 requires the results of c_3 and c_4. Thus, hosts c_1 and c_2 must be visited before the agent visits c_3. Host c_5 has to be visited after c_3 and c_4. The hosts c_1, c_2, and c_4 do not have any constraints. So, the hosts can be visited in the order given in the route or, for example, in the order c_1, c_4, c_2, c_3, c_5. Actually, the component voc can be part of the agent's input control data contained in the component in, but due to the importance of this component for our solution, we present it as a separate agent component.

We can mention the cases of *completely dependent* and *completely independent* computations as special cases of *dependent* and *independent* computations. We say that the agent journey is completely dependent if for $1 < i \leq n$ each c_i requires the results of c_{i-1}. In this case, the hosts have to be visited exactly in the same order as they are prescribed in the route, and the agent cannot fulfill its task when at least one of the required hosts specified in the route is either not available or denies its services to the agent. We say that an agent journey is *completely independent* when there are no dependent computations in it. The properties of being completely dependent and completely independent are not complementary, because on an agent's journey both tasks for dependent and independent computations can be given.

Note that the visiting-order constraints have the property of *transitivity*. If for $1 \leq i < j < k \leq n$ we have $c_i \in voc_j$ and $c_j \in voc_k$, it follows that c_i has to be visited before c_k even if $c_i \notin voc_k$. In the previous example, c_5 depends on c_3 and c_4, while c_3 depends on c_1 and c_2. This implies that c_5 also depends on c_1 and c_2. The transitivity can be exploited to keep voc as small as possible.

The constraints and the route have to be chosen in such a way that the determination of the visiting order is a solvable problem. If there are non-satisfyable dependencies in the visiting-order constraints, this will lead to skipping some hosts in the route in order to avoid unreasonable migrations (cp. the analysis of algorithm *SelectNextHost* in Section 12.5.1), with the consequence that the agent will not bring the desired results to its owner. This could happen even if the route and the visiting-order constraints had been chosen reasonably, but some of the hosts in the route could not be visited because of, e.g., being off-line.

Both cases of hosts being off-line or unreachable due to network problems are not considered to be denial of service. However, there are also other attacks that have the same effects for the agent owner and are similar to denial of service. These comprise altering, replacing, or even deleting, for example, the binary code *bc* of the agent or its data in, out^j. In this scenario, the agent owner may not receive the desired correct results. Strictly considered, these belong to other types of attacks attacks on an agent's integrity. Dealing with integrity attacks — especially concerning the integrity of computation results — is, however, not in the central scope of this chapter. These attacks have already been dealt with in other work (see, e.g., [3]).

Having a mechanism dealing with the problem of malicious hosts performing denial of service attacks would be very useful for the agent owner. The solution presented in this chapter enables the agent owner to identify malicious hosts *a posteriori*. The information an agent owner can obtain by using our solution can help him to obtain better knowledge about the hosts he is dealing with, and allows him to build a personal trust policy potentially containing a black list of mistrusted hosts. He can use such a trust policy for future compositions of agent routes.

12.4 Towards a Solution

In the sequel, we shall present a simplified and rather naive solution of the described problem. Later, we shall point out the shortcomings of this simple approach and tackle them in the next section. Additionally, we shall briefly mention the basics of the solution including cryptographic primitives and infrastructure requirements.

The idea underlying our mechanism is based on the usage of undeniable proofs: When an agent owner does not receive his agent after some waiting time, the suspicion arises that the agent suffered a denial of service by a malicious host. In order to identify the attacking host, the agent owner asks all hosts that were contained in the original route r to show him a proof that they correctly dispatched the agent. The attacking host is surely not able to show such a proof. Undeniable proofs can be realised by the technique of digital signatures (see, e.g., [15]). Using digital signatures assumes the existence of a public key infrastructure (PKI), where each subject in the system — the agent owner and all hosts — has a private key and a corresponding public key that is certified to be authentic by a certification authority. Generally, a party A that uses the public key of another party B is required to trust B's certification authority to issue only correct certificates. This implies for our system that the agent owner must trust the certification authorities of all hosts to be visited, as will become clear later.

We assume now — more or less naively — that all hosts in the system carry out the following rule: upon receiving an agent, each host c_{j+1} must send a confirmation to its predecessor c_j, where $1 \leq j < n$. This confirmation is actually a digital signature $sig_{c_{j+1}}(uid, bc, in, r, voc)$ created by host c_{j+1} on the unique agent identifier uid on its binary code bc, on the route r, and on its input data in and voc. This exchange of messages is depicted in Fig. 12.1. The signature can be used later for the agent owner's investigations to find out who was the last party that saw this specific agent alive. Each host that has received such a confirmation should store it locally in a database. When the agent owner does not receive his agent and starts his investigations, a host c_j having such a confirmation can show it to the owner as evidence that it properly dispatched the agent to the next host c_{j+1}. This evidence also gives the agent owner information which host was visited after c_j. This principle works as long as the hosts being considered did not perform a denial of service.

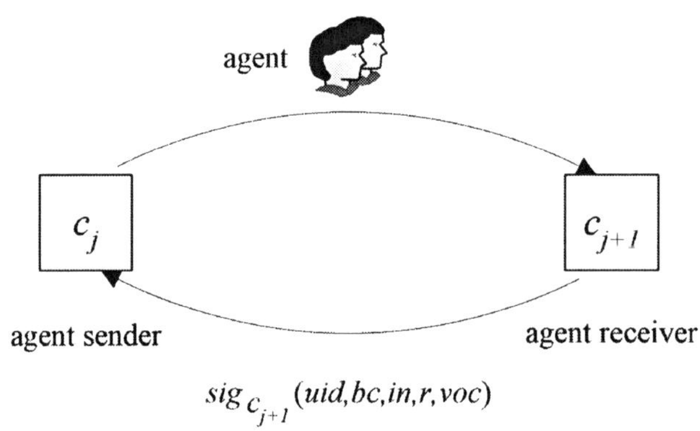

Fig. 12.1: Exchange of agent and confirmation

In all previous considerations, we started from the naive assumption that each host sends a confirmation after having received the agent. Of course, in a real scenario this cannot be guaranteed. Let us consider what happens when host c_{j+1} does not send the confirmation to c_j although it successfully received the agent. Assume that c_{j+1} or any other host after it performed a denial of service, i.e., it annihilated the agent, or did not send it to the next host. When the agent owner starts his investigation procedure, the host c_j cannot show him an evidence that proves its innocence, although it has done its job properly. Thus, the simplified solution presented so far has to be modified in order to be able to handle such cases.

An obvious solution to the problem that c_{j+1} refuses to send a confirmation could be the possibility for c_j to skip c_{j+1} and send the agent to another host, say c_{j+2}. Then, the agent could return to its home properly, and the agent owner could see from the sequence of visited hosts $vc^{\#(c_n)} = vc^{n-1}$ that host c_{j+1} was not visited. The reason for this remains hidden to the agent owner. From his perspective, c_{j+1} could be off-line or refuse to send the confirmation. But this kind of solution could open the doors for another kind of dishonest behaviour, viz., a malicious c_j could intentionally skip c_{j+1} without having even contacted it before. Motivations for c_j to do so could, first, be to damage c_{j+1}'s reputation by making the agent owner believe that c_{j+1} is always off-line or is not willing to follow the rules for exchanging confirmations. Second, c_j could exclude c_{j+1} from providing its services by simply skipping it. The protocol proposed in the next section solves all the problems of the simplified version presented above. In the previous consideration, we have mentioned several times the term *investigation procedure* without explaining how this procedure works. This will also be presented in the next section.

12.5 Detection of Malicious Hosts

The detection of malicious hosts consists of two parts. The first part is the execution of a sender and a receiver protocol, which is presented in Subsection 12.5.1. The second part is an investigation procedure that is executed by the agent owner when the agent does not return to its home. It is presented in Subsection 12.5.4.

12.5.1 The Protocols

In the following, we shall present a *sender protocol*, a *receiver protocol* , and an algorithm *SelectNextHost*. Altogether, they define where and when to create the evidences, and how to deal with the problem of unreachable hosts, hosts not willing to follow the rules, and attacking hosts with respect to our protection goals.

For security reasons, we modify the agent model introduced in Section 12.3. In detail, we assume that agent a is modeled by the following structure:

$$\begin{aligned} a^j = (&uid, bc, in, out^j, r, vc^{\#(c_j)}, voc, m, \\ &sig_h(uid, bc, in, r, voc, m)). \end{aligned} \tag{12.3}$$

with $r = (c_1, ..., c_n)$ containing also a signature of the agent owner h. With this signature, visited hosts are assured about the integrity and the origin of the signed agent components. Note, that in this agent abstraction there is a new parameter $m \in \mathbb{Z}^+$. Parameter m is given by the agent owner, and determines the maximum number of hosts that should try to contact another host not answering properly or not answering at all.

We shall also have some additional signatures in the list of visited hosts. Assume again that on its journey the mobile agent migrates from host c_k to host c_l, $k \in \{1, \ldots, n-1\}$, $l \in \{2, \ldots, n\}$, and $k \neq l$. Then, $vc^{\#(c_l)}$ will be derived from $vc^{\#(c_k)}$ in the following way:

$$vc^{\#(c_l)} = vc^{\#(c_k)}, c_l, sig_{c_l}(vc^{\#(c_k)}, c_l). \tag{12.4}$$

We shall give the reason for this modification after the presentation of algorithm *SelectNextHost* and the protocols.

In the following, we assume that visited hosts will always verify the validity of these signatures, and we shall not mention it again explicitly. Signatures are verified before the hosts start the execution of the protocols presented below. If a host detects an agent with corrupted components, it will forward the agent directly to its owner.

The proposed method combines the mechanism of exchanging confirmations proposed in Section 12.4, and a set of rules for both sending and receiving hosts. The solution consists of a sender and a receiver protocol. The sender uses algorithm *SelectNextHost* as a subroutine. Before we present this algorithm, we introduce some notions used in its definition. For a given element e and set S the predicate $in(e, S)$ returns *true* if $e \in S$, else *false*. The function $card(S)$ returns the number of elements contained in S. The term $\{vc^{\#(c_j)}\}$ describes the set that contains all the hosts visited so far when the agent is at host c_j. For each agent to be processed, a host has an own buffer buf to store contacted host addresses. This is necessary in order not to contact an unreachable host, or a host that did not send a valid confirmation more than once. After forwarding an agent properly, the host can delete the content of buf. The operator $\neg$ represents the negation operator for a given Boolean argument.

Algorithm *SelectNextHost* (at c_j)

```
i = 1
while i < j
    if (¬in(c_i, {vc^#(c_j)}) ∧ ¬in(c_i, buf))
        then
            if (card({c_{i+1}, ..., c_j} ∩ {vc^#(c_j)}) < m)
                then
                    if ( ({vc^#(c_j)} ∩ voc_i) == voc_i )
                        then
                            append c_i to buf
                            NextHost = c_i
                            i = n + 2
                        else
                            i = i + 1
                    endif
```

```
15              else
16                  i = i + 1
17          endif
18      else
19          i = i + 1
20   endif
21 endwhile
22 if (i == j)
23   then
24      i = i + 1
25      while (i ≤ n + 1)
26          if (i == n + 1)
27              then
28                  NextHost = h
29                  i = n + 2
30              else
31                  if (¬in(c_i, {vc^{#(c_j)}}) ∧ ¬in(c_i, buf))
32                      then
33                          if ( ({vc^{#(c_j)}} ∩ voc_i) == voc_i )
34                              then
35                                  append c_i to buf
36                                  NextHost = c_i
37                                  i = n + 2
38                              else
39                                  i = i + 1
40                          endif
41                      else
42                          i = i + 1
43                  endif
44          endif
45      endwhile
46 endif
```

In lines 1–21, the host executing the algorithm checks whether there are still any hosts that should have been visited before itself. If there is such a host that has potentially not been contacted by m other hosts before, it is selected as $NextHost$. The condition in line 7 serves to check if a candidate for $NextHost$ fulfills the visiting-order constraints. If the order of hosts given in the initial route is in accordance with these constraints, line 5 always yields the value *true*. But if there is a contradiction between the initial order and the visiting-order constraints — maybe the agent owner made a mistake in the agent generation — the logic in line 7 prevents the agent from an unreasonable migration. Lines 22–46 describe the case that a host selects others contained in the route behins itself. Here, a host can only be selected if it was not contacted before by the host executing the algorithm, or if the selection is in accordance with the visiting-order constraints.

In the following, we shall present the sender and the receiver protocol. Compared with the confirmation introduced in Section 12.4, the confirmation needed here is different, and also contains the parameters m and $vc^{\#(c_j)}$. If the mobile agent is forwarded from host c_k to host c_l, then the confirmation to be created and replied by c_l is:

$$sig_{c_l}(uid, bc, in, r, voc, m, vc^{\#(c_k)}). \tag{12.5}$$

This confirmation signature and also the agent owner's signature in the agent (see Expression (12.3)) depending on these parameters can be used by the receiving hosts and also by the agent owner in order to detect modifications. The investigation procedure even allows the agent owner to identify the modifying attacker, as will become clear in Subsection 12.5.4. Further reasons for the presence of $vc^{\#(c_k)}$ in the confirmation will be given after presenting the protocols and the illustrating example below.

Sender Protocol. Assume that the agent resides on host c_j. The protocol will be started at host c_j after the agent execution has terminated. The steps of the sender protocol are described in pseudo-code. Note, that the sender protocol is not executed in the agent's first migration when it leaves h. If the result of *SelectNextHost* is $NextHost = h$, then the sender protocol can be stopped immediately.

```
1  NextHost = SelectNextHost(m, voc, vc^{#(c_j)})
2  if (NextHost == h)
3     then stop
4     else
5        store a copy of the agent
6        send agent to NextHost found by SelectNextHost
7        while (no confirmation received and no time-out)
8           wait for confirmation
9        if (confirmation received and valid)
10          then
11             store confirmation in local database
12          else
13             go to step 1
14       endif
15       delete agent copy
16 endif
```

Receiver Protocol. The receiver protocol is the reaction of host, say c_k, to the sender protocol. This protocol starts when the sender contacts the receiver to transfer the agent. This means that when the sender protocol terminates in step 3, the execution of the receiver protocol is not required.

```
receive agent
if ((c_k not in r) or (voc_k not fulfilled))
  then stop
```

```
    else
        create confirmation
        send confirmation
 endif
```

After protocol termination the receiver can do whatever it wants. It can execute the agent and then the sender protocol, or it can be malicious and annihilate the agent. But if it behaves in such a malicious way, then its confirmation stored at its predecessor can be used to identify it as an attacker in the investigation procedure. The copy of the agent at the sender side is stored in order to conserve the agent with its results. Thus, the data out^j collected so far will not be lost regardless of how the receiver will behave later. The *time-out* value can be specified by each host itself and denotes the maximum waiting time for confirmations. The sender is required to verify the validity of the confirmation, i.e., to check that the signature is not forged and the receiver's certificate is valid. If the confirmation is valid, it must be stored for an adequate period in the sender's database. Before we shall discuss the achievements of the presented protocols and of the algorithm *SelectNextHost*, we shall illustrate both with an example.

12.5.2 Example

Consider an owner h that composes an agent with the following parameters:

- $r = (c_1, \ldots, c_6)$,
- $m = 3$,
- $voc = (\emptyset, \emptyset, \{c_1\}, \emptyset, \{c_2, c_3, c_4\}, \{c_3\})$,
- and $vc^0 = \emptyset$ according to the rules.

The agent's journey is depicted in Fig. 12.2. For the sake of the example we assume that c_2 is either off-line, or does not give a response for some reasons — c_2 may even delete the agent after receiving it. In the first migration, h sends the agent to c_1. Here, no confirmation from c_1 is required. Now c_1 updates the list of visited hosts by creating $vc^1 = c_1, sig_{c_1}(c_1)$ and inserts this result in the agent. Then, c_1 starts the sender protocol. With $j = 1$, the algorithm *SelectNextHost* determines $NextHost = c_2$, host c_1 sends the agent to c_2, and waits for confirmation from c_2 that it received the agent. When time-out is reached, c_1 executes algorithm *SelectNextHost* for a second time. With c_2 contained in c_1's buffer and because the necessary results for the computations at c_3 have been collected, it yields $NextHost = c_3$. Assume that c_3 is on-line, but for some reason it does not respond before time-out. Then, the next execution of *SelectNextHost* at c_1 yields $NextHost = c_4$. This follows, since for the computation at c_4 neither the results from c_2 nor the results from c_3 are necessary (line 7 of algorithm *SelectNextHost*).

Now, c_4 starts the receiver protocol, creates the confirmation $sig_{c_4}(uid, bc, in, r, m, voc, vc^1)$, and sends a reply with this confirmation to c_1. Then, c_4 executes the agent, and updates the agent's list of visited hosts by generating

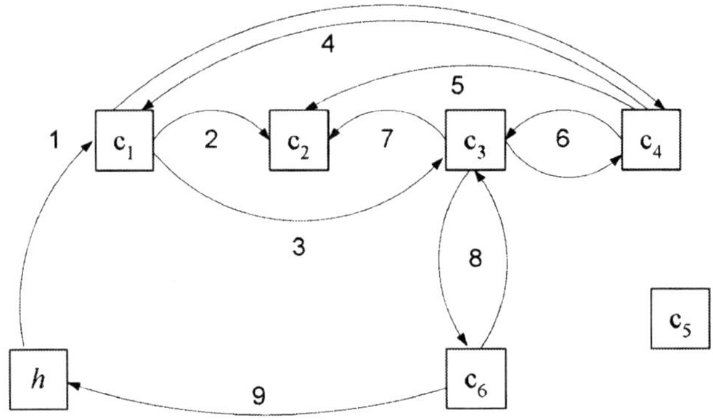

Fig. 12.2: The agent's journey in the example

$vc^{\#(c_4)} = vc^2 = vc^1, c_4, sig_{c_4}(vc^1, c_4)$. The start of the sender protocol leads to the execution of *SelectNextHost* for $j = 4$. At the iteration step $i = 1$, we see that c_1 is already contained in $\{vc^{\#(c_4)}\}$. In iteration step $i = 2$, we obtain $NextHost = c_2$, because c_2 is neither contained in vc^2 nor in c_4's buffer. Furthermore, $card(\{c_3, c_4\}) \cap \{c_1, c_4\}) = 1 < 3$ and $voc_2 = \emptyset$. Now, c_4 contacts host c_2 sending it the agent, but c_2 does not respond with its confirmation before the time-out occurs. Then, c_4 executes algorithm *SelectNextHost* again and it yields $NextHost = c_3$. Now, c_3 will react in time. It receives the agent, creates the confirmation $sig_{c_3}(uid, bc, in, r, m, voc, vc^2)$, and sends a reply with this confirmation to c_4. The update of the list of visited hosts yields $vc^{\#(c_3)} = vc^3 = vc^2, c_3, sig_{c_3}(vc^2, c_3)$. After the execution of the agent, c_3 starts the sender protocol and obtains $NextHost = c_2$. Without having received an answer from c_2, the next execution of *SelectNextHost* could have yielded $NextHost = c_5$ but, as for the computations at c_5 the results from c_2 besides the results from c_3 and c_4 are required, $NextHost$ will not yield c_5. Instead, the result will be $NextHost = c_6$.

Thus, c_3 contacts c_6. After having sent $sig_{c_6}(uid, bc, in, r, m, voc, vc^3)$, host c_6 creates $vc^{\#(c_6)} = vc^4 = vc^3, c_6, sig_{c_6}(vc^3, c_6)$. Now, iteration $i = 2$ of *SelectNextHost* renders c_2, which is neither contained in the list of visited hosts nor in c_6's buffer, but $card(\{c_3, c_4, c_5, c_6\} \cap \{c_1, c_3, c_4, c_6\}) = 3 \not< 3$. Thus, c_2 will not be contacted by c_6. Further execution of *SelectNextHost* will not yield $NextHost = c_5$ — even if c_5 has not been visited — because the visiting-order constraints for c_5 are not fulfilled. Thus, the algorithm gives now $NextHost = h$, and the agent can be sent to its owner without requiring a confirmation.

From the list of the visited hosts in his agent, the owner can see that the hosts c_1, c_4, c_3 and c_6 have been visited, while c_2 and c_5 were not. Of course, it is not clear to the agent owner why c_2 and c_5 have not been visited. But finally, he has received at least some of his desired results. Now, the agent owner can decide to send a copy of the agent on a new journey to c_2 and c_5.

To conclude the example, we note that c_2 was contacted by $m = 3$ parties altogether, i.e., for the first time by c_1, the second by c_4, and the third by c_3.

12.5.3 Properties of the Protocols

Now, we shall discuss the achievements of the protocols and of algorithm *SelectNextHost*. After a host dispatched an agent, it waits for confirmation from the receiver, which is used for potential investigation procedures when a denial of service has occurred. If the next host in the route is not reachable or did not send a valid confirmation, a new destination for the agent is determined by algorithm *SelectNextHost*. According to it, the agent is sent to h if either all hosts in the route have been visited or no host contained in the route still to be visited has given a proper answer. If the protocol and the algorithm are executed correctly, then the hosts given in the route can be visited in many orders, but the constraints voc must be satisfied. Algorithm *SelectNextHost* ensures this (see lines 7 and 33).

If host c_j does not forward the agent even though it received it and confirmed it, then this host can be identified by the confirmation that is stored at c_j's predecessor. Here, the signature in the confirmation also depends on the list of the visited hosts $vc^{\#(c_j)-1}$. The reason for this is described below.

The fact that the solution allows skipping of hosts could be exploited by a malicious host. Such a host could, for example, have the intention to exclude a competitor that is also contained in the route. To do this, it could register the address of the competitor in the list of already visited hosts so that this host will not be contacted anymore on the agent's further journey and is, thus, excluded from the interaction with the agent. Therefore, the list of visited hosts is protected in a sense that this attack is detectable. In our solution, each host obtains as a confirmation to be exchanged in the sender and receiver protocol a signature also depending on the list of visited hosts from its successor. In case of a conflict, this signature can be used as evidence to show that it did not manipulate the list. If one reconstructs the growth of the list, then it can be easily verified if a host entered more addresses than its own to the list of already visited hosts. The successor's signature on uid and the list of visited hosts can be used by an honest host to prove that the list was not manipulated by it in a malicious manner. That was the reason for changing the signature of the confirmation introduced in Section 12.4.

It is possible that a host c_j intends to skip another one being the next host to be visited according to the *SelectNextHost* algorithm. It could claim that this host was not reachable. In such a case it could forward the agent to another host selected by itself out of the set of potential candidates contained in the route. But then, this successor of c_j, if honest, would send the agent to the skipped host after having executed it. The only way for c_j to exclude a host is to initiate a collusion, e.g., by bribing. In the context of our solution this would mean for c_j to convince the next $m-1$ consecutive hosts to collude. If only one of these hosts is not willing to collude then the attack does not work.

The only possibility would be to also skip the host refusing to collude. But then, the process starts all over again, i.e., the attacker requires new $m-1$ hosts following the refusing one so that it can be skipped. Thus, besides a special case to be explained shortly, attacks can only have success if the size of a collusion is at least m — one initiator and $m-1$ partners.

Of course, all this reasoning about collusion size only holds as far as there are at least $m-1$ hosts that can be visited. Consider, for example, the case that we have n route entries and let the agent reside at host c_{n-2}. Assume that c_{n-2} decides to skip c_{n-1}. In this case, c_{n-2} only has to collude with c_n if all other hosts have been visited before. This means, that here the collusion size is 2, which can be smaller than m. In the case of a malicious c_{n-1}, the skipping of c_n works even without initiating a collusion. Thus, if some hosts are contained near to the end of the route and have not been visited, the agent owner can decide to contact these hosts in a new agent journey. We see that a manipulation of $NextHost$ does not lead to denial of service (step 1).

We now briefly discuss the implications of improper executions of the other steps, assuming that all sender and receiver protocols have been run properly before the sender protocol is executed by c_j.

If c_j does not store the agent copy (step 5), or deletes the copy too early, then c_j can be identified as a denial of service attacker. This is, e.g., possible when c_j does not have the agent copy anymore when it receives a forged signature from c_{j+1}. Then, c_{j+1} could delete the agent, but the investigation procedure will identify c_j as an attacker. Thus, c_j is interested in having an agent copy for the required time.

If c_j does not verify the signature obtained from c_{j+1} in a correct way (step 9), then it has no valid evidence in a possible investigation procedure. This would mean that, e.g., c_{j+1} could perform denial of service, but c_j would be identified as the attacker, since it cannot prove that it sent the agent to its successor even though it did. Thus, c_j has a natural interest to verify received signatures. The same argument holds for the case when c_j does not store the received confirmation in its database.

Let us consider the receiver protocol. Here, it can also be detected if the receiver c_{j+1} does not follow the rules. The case when c_{j+1} does not send a valid confirmation to c_j is already included in the sender protocol and was discussed above — c_j will send the agent to its home. Even if c_{j+1} decides to forward the received agent on its original route without having given a valid confirmation to c_j, it would have no benefit. It can be assumed that it would receive another copy of the agent from the agent owner. The situation is also clear if c_{j+1} sends a valid confirmation to c_j and then performs denial of service.

12.5.4 Investigation Procedure

When the owner becomes suspicious that some malicious party could have deleted his agent, he starts the investigation procedure. It basically consists of consecutively applying an investigation protocol consisting of two steps: The agent owner's request, and the answer in which a host shows its evidence. In the investigation procedure, the hosts must not necessarily be queried in the same order as they are given in the route r, because the real order of visits can differ from the one prescribed in r. In fact, the investigation must follow the actual order in which the hosts were visited. This is possible, since all hosts — except the attacker — can provide the agent owner with the identity of their successor by presenting the confirmation. Thereby, the agent owner can easily derive the real route step by step.

The owner starts the investigation procedure by requesting the first host in the agent route r, say c_1. If c_1 has done its job properly, it can present its successor's confirmation, which includes the signature on vc^1 and other agent components. Thus, after a positive check of the confirmation presented by c_1, the owner is convinced that c_1 has dispatched the agent properly. Furthermore, he knows c_1's successor to which he will send his next request. Then, the sending of requests and collecting confirmations is repeated according to the same known principle: If the confirmation presented by a host c_i is valid, send the next request is sent to the host found in $vc^{\#(c_i)+1}$. If the contacted host cannot present a valid confirmation, then the attacker is identified and the investigation procedure ends.

Example. Consider the example presented in Section 12.5.1 again, but now assume that c_3 behaves differently: It receives the agent from c_4 and sends a valid confirmation to c_4; but after that it performs a denial of service attack, i.e., it deletes the agent and does not run the sender protocol (Fig. 12.3a). After the agent owner has waited for some predefined time t for his agent, he becomes suspicious that the agent suffered a denial of service and starts the investigation procedure (Fig. 12.3b) by requesting the first host in r — here, this is host c_1. Since c_1 has forwarded the agent to c_4, it is able to show to the agent owner the evidence for this, namely, the confirmation $sig_{c_4}(uid, bc, in, r, m, voc, vc^{\#(c_1)}) = sig_{c_4}(uid, bc, in, r, m, voc, vc^1)$ that it received from c_4. After a positive check of this confirmation, the owner is convinced that c_1 has dispatched the agent properly, and he learns that c_4 was c_1's successor. So, the owner will now request an evidence from c_4. Host c_4 is able to prove its innocence, since it has the confirmation $sig_{c_3}(uid, bc, in, r, m, voc, vc^{\#(c_4)}) = sig_{c_3}(uid, bc, in, r, m, voc, vc^2)$ from c_3 for having forwarded the agent. After checking it, the owner is convinced that c_4 has dispatched the agent properly as well, and learns that the next visited host was c_3. So, the owner sends a request to c_3. But c_3 is not able to show any confirmation, because it has not forwarded the agent according to the protocol rules. Thus, the owner identifies c_3 as an attacker. He can use this information to modify his trust policy regarding the attacking host c_3. □

The investigation procedure also works for identifying attackers that created new agents by using or modifying an original agent's components, like uid, bc, in, r, and m. Since exchanged signatures depend on all these components, the party showing an invalid signature for them will be identified to

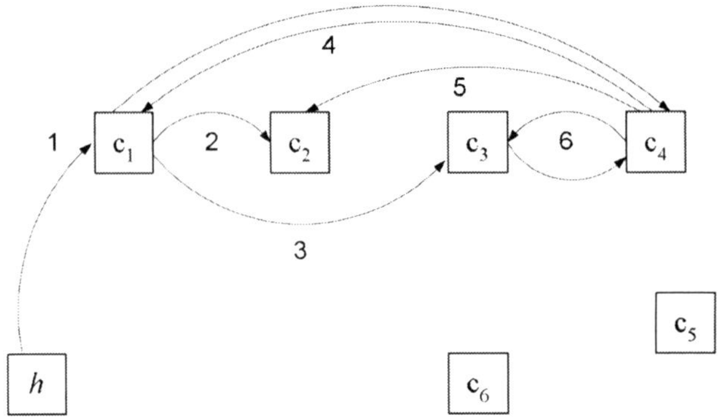

(a)

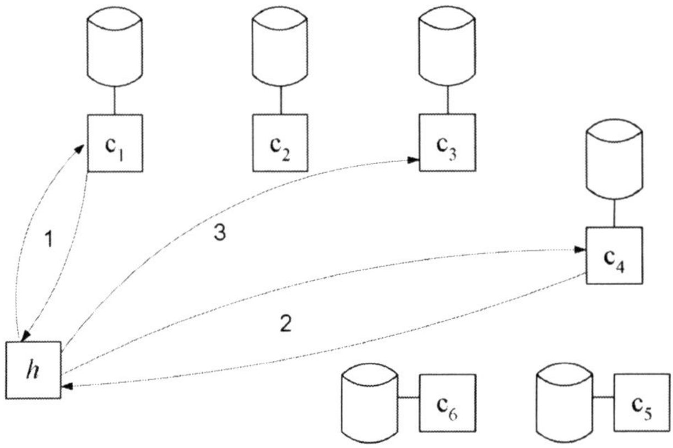

(b)

Fig. 12.3: An agent's journey with DoS from host c_3 (a), and the investigation procedure (b)

be the attacker. If we require that all exchanged confirmations are checked by the hosts on receipt, then there is no possibility for a sending host to deny that it dispatched a modified agent. If it receives an invalid signature even though it dispatched the agent correctly, the rules of the game — as given above — prescribe what to do.

12.6 Conclusion

In this chapter we treated the problem of denial of service attacks in mobile agent systems performed by malicious hosts, i.e., that a host annihilates an agent or is not willing to let it continue its route. We proposed a method realised with a combination of cryptographic techniques and a set of behaviour rules that enable the agent owner to identify the attacking host. Our solution for attacker identification works both for the case of dependent and independent agent computations, because the algorithm for the determination of the next host takes care of the computations' dependency.

A malicious host is identified by an investigation procedure, which will be executed by the agent owner in the case that his agent does not return. The presented solution does not strongly prevent the deletion of mobile agents. But it can be assumed that it has some preventive power to a certain extent, because an attacker can be uniquely identified. The outcome of the investigation procedure can be used by an agent owner when composing future routes of his agents.

A problem to be handled in the future is the case of dynamic routes, i.e., migrations to hosts initiated by hosts or agents that were not contained in the route composed by the agent owner.

Acknowledgement

We are grateful to Professor F. Kaderali holding the chair of Communication Systems at FernUniversität in Hagen for his support and critical review of this work.

References

1. Ametller J, Robles S, Ortega-Ruiz JA (2004) Self-protected Mobile Agents. In: Proc AAMAS. ACM
2. Baumann J, Hohl F, Rothermel K (1997) Mole — Concepts of a Mobile Agent System. Technical Report TR-1997-15
3. Corradi A, Cremonini M, Montanari R, Stefanelli C (1999) Mobile Agents Integrity for Electronic Commerce Applications. Information Systems 24 (6), Elsevier Science
4. Cubaleska B, Qiu W, Schneider M (2002) How to Play Sherlock Holmes in the World of Mobile Agents. In: Proc 7th Australasian Conf on Information Security and Privacy, LNCS 2384, Springer
5. Dadon-Elichai A (2004) RDS: Remote Distributed Scheme for Protecting Mobile Agents. In: Proc AAMAS, ACM
6. Dobrev S, Flocchini P, Prencipe G, Santoro N (2001) Mobile Search for a Black Hole in an Anonymous Ring. In: Proc DISC 2001, LNCS 2180, Springer
7. Ferreira LC, Dahab R (2002) Blinded-key Signatures: Securing Private Keys Embedded in Mobile Agents. In: Proc ACM Symp on Applied Computing
8. Ferreira LC, Uto N, Dahab R (2003) Combining Techniques for Protecting Mobile Agents. In: Actas del Secondo Congreso Iberoamericano de Seguridad Informatica, Mexico
9. Hohl F (1998) Time Limited Blackbox Security: Protecting Mobile Agents from Malicious Hosts. In: Mobile Agents and Security, LNCS 1419, Springer

10. Karjoth G, Asokan N, Gülcü C (1998) Protecting the Computation Results of Free-roaming agents. In: Rothermel K, Hohl F (eds) 2nd Int Workshop on Mobile Agents, LNCS 1477, Springer, Berlin
11. Karnik N, Tripathi A (1998) Agent Server Architecture for the Ajanta Mobile-Agent System. In: Proc Int Conf on Parallel and Distributed Processing Techniques and Applications, Las Vegas, NV
12. Lange DB, Oshima M (1998) Java Agent API: Programming and Deploying Aglets with Java. Addison-Wesley
13. Chun Lee D, Kim JG (2001) Adaptive Migration Strategy for Mobile Agents on Internet. In: Proc 2nd Int Workshop Technologies for E-Services, LNCS 2193, Springer
14. Lee H, Alves-Foss J, Harrison S (2004) The Use of Encrypted Functions for Mobile agent Security. In: Proc 37th Hawaii Int Conf on System Sciences
15. Menezes AJ, van Oorschot PC, Vanstone SA (1997) Handbook of Applied Cryptography. CRC Press, Boca Raton, FL
16. Sander T, Tschudin CF (1998) Towards Mobile Cryptography. In: IEEE Symp on Security and Privacy
17. Schneider FB (1997) Towards Fault-tolerant and Secure Agentry. In: Proc 11th Int Workshop on Distributed Algorithms, LNCS 1320, Springer
18. Tripathi A, Karnik N, Vora M, Ahmed T (1998) Ajanta — A System for Mobile-Agent Programming. Technical Report TR98-016
19. Vigna G (1998) Cryptographic Traces for Mobile Agents. In: Mobile Agents and Security, LNCS 1419, Springer
20. Westhoff D, Schneider M, Unger C, Kaderali F (1999) Methods for Protecting a Mobile Agent's Route. In: Proc Information Security, 2nd Int Workshop, LNCS 1729, Springer
21. Westhoff D, Schneider M, Unger C, Kaderali F (2000) Protecting a Mobile Agent's Route Against Collusions. In: Proc Selected Areas in Cryptography, 6th Annual Int Workshop, LNCS 1758, Springer
22. Yee BS (1999) A Sanctuary for Mobile Agents. In: Vitek J, Jensen CD (eds) Secure Internet Programming, LNCS 1603, Springer

13

Fighting Internet Congestion by Means of Chaos Control

Zhong Li

Faculty of Electrical and Computer Engineering
FernUniversität in Hagen, 58084 Hagen, Germany
zhong.li@fernuni-hagen.de

Summary. The end-to-end congestion control of the Transmission Control Protocol (TCP) is modeled together with the Random Early Detection (RED) router queue management as a discrete-time dynamical system, which can exhibit complex bifurcating and chaotic behaviour. Then, the chaotic characteristics of the TCP-RED model are analysed. To control bifurcation and chaos in the congestion control system, a time-delayed feedback control method is proposed. Finally, a simulation example is presented to illustrate the effectiveness of the proposed congestion control method.

Key words: Internet congestion, TCP-RED, bifurcation, chaos, chaos control, time-delayed feedback control.

13.1 Introduction

The congestion control mechanisms in today's Internet already represent one of the largest artificial feedback systems deployed. As the Internet continues to expand in size, diversity and reach, playing an ever-increasing rôle in human lifes and for the integration of other networks indispensable for modern civilisation (e.g., in transportation or finance), it becomes ever more crucial to have a solid understanding of how this fundamental resource is controlled. Given the scale and complexity of the network, however, and the heuristic, intricate nature of many deployed control mechanisms, until recently this problem appeared to be well beyond the reach of analytical modeling and feedback control theory. Hence, a reliable and accurate traffic model is essential which involves time-delays inevitably being introduced in communication networks, both due to limited bandwidth, but also due to overhead in the communication nodes and in the network. Such a traffic model is needed in order to evaluate the performance of the Internet's networks, and for use in network planning and designs.

The Transmission Control Protocol (TCP) is to ensure end-to-end control of the data transmissions over the Internet. Since most of today's Internet

Z. Li: *Fighting Internet Congestion by Means of Chaos Control*, Studies in Computational Intelligence (SCI) **42**, 269–283 (2007)
www.springerlink.com © Springer-Verlag Berlin Heidelberg 2007

traffic is generated by data transfer applications (such as web sessions) using TCP, the Transport Control Protocol/Internet Protocol (TCP/IP) is becoming the most important data transmission standard. It is, therefore, a major challenge to understand its dynamics and to evaluate its performance, in order to improve the efficiency of the network resources' utilisation.

However, an uncontrolled network may suffer from severe congestion, which can cause high packet loss rates and increasing delays, and can even let the whole system break down in form of a *congestion collapse* (or Internet meltdown). This is the state in which any increase in the imposed load leads to a decrease of the useful work done by the network.

While available bandwidth is increasing, the demand for that bandwidth is growing even faster. The problem of congestion cannot be solved by introducing almost "infinite" buffer space inside the network. In fact, too much buffer space in the routers can be more harmful than too little, because packets will have to be dropped only after they have consumed valuable network resources.

Given the proliferation of applications based on the Internet and the expected growth of the associated traffic in the years to come, the rising volume of Internet traffic is now causing massive problems. It was reported in [9], that with two TCP connections the behaviour of the aggregated traffic could be varied according to the system parameters. For most settings, the aggregated traffic has a simple and synchronised behaviour. For some settings, however, the traffic characteristic show some very complex behavioural patterns. Using simulations with 30 TCP connections it was demonstrated that the aggregated traffic behaves chaotically. Due to the unpredictability of a chaotic system, the network capacity is significantly affected and its optimisation becomes impossible. For system designers and network operators it is essential to avoid such undesirable situations.

It was also reported in [7], that Internet traffic jams are chaotic, not random. This is a crucial distinction, because chaos can be controlled. Chaotic systems can occupy a wide range of unstable oscillation states. A typical system will sit in one state for a short period of time and, then, jump randomly to another state. Although it is impossible to predict where a chaotic system will jump, we can observe them and identify the range of characteristic frequencies that the system is likely to occupy. Then, we can use this information to trap the system.

Control problems in communication networks are inherently high-dimensional, non-linear, dynamic and complex. The stability of the Internet so far is, to a large extent, due to the congestion control and avoidance algorithms implemented in its transmission control protocol (TCP), which employs the additive-increase/multiplicative-decrease (AIMD) rate control mechanism. Remarkable progress has been made only recently towards a theoretical understanding of TCP congestion control, both for its equilibrium characterisation using optimisation theory, and its dynamics using control theory [10].

It has become clear that the TCP congestion control and avoidance mechanisms, though necessary and powerful, are not sufficient to provide good services under all circumstances. Basically, there is a limit to how much control can be accomplished from the edges of the network. Some mechanisms are needed in the routers to implement endpoint congestion avoidance mechanisms. Traditionally, the so-called drop-tail mechanism has served the Internet well for years, but one of the main shortcomings of this TCP congestion control algorithm is that the sending sources reduce their transmission rates only after detecting packet loss due to queue overflow. This is a problem, since a considerable amount of time may elapse between the dropping of a packet at a router and the actual detection of the loss by the source. In the meantime, a large number of further packets may be dropped as the source continues to transmit at a rate that the network cannot support. Thus, drop-tail buffer management forces network operators to choose between high utilisation (requiring large buffers) and low delay (requiring small buffers). One solution to this problem is for the routers to drop (or mark) packets before a queue becomes full, so that end nodes can respond to congestion before buffers overflow. This approach is called "Active Queue Management (AQM)". One form of AQM for deployment in the Internet is Random Early Detection (RED).

However, it is very difficult to parameterise RED in order to attain good performance under different congestion scenarios. In almost all studies the parameter settings are based on heuristics, and proposed configurations are suitable for the particular traffic conditions studied, only. The method RED tries to stabilise the average queue length at a target value, but its current version does not succeed in achieving this objective, because the equilibrium average queue length strongly depends on traffic load patterns and parameter settings. Instability in TCP-RED often leads to oscillations of average queue length.

Recently, it has been shown that TCP-RED can approximately be modeled as a first-order non-linear map, and exhibits a rich variety of irregular behaviours such as bifurcation and chaos.

In this chapter, first the TCP-RED mechanism is introduced from a control system point of view, and its discrete dynamical system model is derived. Then, chaotic properties of the TCP-RED map are analysed, and its chaotic behaviour is shown under some parameter conditions. Finally, a time-delayed feedback control method is proposed to stabilise the average queue length at a desired value, whose effectiveness is shown by a simulation example.

13.2 Modeling the TCP-RED Congestion Control System

A discrete-time dynamical model of TCP congestion control at the hosts coupled with RED active queue management at the routers is to be derived.

Upon detection of congestion, the sources should inject their packets into the network more slowly. In order for a host to be able to detect congestion, the routers must be able to provide *feedback* information that the network is currently (or is about to become) overloaded. Packet drops were, and to great extent are still, the only means for a router to fight congestion. When detecting packet drops, the sources interpret this as a fed-back congestion indication and, then, reduce their rates.

The feedback from the network and the response by the sources are the foundations of Internet congestion control and are very important as they facilitate decentralised resource allocation. Up to now, congestion control mechanisms were mainly implemented in the end hosts. However, with decisions made at the end hosts and treatment of routers as black boxes simply dropping packets, there is clearly a limit to how much control can be achieved over the allocation of network resources. This also limits the range of services a network is capable to offer.

Active Queue Management can be allocated in routers as they are exactly aware of congestions. One of the most prominent AQM mechanism is RED, which controls congestion by randomly dropping packets with a probability associated with the average queue length in the router buffers.

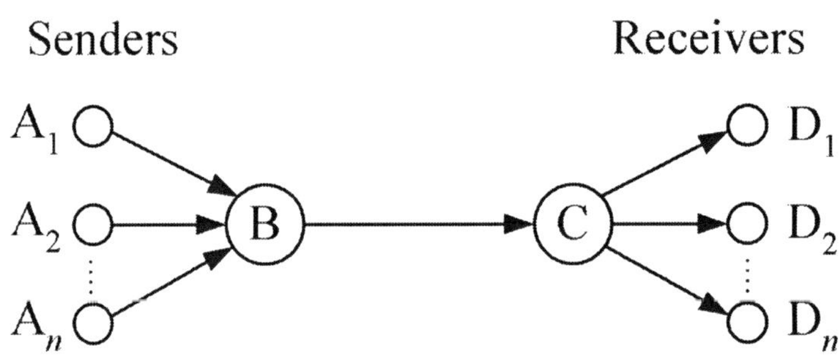

Fig. 13.1: System of n TCP flows passing through a common link

To derive a discrete-time, analytical model of the TCP-RED congestion control system, we consider a system of n TCP flows passing through a common link with capacity c as shown in Fig. 13.1. Assume that all access links, $A_i - B$ and $C - D_i$, have enough capacity so that $B - C$ is the sole bottleneck link, i.e., the only link where the rate of incoming traffic can surpass the link capacity. Further assume that each TCP flow has the same Round Trip Time (RTT) R. The actual queue size in the router is sampled every $\Delta t = R$ units of time, and the RED controller provides a new value of the drop probability p every Δt units of time. Thus, the whole TCP-RED congestion control system can be modeled as a discrete-time feedback dynamical system, whose diagram is shown in Fig. 13.2.

Suppose that at time t_k the actual queue length is q_k. Due to bursts of network traffic and other perturbations, the actual queue length is fluctuating

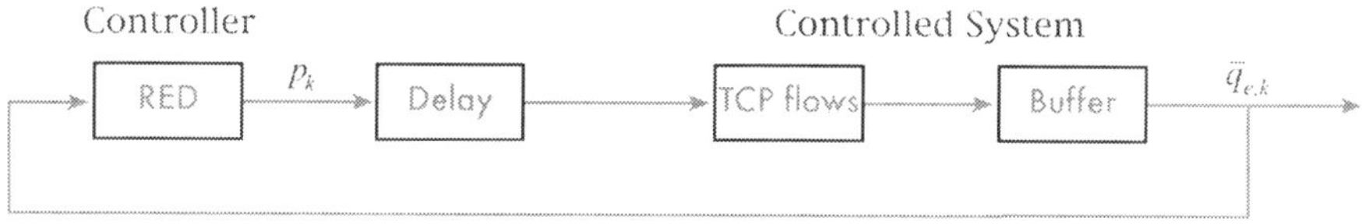

Fig. 13.2: TCP-RED as a discrete-time feedback control system

considerably, so that a low-pass filter A is required. The filtered average queue length is in the form of

$$\bar{q}_{e,k} = A(\bar{q}_{e,k-1}, q_k). \tag{13.1}$$

RED manages the queue length by randomly dropping packets with a probability p_k that is an increasing function H of the average queue length:

$$p_k = H(\bar{q}_{e,k}). \tag{13.2}$$

At time $t_{k+1} = t_k + \Delta t$, the TCP hosts adjust the sending rates according to the drop probability p_k, which is described as:

$$q_{k+1} = G(p_k). \tag{13.3}$$

Combining Eqs. (13.1)–(13.3) yields the following discrete-time dynamical system model for the average queue length:

$$\bar{q}_{e,k+1} = A(\bar{q}_{e,k}, G(H(\bar{q}_{e,k}))) \equiv f(\bar{q}_{e,k}). \tag{13.4}$$

Owing to the complexity of the Internet, it is impossible to derive an exact expression of the map f, but a simplified analytical formula for the map f can be given to compute the average queue length as:

$$\bar{q}_{e,k+1} = A(\bar{q}_{e,k-1}, q_k) = (1-\omega)\bar{q}_{e,k-1} + \omega q_k, 0 < \omega < 1, \tag{13.5}$$

which appears as an exponential weighted moving average of the queue length and can be recast as

$$\bar{q}_{e,k} = \omega \sum_{i=0}^{k} (1-\omega)^i q_{k-i}, \text{if } \bar{q}_{e,0} = 0. \tag{13.6}$$

The RED algorithm manages the queue length by randomly dropping packets with a probability p_k. Assume that the packet dropping probability increases linearly from 0, when the average queue length is at the RED parameter *minthresh* $q_{\min}$, to a drop rate $p_{\max}$, when the average queue length reaches the parameter *maxthresh* $q_{\max}$; then it increases linearly from $p_{\max}$ to 1 as the average length varies from $q_{\max}$ to $2q_{\max}$, as shown in Fig. 13.3.

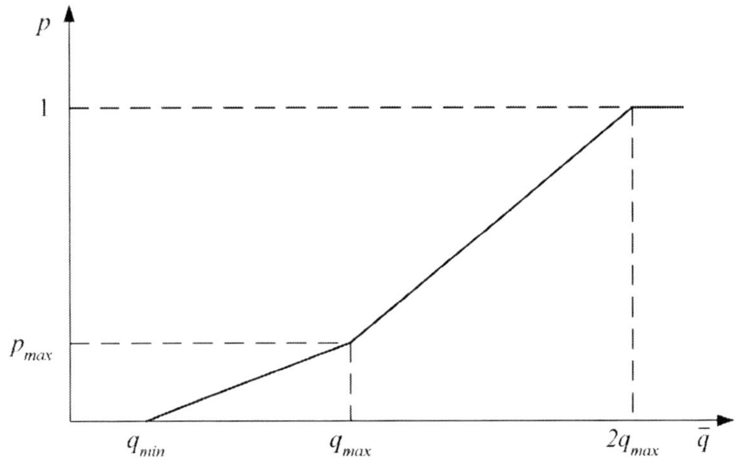

Fig. 13.3: Packet dropping probability of the RED algorithm

More precisely, the dropping probability p_k is formulated as:

$$p_k = H(\bar{q}_{e,k}) = \begin{cases} 0, & 0 \leq \bar{q}_{e,k} < q_{\min}, \\ \dfrac{\bar{q}_{e,k} - q_{\min}}{q_{\max} - q_{\min}} p_{\max}, & q_{\min} \leq \bar{q}_{e,k} < q_{\max}, \\ p_{\max} + \dfrac{1 - p_{\max}}{q_{\max}} (\bar{q}_{e,k} - q_{\max}), & q_{\max} \leq \bar{q}_{e,k} \leq 2q_{\max}, \\ 1, & 2q_{\max} \leq \bar{q}_{e,k} \leq B, \end{cases} \tag{13.7}$$

where B is the buffer size.

According to the procedure given in [3], the map G in (13.3) can be expressed as:

$$G(p) = \begin{cases} \min(B, \dfrac{c}{M}(T_R^{-1}(p, \dfrac{c}{n}) - R_0)), & p \leq p_0, \\ 0, & otherwise, \end{cases} \tag{13.8}$$

where T is the throughput of the TCP flow (in bits/sec), M the packet size (in bits), R_0 the propagation and transmission time, T_R^{-1} and T_p^{-1} the inverse of $T(p, R)$ in R and p, respectively, and $p_0 = T_p^{-1}(\frac{c}{n}, R_0)$ the maximum dropping probability for which the system is fully utilised, i.e., for $p > p_0$, senders will have their rates too small to keep the link fully utilised.

The throughput function is described as [6]:

$$T(p, R) = \frac{M}{R} \frac{K}{\sqrt{p}}, \tag{13.9}$$

where K is a constant satisfying $1 \leq K \leq \sqrt{8/3}$. The map G writes as:

$$G(p) = \begin{cases} \min\left(B, \frac{c}{M}\left(\frac{MK}{\frac{c}{n}\sqrt{p}} - R_0\right)\right), p \leq p_0, \\ 0, \qquad\qquad\qquad\qquad\qquad\qquad otherwise, \end{cases} \tag{13.10}$$

where

$$p_0 = \left(\frac{nKM}{R_0 c}\right)^2.$$

Given the maps A, H and G in (13.5)–(13.7), the TCP-RED congestion control system can, thus, be modeled as a one-dimensional map for the average queue length:

$$\bar{q}_{e,k+1} = f(\bar{q}_{e,k}, \rho) = (1-\omega)\bar{q}_{e,k} + \omega G(H(\bar{q}_{e,k})), \tag{13.11}$$

where ρ is a vector of system parameters.

Consider the case $p_0 > p_{\max}$.

1. $G(H(\bar{q}_{e,k})) \equiv 0$, if

$$H(\bar{q}_{e,k}) = p_{\max} + \frac{1-p_{\max}}{q_{\max}}(\bar{q}_{e,k} - q_{\max}) > p_0,$$

or equivalently,

$$\bar{q}_{e,k} > b_1 \equiv \frac{q_{\max}(p_0 - p_{\max})}{1 - p_{\max}} + q_{\max}.$$

Thus, one has

$$\bar{q}_{e,k+1} = (1-\omega)\bar{q}_{e,k}, \quad \bar{q}_{e,k} > b_1.$$

2. $G(H(\bar{q}_{e,k})) \equiv B$, if

$$\begin{cases} H(\bar{q}_{e,k}) \leq p_0, \\ B \leq \frac{c}{M}\left(\frac{MK}{\frac{c}{n}\sqrt{H(\bar{q}_{e,k}}} - R_0\right). \end{cases} \Rightarrow H(\bar{q}_{e,k}) \leq \left(\frac{nK}{B + \frac{R_0 c}{M}}\right)^2.$$

a) If $p_0 > \frac{nK}{B + \frac{R_0 c}{M}} \geq p_{\max}$,

$$H(\bar{q}_{e,k}) = p_{\max} + \frac{1-p_{\max}}{q_{\max}}(\bar{q}_{e,k} - q_{\max}) \leq \left(\frac{nK}{B + \frac{R_0 c}{M}}\right)^2,$$

or equivalently,

$$\bar{q}_{e,k} \leq B_2 \equiv \frac{q_{\max}}{1-p_{\max}}\left(\left(\frac{nK}{B + \frac{R_0 c}{M}}\right)^2 - p_{\max}\right) + q_{\max}.$$

b) If $p_0 > p_{\max} > \dfrac{nK}{B + \frac{R_0 c}{M}}$,

$$H(\bar{q}_{e,k}) = \frac{\bar{q}_{e,k} - q_{\min}}{q_{\max} - q_{\min}} p_{\max} \le \left(\frac{nK}{B + \frac{R_0 c}{M}} \right)^2,$$

or equivalently,

$$\bar{q}_{e,k} \le b_3 \equiv \frac{q_{\max} - q_{\min}}{p_{\max}} \left(\frac{nK}{B + \frac{R_0 c}{M}} \right)^2 + q_{\min}.$$

Thus, one has

$$\bar{q}_{e,k+1} = (1-\omega)\bar{q}_{e,k} + \omega B, \quad \bar{q}_{e,k} \le b_2 (\text{ or } \bar{q}_{e,k} \le b_3). \tag{13.12}$$

3. $G(H(\bar{q}_{e,k})) \equiv \dfrac{c}{M} \left(\dfrac{MK}{\frac{c}{n}\sqrt{H(\bar{q}_{e,k})} - R_0} \right)$, if

$$\begin{cases} H(\bar{q}_{e,k}) \le p_0, \\ B \ge \dfrac{c}{M} \left(\dfrac{MK}{\frac{c}{n}\sqrt{H(\bar{q}_{e,k})}} - R_0 \right), \end{cases} \Rightarrow H(\bar{q}_{e,k}) \ge \left(\frac{nK}{B + \frac{R_0 c}{M}} \right)^2.$$

a) For $q_{\max} < \bar{q}_{e,k} \le b_1$,

$$H(\bar{q}_{e,k}) \equiv p_{\max} + \frac{1 - p_{\max}}{q_{\max}} (\bar{q}_{e,k} - q_{\max}),$$

and

$$\begin{aligned} \bar{q}_{e,k} &\equiv \tilde{f}(\bar{q}_{e,k}, \rho) \\ &= (1-\omega)\bar{q}_{e,k} + \omega \left(\frac{nK}{\sqrt{p_{\max} + \frac{1-p_{\max}}{q_{\max}} (\bar{q}_{e,k} - q_{\max})}} - \frac{R_0 c}{M} \right). \end{aligned}$$

b) For $b_3 < \bar{q}_{e,k} \le q_{\max}$,

$$H(\bar{q}_{e,k}) \equiv \frac{\bar{q}_{e,k} - q_{\min}}{q_{\max} - q_{\min}} p_{\max},$$

and

$$\begin{aligned} \bar{q}_{e,k+1} &\equiv \tilde{f}(\bar{q}_{e,k}, \rho) \\ &= (1-\omega)\bar{q}_{e,k} + \omega \left(\frac{nK}{\sqrt{\frac{\bar{q}_{e,k} - q_{\min}}{q_{\max} - q_{\min}} p_{\max}}} - \frac{R_0 c}{M} \right). \end{aligned}$$

In summary, if $p_0 > \dfrac{nK}{B+\frac{R_0 c}{M}} \geq p_{\max}$, the TCP-RED model (13.11) can be written as:

$$
\begin{aligned}
\bar{q}_{e,k+1} &= f(\bar{q}_{e,k},\rho) \\
&= \begin{cases} (1-\omega)\bar{q}_{e,k}, & \bar{q}_{e,k} > b_1, \\ \tilde{f}(\bar{q}_{e,k},\rho), & b_2 < \bar{q}_{e,k} \leq b_1, \\ (1-\omega)\bar{q}_{e,k} + \omega B, & \bar{q}_{e,k} \leq b_2; \end{cases}
\end{aligned}
\tag{13.13}
$$

if $p_0 > p_{\max} > \dfrac{nK}{B+\frac{R_0 c}{M}}$, it reads as:

$$
\begin{aligned}
\bar{q}_{e,k+1} &= f(\bar{q}_{e,k},\rho) \\
&= \begin{cases} (1-\omega)q_{e,k}, & q_{e,k} > b_1, \\ \tilde{f}_1(\bar{q}_{e,k},\rho), & q_{\max} < \bar{q}_{e,k} \leq b_1, \\ \tilde{f}_2(\bar{q}_{e,k},\rho), & b_3 < \bar{q}_{e,k} \leq q_{\max}, \\ (1-\omega)\bar{q}_{e,k} + \omega B, & \bar{q}_{e,k} \leq b_3. \end{cases}
\end{aligned}
\tag{13.14}
$$

It is obvious that $b_3 < q_{\max} < b_2 < b_1$. In practice, the average queue length should be avoided to be greater than $q_{\max}$, which results in too small TCP sending rates. It is shown that if $\bar{q}_{e,k} \leq b_2$, map (13.13) becomes the linear one (13.12), which has a unique fixed point B. This fixed point is not desirable. Similarly, in map (13.14), only $b_3 < \bar{q}_{e,k} \leq q_{\max}$ is appropriate. Therefore, it is necessary to introduce the restriction $p_0 \leq p_{\max}$ for a good performance of RED.

Now, consider the case $p_0 \leq p_{\max}$.

1. $G(H(\bar{q}_{e,k})) \equiv 0$, if

$$
H(\bar{q}_{e,k}) \equiv \frac{\bar{q}_{e,k} - q_{\min}}{q_{\max} - q_{\min}} p_{\max} > p_0,
$$

or equivalently,

$$
\bar{q}_{e,k} > b_4 \equiv \frac{p_0(q_{\max} - q_{\min})}{p_{\max}} + q_{\min}.
$$

Thus, one has

$$
\bar{q}_{e,k} = (1-\omega)\bar{q}_{e,k}, \quad \bar{q}_{e,k} > b_4.
$$

It is obvious that when $p_0 = p_{\max}$, $b_4 = q_{\max}$.

2. $G(H(\bar{q}_{e,k})) \equiv B$, if

$$
\begin{cases} H(\bar{q}_{e,k}) \leq p_0, \\ B \leq \frac{c}{M}\left(\frac{MK}{\frac{c}{n}\sqrt{H(\bar{q}_{e,k}}} - R_0\right), \end{cases} \Rightarrow H(\bar{q}_{e,k}) \leq \left(\frac{nK}{B+\frac{R_0 c}{M}}\right)^2,
$$

or equivalently,

$$\bar{q}_{e,k} \le b_3 \equiv \frac{q_{\max} - q_{\min}}{p_{\max}} \left(\frac{nK}{B + \frac{R_0 c}{M}} \right)^2 + q_{\min}.$$

Thus, one has

$$\bar{q}_{e,k+1} = (1-\omega)\bar{q}_{e,k} + \omega B, \quad \bar{q}_{e,k} \le b_3.$$

3. For $b_3 < \bar{q}_{e,k} \le b_4$, one has

$$\begin{aligned} \bar{q}_{e,k+1} &\equiv \tilde{f}(\bar{q}_{e,k}, \rho) \\ &= (1-\omega)\bar{q}_{e,k} + \omega \left(\frac{nK}{\sqrt{\frac{\bar{q}_{e,k} - q_{\min}}{q_{\max} - q_{\min}} p_{\max}}} - \frac{R_0 c}{M} \right). \end{aligned}$$

In summary, if $p_0 \le p_{\max}$, the TCP-RED map (13.11) is written as:

$$\begin{aligned} \bar{q}_{e,k+1} &= f(\bar{q}_{e,k}, \rho) \\ &= \begin{cases} (1-\omega)\bar{q}_{e,k}, & \bar{q}_{e,k} > b_4, \\ \tilde{f}(\bar{q}_{e,k}, \rho), & b_3 < \bar{q}_{e,k} \le b_4, \\ (1-\omega)\bar{q}_{e,k} + \omega B, & \bar{q}_{e,k} \le b_3. \end{cases} \end{aligned} \tag{13.15}$$

It is remarked that an exact mathematical model of the TCP-RED congestion control system is very difficult, if not impossible, to be derived; however, Eq. (13.15) characterises, at least to some extent, some important features of TCP-RED for the purpose of improving the performance of Internet congestion control. Here and throughout, only the case $p_0 \le p_{\max}$ is considered.

13.3 Chaotic Dynamics of the TCP-RED System

It is clear that if $\bar{q}_{e,k} \ge b_4$, the TCP sending rates will be too small to keep the link highly utilised; on the other hand, if $\bar{q}_{e,k} \le b_3$, the TCP sending rates will be too large to keep delays low. Therefore, it is appropriate to assume that $b_3 < \bar{q}_{e,k} < b_4$, within which the system is indeed the strongly nonlinear map $\tilde{f}$, whose fixed point $\bar{q}_e^* \in (b_3, b_4)$ is a real solution of the equation:

$$(\bar{q}_e^* - q_{\min})(\bar{q}_e^* + \frac{R_0 c}{M})^2 = \frac{(nK)^2}{p_{\max}}(q_{\max} - q_{\min}).$$

The eigenvalue of the system at the fixed point is

$$\begin{aligned}\alpha &\equiv \frac{\partial f(\bar{q}_e^*, \rho)}{\partial \bar{q}_e} \\ &= 1 - \omega - \frac{\omega n K}{2(\bar{q}_e^* - q_{\min})^{\frac{3}{2}}} \sqrt{\frac{q_{\max} - q_{\min}}{p_{\max}}} \\ &< 1.\end{aligned} \tag{13.16}$$

Since $\bar{q}_e^* \leq b_4$, one has

$$\begin{aligned}\alpha &> 1 - \omega - \frac{\omega n K}{2(b_4 - q_{\min})^{3/2}} \sqrt{\frac{q_{\max} - q_{\min}}{p_{\max}}} \\ &= 1 - \omega - \frac{\omega \left(B + \frac{R_0 c}{M}\right)^3}{2(nK)^2} \sqrt{\frac{q_{\max} - q_{\min}}{p_{\max}}} \\ &\equiv \bar{\alpha}.\end{aligned} \tag{13.17}$$

It is clear that $\bar{\alpha}$ is a decreasing function of parameters ω and $p_{\max}$.

If $-1 < \alpha < 1$, the fixed point $\bar{q}_e^*$ is asymptotically stable. However, it is shown that the TCP-RED system can exhibit complex oscillating behaviour, contradictorily implying the instability of the fixed point.

Suppose that the system setting is as follows, here and throughout [6]:

$$\begin{aligned}&B = 300 \text{ packets}, n = 20, K = \sqrt{\frac{8}{3}}, R_0 = 0.1 \text{ sec} \\ &c = 1500 \text{ kbps}, M = 0.5 \text{ kb}, q_{\max} = 100, p_{\max} = 0.5.\end{aligned}$$

Taking the exponential averaging weight ω or the RED parameter $q_{\min}$ as bifurcation parameter, the TCP-RED model can exhibit bifurcation behaviour as illustrated in Figs. 13.4 and 13.5. It is shown in Fig. 13.4 that the system exhibits a period-doubling bifurcation at $\omega \approx 0.047$. For $\omega < 0.047$, the average queue length $\bar{q}_{e,k}$ stabilises on the fixed point $\bar{q}_e^*$. For $0.047 < \omega < 0.049$, the system exhibits a period-2 oscillation. This oscillation lies in the safe region (b_3, b_4) which is, thus, considered as a "benign" oscillation, since it does not significantly affect either throughput or delay. Further increasing ω leads to a two-bands chaotic phenomenon with the upper-band chaotic oscillation colliding with the upper border b_4. For $\omega > 0.055$, the two-band chaotic oscillations become a single-band chaotic oscillation. Such a border-collision-induced chaotic oscillation is "malignant" and is, thus, to be avoided, since it can result in significantly decreasing throughput and increasing dropping rates.

Similarly, Fig. 13.5 shows the bifurcation diagram of the average queue length with respect to the parameter $q_{\min}$ at $\omega = 2^{-6}$.

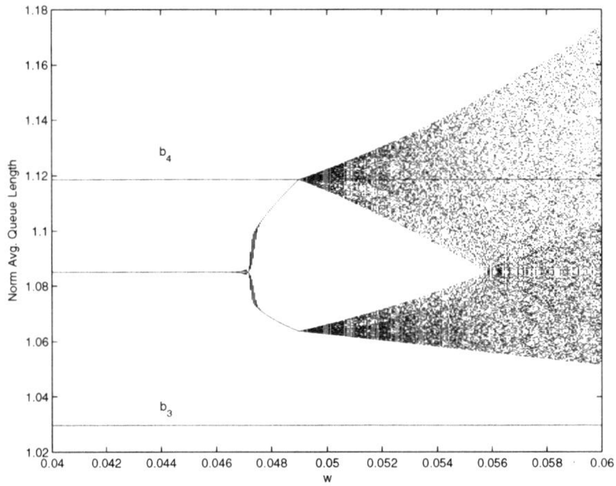

Fig. 13.4: Bifurcation diagram of the normalised average queue length (dividing $q_{\min}$) w.r.t. the weight ω; the upper line represents b_4 and the lower one b_3

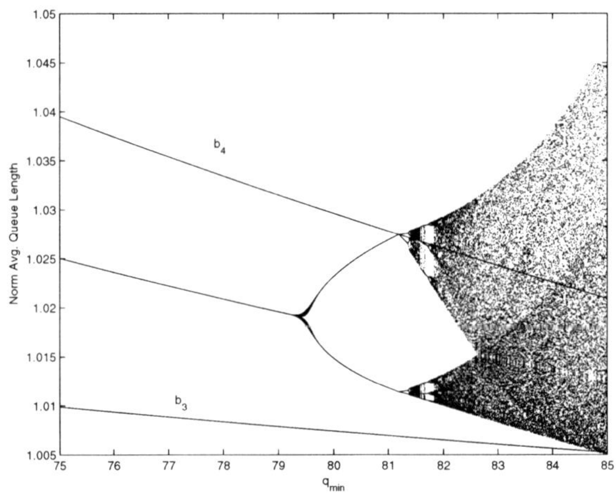

Fig. 13.5: Bifurcation diagram of the normalised average queue length (dividing $q_{\min}$) w.r.t. the weight $q_{\min}$; the upper line represents b_4 and the lower one b_3

13.4 Controlling Chaos in the TCP-RED System

A large number of bifurcation and chaos control approaches are currently available [2, 1, 5, 8, 4]. Many of them, however, cannot directly be applied to the control of oscillations in the TCP-RED system, due to the complexity of the network structure and physical limitations on allowable control. Here, we apply the time-delayed feedback control (TDFC) method to stabilise the

TCP-RED map. A simulation example will be given to show the effectiveness of the proposed algorithm to control Internet congestions.

Consider the following control system:

$$\bar{q}_{e,k+1} = f(\bar{q}_{e,k}, \rho) + u_k, \tag{13.18}$$

where the control input was proposed in [8] of the form:

$$u_k = h(\bar{q}_{e,k} - (1-r)\sum_{j=1}^{\infty} r^{j-1}\bar{q}_{e,k-j}), \tag{13.19}$$

in which h is the feedback gain, $r \in [0,1]$. It is noted that when $r = 0$ the control law corresponds to the time-delayed feedback control (TDFC) proposed in [5].

Theorem 1. *Consider the discrete-time control system* (13.15), (13.18), *and* (13.19). *The fixed point* $\bar{q}_e^*$ *is asymptotically stable iff*

$$\frac{r+3}{r-1} < \alpha < 1,$$

and

$$-\frac{1}{2}(1+r)(1+\alpha) < h < 1 - \alpha r,$$

where $\alpha \equiv \dfrac{\partial f(\bar{q}_e^*, \rho)}{\partial \bar{q}_e}$.

Proof. Denote $\delta\bar{q}_{e,k} = \bar{q}_{e,k} - \bar{q}_e^*$ and $e_k = \delta\bar{q}_{e,k} - \delta\bar{q}_{e,k-1}$. Linearising system (13.18) about the fixed point gives

$$\delta\bar{q}_{e,k+1} = \alpha\delta\bar{q}_{e,k} + u_k. \tag{13.20}$$

Substituting (13.19) into (13.20) yields the following second-order linear difference equation:

$$\begin{bmatrix} \delta\bar{q}_{e,k+1} \\ e_{k+1} \end{bmatrix} = \begin{bmatrix} \alpha + r(1-\alpha) & \alpha r \\ (1-4)(\alpha-1) & \alpha r + h \end{bmatrix} \begin{bmatrix} \delta\bar{q}_{e,k} \\ e_k \end{bmatrix} \tag{13.21}$$

The stability of (13.21) is governed by the characteristic equation:

$$\lambda^2 - (\alpha + h + r)\lambda + \alpha r + h = 0. \tag{13.22}$$

Condition $|\lambda| < 1$ guarantees the stability of the fixed point, which results in the stability conditions of the theorem. Thus, the proof is completed.

To achieve chaos control with small perturbation, the control law is taken in the form:

$$u_k = \begin{cases} h(\bar{q}_{e,k} - (1-r)\sum_{j=1}^{\infty} r^{j-1}\bar{q}_{e,k-j}), & |u| < \varepsilon, \\ 0, & \text{otherwise}, \end{cases} \tag{13.23}$$

where ε is a small positive number.

Choosing the above given parameter settings, together with $\omega = 0.07$, $p_{\max} = 0.1$ and the initial state $\bar{q}_{e,0} = 60$, the control law (13.23) is applied to control chaos in the TCP-RED system at $r = 0$ (corresponding to TDFC) and $r = 0.2$ (corresponding to the proposed one), respectively.

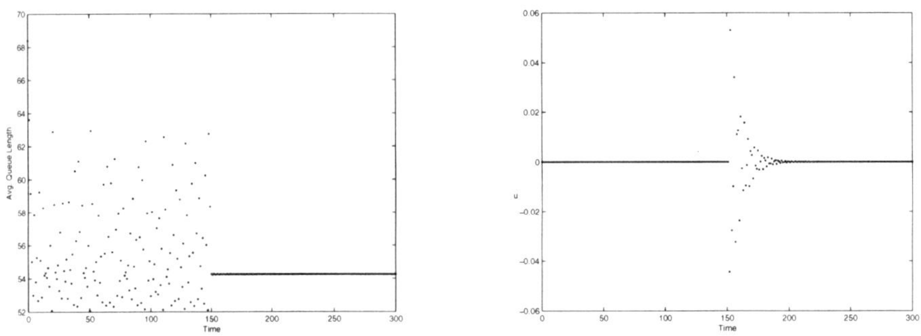

(a) (b)

Fig. 13.6: Time-delayed feedback control of the TCP-RED system with $h = 0.8$ and $c = 0.5$ at $r = 0$: (a) temporal evolution of the average queue length $\bar{q}_{c,k}$; (b) temporal evolution of the control force u_k

It is shown in Fig. 13.6 and Fig. 13.7 that, by the proposed control law, the system is stabilised on the fixed point $\bar{q}_e^*$, and thereafter the control input becomes zero, while by TDFC, the control performance is similar, but the control input keeps oscillating with a small amplitude.

It should be noted that the physical meaning of this control method is not clear, yet, and it is not easy to be implemented in routers. In addition to this time-delayed feedback control method, some other chaos control methods can also be considered, such as the adaptive control method.

13.5 Conclusions

Internet congestion control is of fundamental significance for today's ubiquitous network traffic. In this chapter, a rough discrete-time dynamical model of the TCP-RED congestion control system was derived, with which some

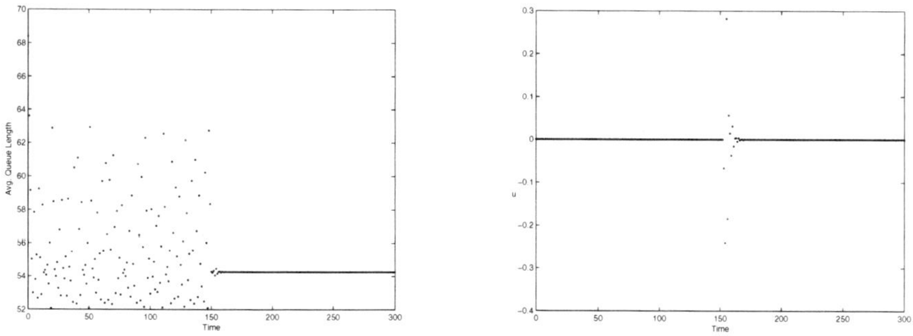

(a) (b)

Fig. 13.7: Time-delayed feedback control of the TCP-RED system with $h = 0.8$ and $\varepsilon = 0.5$ at $r = 0.2$: (a) temporal evolution of the average queue length $\bar{q}_{e,k}$; (b) temporal evolution of the control force u_k

fundamental characteristics of the system can be investigated to the end of improving its performance. It was shown that this system can exhibit complex bifurcating and chaotic behaviour. Thus, chaos control methods can be employed to stabilise chaotic congestion control systems. The problem of how to implement chaos control methods in routers, however, is still open and to be studied further.

References

1. Chen G (1999) Controlling Chaos and Bifurcations in Engineering Systems. CRC Press, Boca Raton, FL
2. Chen G, Dong X (1998) From Chaos to Order: Perspectives, Methodologies, and Applications. World Scientific, Singapore
3. Firoiu V, Borden M (2000) A Study of Active Queue Management for Congesiton Control. In: Proc IEEE INFOCOM'2000. `http://www.ieee.infocom.org/2000`
4. Konishi K, Kokame H (1998) Phys Lett A 248: 359–368
5. Pyragas K (1992) Phys Lett A 170: 421–428
6. Ranjan P, Abed E (2002) Nonlinear Instabilities in TCP-RED. In: Proc IEEE INFOCOM'2002. `http://www.ieee-infocom.org/2002`
7. Sincell M (2001) Physicists Use Chaos to Calm the Web. Physics World Digest 7
8. Socolar J, Sukow D, Gauthier D (1994) Phys Rev E 50: 3245–3248
9. Veres A, Boda M (2000) The Chaotic Nature of TCP Congestion Control. In: Proc IEEE INFOCOM'2000, pp 1715–1723
10. Wang X, Chen G, Ko K (2001) A Stability Theorem for Internet Congestion Control. Systems and Control Letters 45: 81–85

14

Securing Communication by Chaos-based Encryption

Ping Li[1], Zhong Li[1], Wolfgang A. Halang[1], and Guanrong Chen[2]

[1] Faculty of Electrical and Computer Engineering
FernUniversität in Hagen, 58084 Hagen, Germany
{ping.li|zhong.li|wolfgang.halang}@fernuni-hagen.de

[2] Dept. of Electronic Engineering
City University of Hong Kong, Kowloon, Hong Kong SAR, P.R. China
gchen@ee.cityu.edu.hk

Summary. As a prerequisite for data encryption, a special spatiotemporal chaotic system, viz., a coupled map lattice (CML), is employed to design a multiple-output pseudo-random bit generator (PRBG). The CML can exhibit complicated chaotic phenomena both in space and in time, which is highly desirable for the PRBG to be applicable for encryption purposes as it guarantees random behaviour of the generated bit sequences. Cryptographic properties such as probability distribution, auto- and cross-correlation, and FIPS 140-2 tests of the PRBG are numerically investigated to determine certain parameter ranges where the PRBG has good cryptographic properties. Based on this PRBG, a stream cipher is presented. Its security is assessed by analysing diffusion and confusion, using various typical attacks. In addition, the encryption speed of the cipher is evaluated. As an example, an image is encrypted with the cipher. It is shown that the cipher has both low computation expense and high security and efficiency.

Key words: Pseudo-random bit generator, coupled map lattice, r-adic map, cryptography, cryptographic properties, cipher, security.

14.1 Introduction

Chaos-based encryption has attracted much interest in the last two decades. The main reason for it lies in that chaotic systems are characterised by high sensitivity to initial conditions and control parameters, random-like behaviour, and unstable periodic orbits with long periods, which are quite advantageous to ciphers [7, 26, 13]. Until now, a number of chaos-based ciphers have been proposed. Many of them, however, turned out unpromising in practice due to the following reasons.

P. Li et al.: *Securing Communication by Chaos-based Encryption*, Studies in Computational Intelligence (SCI) **42**, 285–306 (2007)
www.springerlink.com © Springer-Verlag Berlin Heidelberg 2007

1. For ciphers where the orbits of chaotic systems with simple structures are directly used to encrypt plaintext, useful information can be extracted from the chaotic orbits to break them [22, 27].
2. There exists dynamical degradation of chaotic systems when realised with digital computers, which threatens the security of the ciphers based on these chaotic systems [25, 15].
3. Some chaos-based ciphers exhibit low processing speed [1], which makes them infeasible in practice.

To overcome these drawbacks, multiple chaotic systems [20, 14], high-dimensional chaotic systems [4], multiple iterations of chaotic systems [29], and other techniques [21] have been proposed to improve chaos-based ciphers. Especially, using spatiotemporal chaotic systems [16, 10, 11, 12] constitutes a significant advance in this aspect due to the following special features inherent to spatiotemporal chaos:

1. The orbit of a spatiotemporal chaotic system has a long period even with dynamical degradation of digital chaos [24].
2. The randomness of a spatiotemporal system's orbit is guaranteed by the complex dynamics with a large number of positive Lyapunov exponents.
3. There are multiple sites in a spatiotemporal chaotic system, which can generate independent keystreams simultaneously.

In this chapter, a spatiotemporal chaotic map is employed to construct a multiple-output PRBG. To make the PRBG effective, a simple coupled map lattice with the essential features of spatiotemporal chaos, which consists of r-adic maps coupled by a one-way rule, is adopted as spatiotemporal chaotic map. Chaotic behaviour of the CML is desired to construct the PRBG, and analysed via the Lyapunov exponent spectrum. Moreover, the cryptographic properties of the PRBG such as period, probability distribution, auto-correlation and cross-correlation are investigated, and the FIPS 140-2 test executed for various parameters. Thus, the ranges of the parameters within which the PRBG has perfect cryptographic properties are determined.

Moreover, a simple stream cipher is designed. Keystreams are generated from the PRBG by using a part of the multiple output, to encrypt plaintext by the XOR function. The cryptographic properties of the keystreams such as probability distribution, auto-correlation and run are investigated numerically. The results show that the properties are satisfactory. The security of the cipher is evaluated by analysing its confusion and diffusion, and applying error function attack, known-plaintext attack, differential attack, brute-force attack, and chosen-plaintext/ciphertext attack. It is shown that the security level of the cipher has a power relation with the size of the CML. Moreover, even for small sizes of the CML, the cipher can easily reach high security levels. In addition, efficient parallel operation of the cipher is guaranteed by close-to-zero cross-correlation among multiple keystreams, and the encryption speed is high even if the size of the CML in the cipher is small. Finally, an image is encrypted with the cipher to illustrate its effectiveness in practice. As compared to the cipher based on a spatiotemporal chaotic system with much more complicated configuration presented in [16], the proposed cipher exhibits higher security, higher encryption speed, and lower computation cost.

This chapter is organised as follows. In Section 14.2, a multiple-output PRBG based on a CML is constructed and its Lyapunov exponents are analysed. The cryptographic properties of the PRBG for various parameters are investigated in Section 14.3. A cipher based on the PRBG is proposed in Section 14.4. The statistical properties, the security, and the efficiency of the cipher are analysed in Section 14.5 and, finally, conclusions are drawn.

14.2 A CML-based Multiple-Output PRBG

14.2.1 One-way CML based on r-adic Maps

A spatiotemporal chaotic system is a spatially extended system, which can exhibit chaos in both space and time. A CML [6], which has been widely employed to study spatiotemporal chaos, is used as spatiotemporal chaotic system here. A CML is a dynamical system with discrete-time, discrete-space, and continuous states. It consists of nonlinear maps located on the lattice sites, named local maps. Each local map is coupled with other local maps in terms of certain coupling rules. Because of the intrinsic nonlinear dynamics of each local map, and the diffusion due to the spatial coupling among local maps, a CML can exhibit spatiotemporal chaos. Further, by adopting various local maps and coupling methods [3], various CMLs can be constructed.

In the literature, the logistic map is often used in chaos-based PRBGs. However, since the invariable density function of the logistic map is equal to $\frac{1}{\pi\sqrt{x(1-x)}}$ [9], the logistic map may be improper to be used for designing good PRBGs. Whereas, the piecewise linear chaotic map $f : I \mapsto I$, $I = [a, b] \subset \Re$, has uniform invariable density function, namely, $\frac{1}{b-a}$ [2]. Therefore, the piecewise linear chaotic maps, such as the r-adic map [3], the skew tent map [5, 18] and the generalised four-segments tent map [29], may be good candidates to design chaos-based PRBGs, since they have simple structures and good statistical properties. A r-adic map is used here as local map. Additionally, to let the PRBG have low computational expense, the simplest coupling method, i.e., one-way coupling is employed.

A one-way CML based on r-adic maps is constructed as

$$x_{n+1}^{j} = (1-\epsilon)f(x_n^j) + \epsilon f(x_n^{j-1}), \tag{14.1}$$

$$f(x, r) = rx \mod 1, \tag{14.2}$$

where x_n^j represents the state variable of the jth site ($j = 1, 2, ..., L$, $L \in N$, L is the number of the sites in the CML) at time n ($n = 1, 2, ...$), $\epsilon \in (0, 1)$ is a coupling strength, $r > 1$ is the parameter of the r-adic map. The periodic boundary condition, $x_n^0 = x_n^L$, is used in the CML.

14.2.2 Lyapunov Exponents of Spatiotemporal Chaotic Systems

For design of a cipher, it is desirable that a CML exhibits spatiotemporal chaos. This is analysed here by the Lyapunov exponent spectrum.

The ith Lyapunov exponent (LE) is obtained as

$$\lambda_i = \lim_{n\to\infty} \frac{1}{n} \ln[\mathrm{eig}(\prod_{k=1}^{n} J_k)]_i, \tag{14.3}$$

For $f'(x_n^i) = r(i \in [1, L], n \in N)$ one has $J_n = rM$, with M being the constant matrix

$$M = \begin{bmatrix} 1-\epsilon & 0 & \cdots & 0 & \epsilon \\ \epsilon & 1-\epsilon & 0 & \cdots & 0 \\ 0 & \epsilon & 1-\epsilon & \cdots & 0 \\ \vdots & \ddots & \ddots & \ddots & \vdots \\ 0 & \cdots & \epsilon & 1-\epsilon & 0 \\ 0 & 0 & \cdots & \epsilon & 1-\epsilon \end{bmatrix}. \tag{14.4}$$

Since the matrix M is symmetric and circulant [17], its eigenvalues $\beta_j (j = 1, 2, ..., L)$ are given by

$$\beta_j = 1 - \epsilon + \epsilon e^{\frac{-2\pi j}{L} i}. \tag{14.5}$$

The LEs are then derived as

$$\lambda_i = \ln \mathrm{r} + \ln|\phi_\mathrm{i}(\epsilon, \mathrm{L})|, \tag{14.6}$$

$$\phi_i(\epsilon, L) = \sqrt{1 - 2\epsilon(1-\epsilon)(1 - \cos\frac{2\pi}{\mathrm{L}})}.$$

The Lyapunov exponent spectrum, i.e., the LEs in decreasing order, is given by

$$\lambda_i = \begin{cases} \ln \mathrm{r} + \ln|\phi_{\mathrm{i}/2}(\epsilon, \mathrm{L})| & i \;\text{ even,} \\ \ln \mathrm{r} + \ln|\phi_{(\mathrm{i}-1)/2}(\epsilon, \mathrm{L})| & i \;\text{ odd.} \end{cases} (i = 1, 2, ..., L) \tag{14.7}$$

That is, $\lambda_1 = \ln \mathrm{r} > \lambda_2 > ... > \lambda_\mathrm{L} = \ln \mathrm{r}|1 - 2\epsilon|$. To make all LEs positive, i.e., $\ln \mathrm{r}|1 - 2\epsilon|$ is positive, the following inequality should hold

$$r|1 - 2\epsilon| > 1. \tag{14.8}$$

14.2.3 Generation of Multiple PRBSs

When the CML (14.1) behaves chaotically, x_n^i can be regarded as a pseudo-random number, which means that $\{x_n^i\}$ $(n = 1, 2, ...)$ can be used as a pseudo-random number sequence (PRNS), denoted by $\mathrm{PRNS_i}$. Therefore, L PRNSs can be generated simultaneously from the CML with L sites. The CML can generate multiple PRBSs in the following two steps: (1) one CML with multiple sites generates multiple PRNSs, and (2) each PRNS generates multiple PRBSs. To illustrate the mechanism clearly, a diagram is shown in Fig. 14.1, where "LM" is the abbreviation for local map.

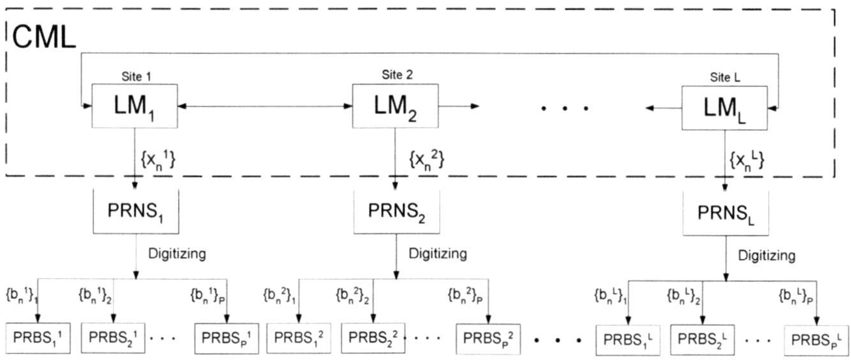

Fig. 14.1: Multiple PRBSs generated from the CML

The digitising process is described in detail as follows. x_n^i can be represented as a binary sequence [8, 21, 28]

$$x_n^i = 0.b_{n1}^i, b_{n2}^i, ..., b_{nP}^i, \tag{14.9}$$

where P stands for a certain precision. Here, P is assumed as 52, which is the maximum precision in standard computer realisation, in iterating the CML. Based on the binary representation, the digitisation method is shown in the Fig. 14.2.

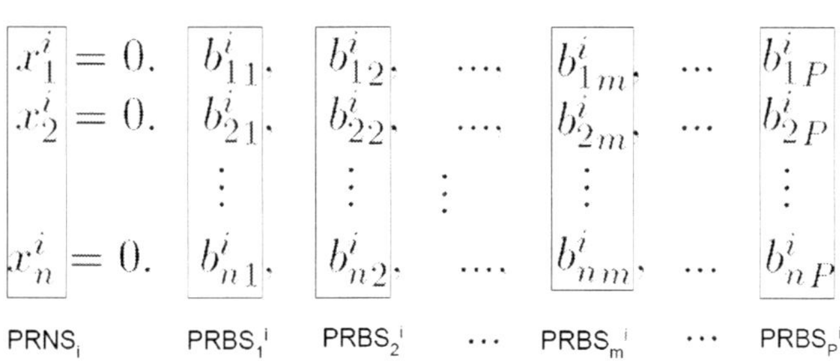

Fig. 14.2: Digitisation method

It is shown that the mth bits of the binary representations of $\mathrm{PRNS_i}$ comprise a PRBS, i.e., $\{b^i_{nm}\}$ $(m = 1, 2, ..., 52)$, denoted by $\mathrm{PRBS^i_m}$. Thus, 52 PRBSs are generated from one PRNS by using this digitisation method. In total, $52L$ PRBSs can be generated at one time from the CML.

14.3 Cryptographic Properties of a CML-based Multiple-Output PRBG

For a PRBG suitable for encryptions, the following cryptographic properties should be satisfied: (1) long period, (2) balance on $\{0, 1\}$, (3) δ-like autocorrelation, and (4) close-to-zero cross-correlation. In this section, these properties of the PRBG are investigated for various sets of parameters.

14.3.1 Long Period

Generally speaking, if chaotic maps are realised in a computer with a finite precision, there exist short periods of the chaotic orbits. This problem can, however, be avoided in spatiotemporal chaotic systems. The period of the CML with L sites is derived numerically to be around 10^{7L} [24]. Since the CML has a symmetric configuration, the periods of all PRNSs generated from the CML are the same as that of the CML. Therefore, 52 PRBSs generated from the PRNS have the period $O(10^{7L})$, which is long enough for cryptographic applications.

14.3.2 Balance

A PRBS, $\{b_n\}$ $(n = 0, ..., N-1)$, to be balanced on $\{0, 1\}$ means that it has a uniform probability distribution, i.e., $P\{b_n = 0\} = P\{b_n = 1\}$. In this section, the probability distributions of 52 PRBSs generated from one arbitrary PRNS of the PRBG with various r, ϵ and L are analysed by computing a scaled difference ΔP between $P\{(b_n) = 0\}$ and $P\{(b_n) = 1\}$ of each PRBS, i.e., $\Delta P = \frac{N_1 - N_0}{N/2}$ (N_1 and N_0, N are the numbers of "1" and "0", and the length of the PRBS, respectively). The lengths of the 52 PRBSs are assumed as 10^4, and the parameters are assumed here and throughout in the following way: first, increase r from 2 to 6 by 0.1 each time, while fixing ϵ as 0.95 and L as 8; then, increase ϵ from 0.01 to 0.99 by 0.02 each time, while fixing r as 4 and L as 8; finally, increase L from 8 to 64 by 1 each time with fixing r as 4 and ϵ as 0.95.

Fig. 3(a) shows ΔP of these 52 PRBS outputs from the PRBG for various r, where the x-axis is the index of 52 PRBSs, denoted by "i", the y-axis denotes the various r, and the z-axis stands for ΔP. Additionally, by setting a threshold for ΔP as 0.06, Figs. 3(d) and 3(f) are obtained in the following way. If ΔP of a PRBS is smaller than the threshold, the point corresponding

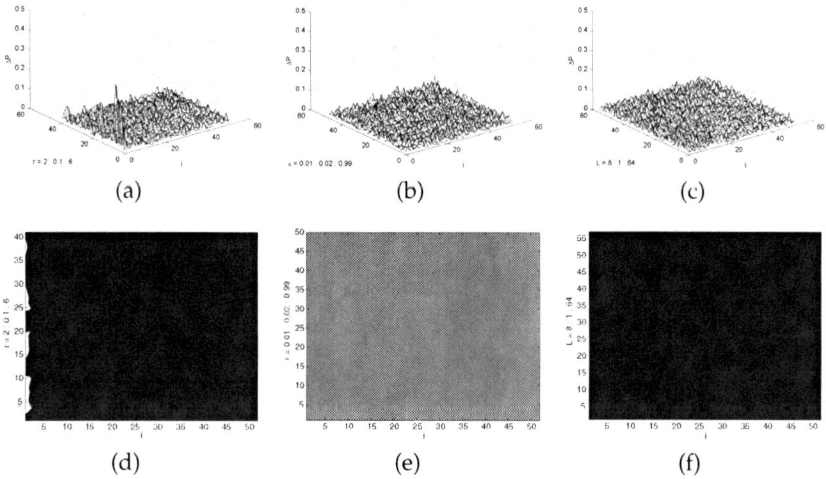

Fig. 14.3: ΔP of the PRBG for various parameters

to the index of the PRBS and r of the PRBG from which the PRBS outputs is drawn black, otherwise the point is drawn white. In the same way, ΔP of 52 PRBSs from the PRBG for various ϵ and various L are plotted in Figs. 3(b) and 3(e) as well as in Figs. 3(c) and 3(f), respectively. It is shown that only the first four PRBSs are not balanced, if r is equal to certain values, and that the rest PRBSs are balanced whatever r is. Moreover, all 52 PRBSs are balanced whatever ϵ and L are.

14.3.3 δ-like Auto-correlation

Auto-covariance, which equals mean-removed auto-correlation, is a special case of cross-covariance. The auto-covariance of a PRBS is described as

$$\begin{aligned}
&C_{ii}(\tau) = \hat{C}_{ii}(\tau)/\hat{C}_{ii}(0),\\
&\hat{C}_{ii}(\tau) = \tfrac{1}{N}\textstyle\sum_{n=1}^{N}(b_n - \bar{b}_n)(b_{n+|\tau|} - \tfrac{1}{N}\sum_{k=1}^{N} b_k),\\
&\bar{b}_n = \tfrac{1}{N}\textstyle\sum_{k=1}^{N} b_k,\\
&|\tau| = 0, 1, ..., N-1.
\end{aligned} \qquad (14.10)$$

δ-like auto-correlation means $C_{ii}(0) = 1$ and $\{C_{ii}(\tau)\}(|\tau| = 1, 2, ..., N-1)$ or the maximum of $\{C_{ii}(\tau)\}(|\tau| = 1, 2, ..., N-1)$ is close to zero. The maximum auto-covariances of 52 PRBSs generated from one arbitrary PRNS of the PRBG for various parameters are computed and shown in Fig. 14.4. It is indicated that the maximum auto-covariances of the first four PRBSs are quite far from zero and the rest except for the PRBSs from the PRBG with r close to 0.99 which are close to zero.

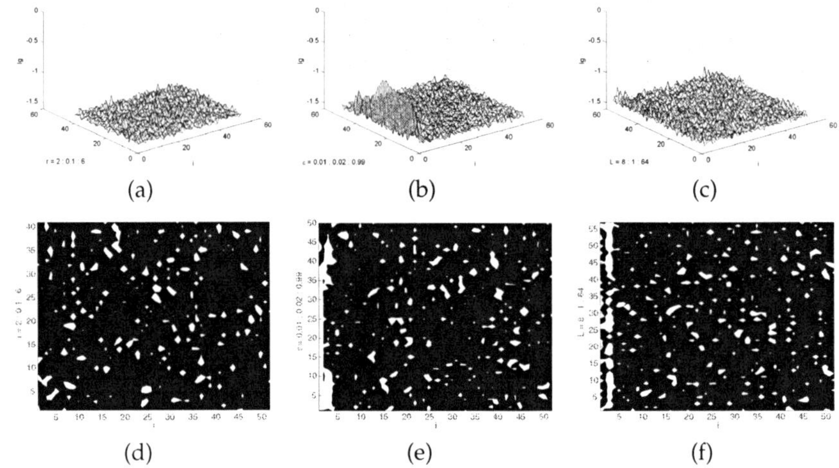

Fig. 14.4: Maximum auto-covariances of the PRBG for various parameters

Additionally, we set the threshold of the maxima as 0.038, and obtain Figs. 4(d)–4(f) in the same way as those of the previous section. It is shown that the maximum auto-covariances of some PRBSs output by the PRBG for certain parameter values are larger than 0.038. We set another threshold of the maximum auto-covariance as 0.055 and find that the maximum auto-covariances of all the 5th–52th PRBSs output by the PRBG for all parameters values in the Figures are smaller than 0.055. Therefore, the auto-correlation of the PRBG discarding the first four PRBSs satisfies the requirements of cryptography.

14.3.4 Close-to-zero Cross-correlation

In order that the 5th–52th PRBSs output by the PRBG can be applied in parallel, the cross-covariance between arbitrary two of them should be close to zero. The cross-covariance, i.e., the mean-removed cross-correlation, between $\mathrm{PRBS_i}$ and $\mathrm{PRBS_j}$ is described as

$$\begin{array}{l} C_{ij}(\tau) = \hat{C}_{ij}(\tau)/\sqrt{\hat{C}_{ii}(0)\hat{C}_{jj}(0)}, \\ \hat{C}_{ii}(\tau) = \frac{1}{N}\sum_{n=1}^{N}(b_n^i - \bar{b}_n^i)(b_{n+|\tau|}^i - \bar{b}_n^i), \\ \hat{C}_{ij}(\tau) = \frac{1}{N}\sum_{n=1}^{N}(b_n^i - \bar{b}_n^i)(x_{n+|\tau|}^j - \bar{b}_n^j), \\ \bar{b}_n^i = \frac{1}{N}\sum_{k=1}^{N} b_k^i, \\ |\tau| = 0, 1, ..., N-1. \end{array} \tag{14.11}$$

A maximum cross-covariance, denoted by $C_{i_1 i_2}$, between $\mathrm{PRBS^i_{m_1}}$ and $\mathrm{PRBS^i_{m_2}}$ $(i, m_1, m_2 \in N,\ i \in [1, L],\ m_1, m_2 \in [5, 52],\ m_1 \neq m_2)$ output by the PRBG for various parameters are computed, which is shown in Fig. 14.5. Additionally, another maximum cross-covariance, denoted by $C_{i_1 j_2}$, between $\mathrm{PRBS}^{\mathrm{i}}_{\mathrm{m}_1}$ and $\mathrm{PRBS}^{\mathrm{j}}_{\mathrm{m}_2}$ $(i, j, m_1, m_2 \in N,\ i, j \in [1, L],\ i \neq j,\ m_1, m_2 \in [5, 52])$ of the PRBG for various parameters is shown in Fig. 14.6.

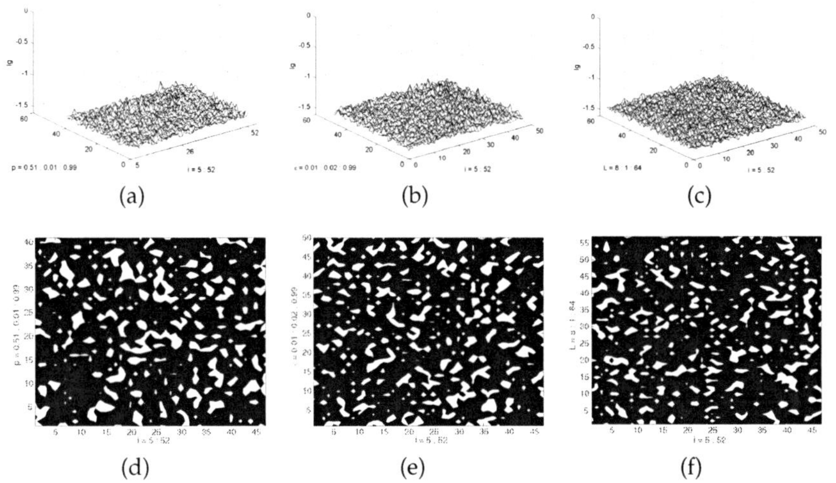

Fig. 14.5: Maximum $C_{i_1 i_2}$ of the PRBG for various parameters

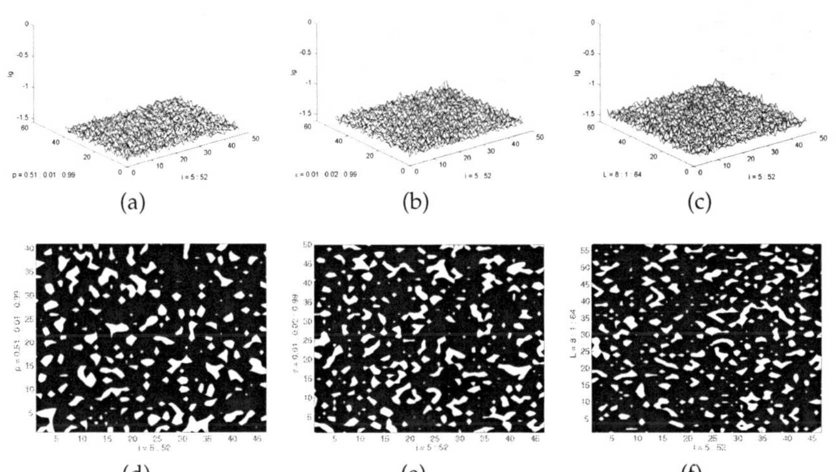

Fig. 14.6: Maximum $C_{i_1 j_2}$ of the PRBG for various parameters

We set the threshold for the maximum cross-covariance as 0.038, and obtain Figs. 5(d)–5(f) and 6(d)–6(f). It is indicated that the maximum cross-covariance between some pairs of PRBSs output by the PRBG for certain parameter values is larger than 0.038. We set another threshold of the maximum cross-covariance as 0.055 and obtain the result that the maximum cross-covariances between all pairs of PRBSs output by the PRBG for all parameters values are smaller than 0.055. Therefore, the cross-correlation of the PRBG discarding the first four PRBSs is acceptable from the cryptographic point of view.

14.3.5 Statistical Tests

In practice, statistical tests are employed to investigate the randomness of PRBGs and, thus, provide additional confidence that PRBGs are acceptable from the statistical point of view. In this chapter, the FIPS 140-2 tests [19] are used to evaluate the randomness of the proposed PRBG. FIPS 140-2 specifies four statistical tests, i.e., monobit test, poker test, run test, and long run test, all of which should be passed for a PRBS to pass FIPS 140-2.

The 52 PRBSs generated from one arbitrary PRNS of the PRBG for various parameters are subjected to FIPS 140-2. The results are shown in Fig. 14.7–14.9, where a black point corresponds to the index of a PRBS which

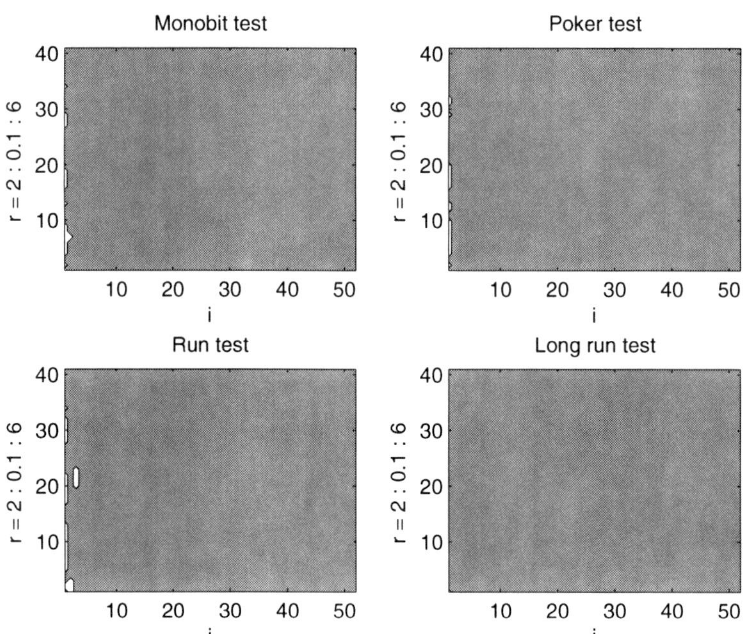

Fig. 14.7: FIPS 140-2 test of the PRBG for various r

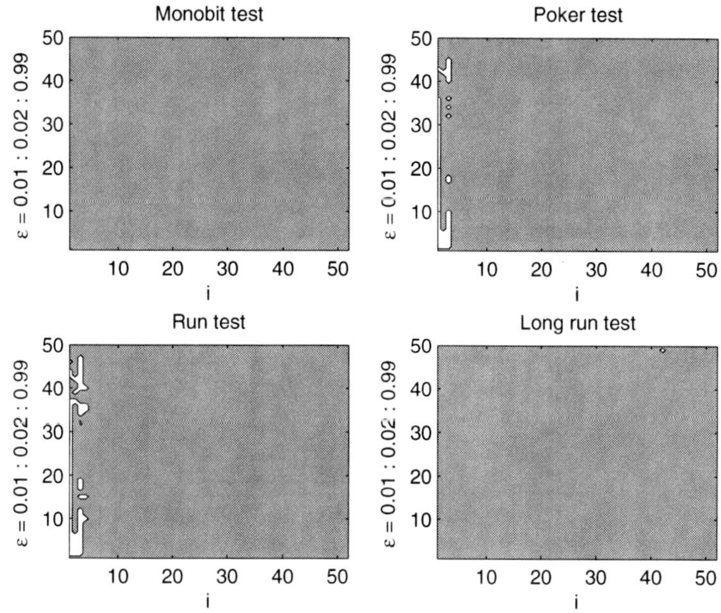

Fig. 14.8: FIPS 140-2 test of the PRBG for various ϵ

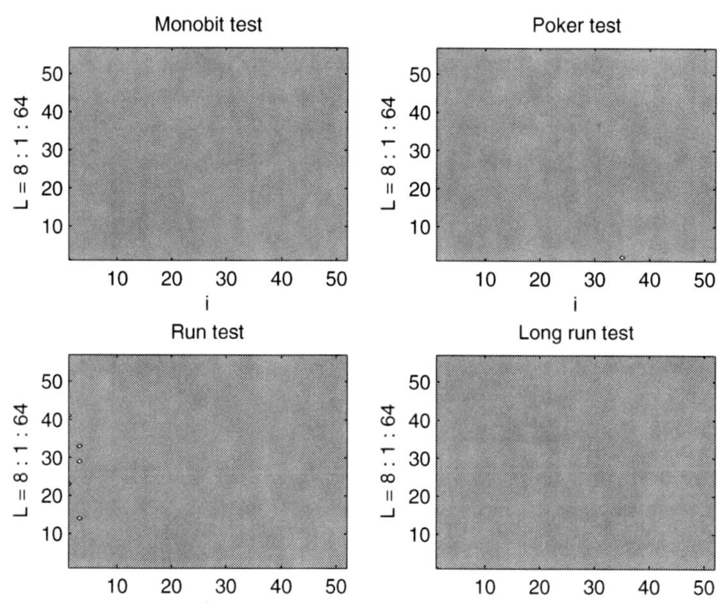

Fig. 14.9: FIPS 140-2 test of the PRBG for various L

passes the test and the parameter of the PRBG from which the PRBS outputs. It is shown that the 5th–52th PRBSs have good randomness, while the first four PRBSs have not, which should therefore be discarded from a PRBG.

14.4 A Stream Cipher based on a PRBG

In this section, a stream cipher based on the proposed PRBG is designed. Since the first 4 PRBSs from the PRBG should be discarded, only the remaining multiple PRBSs are used in the cipher. The encryption is described by

$$\begin{aligned} x_{n+1}^j &= (1-\epsilon) f(x_n^j, a_j) + \epsilon f(x_n^{j-1}, a_{j-1}), \\ f(x_n^j, a_j) &= (4 + 0.01 a_j) x_n^j \mod 1, \\ K_n^j &= \text{int}(x_n^j \times 2^u) \mod 2^v, \\ C_n^j &= P_n^j \oplus K_n^j, \end{aligned} \tag{14.12}$$

where $u, v \in N$, K_n^j, P_n^j and C_n^j are keystream, plaintext and ciphertext, respectively, and $\oplus$ means bitwise exclusive or. Actually, the CML is used as a pseudo-random number generator to produce L keystreams by imposing the algebraic *int* and *mod* operations on the outputs of the CML. The plaintext is subjected bit by bit to the antivalence operation with the keystream to produce the ciphertext. Encryption keys are assumed as $a_j \in [0, 1)$, denoted by a $= \{a_1, a_2, ..., a_L\}$.

The decryption, whose configuration and parameters are the same as those of the encryption, is described as

$$\begin{aligned} y_{n+1}^j &= (1-\epsilon) f(y_n^j, a'_j) + \epsilon f(y_n^{j-1}, a'_{j-1}), \\ f(y_n^j, a'_j) &= (4 + 0.01 a'_j) y_n^j \mod 1, \\ {K'}_n^j &= \text{int}(y_n^j \times 2^u) \mod 2^v, \\ {P'}_n^j &= C_n^j \oplus {K'}_n^j, \end{aligned} \tag{14.13}$$

where $a'_j \in [0, 1)$, denoted by a$' = \{a'_1, a'_2, ..., a'_L\}$, are decryption keys. When a$'$ = a and $y_0^j = x_0^j$, these two CMLs are synchronised, i.e., $y_n^j = x_n^j$, thus produce identical keystreams, i.e., ${K'}_n^j = K_n^j$. As a result, the plaintext is decrypted, i.e., ${P'}_n^j = P_n^j$.

Remarks:

1. r, e are assumed to be in the range $[4, 4.01)$ and as 0.95, respectively, to obtain the keystream with good statistical properties in terms of the inequality. (14.8) and the results numerically analysing the statistical properties of the PRBG were given in Section 14.3.
2. u is fixed as 52, because the number of significant bits of the standard binary representation of a double-precision floating point number in a computer is 52. v is assumed as 32, i.e., only the lowest 32 significant bits are extracted from the 52 bits due to the following reasons. First, our

numerical investigations revealed that only the lowest 48 significant bits can be used to generate keystreams with satisfactory statistical properties. Secondly, the smaller v is, the harder it is to break the cipher with the known-plaintext attack, as will be indicated in Section 14.5.2. However, from the implementation point of view, the larger v is, the more efficient the cipher will be. Therefore, by considering these factors together, v is fixed as 32.

3. The determination of L depends on the following considerations. Numerical investigations show that the period of the keystreams generated from the CML with $L > 3$ is larger than 2^{100}, which is acceptable for cryptography, and that L has no evident effects on other statistical properties of the keystreams. Meanwhile, it does not influence the encryption speed, too, which will be indicated in Section 14.5.3. In additiony, the cost of breaking the cipher is about 2^{40L}, which will be analysed in detail in Section 14.5.2. Therefore, in investigating a concrete cipher thereafter, L is assumed as 4 to meet the two requirements.

A keyspace is defined as a set of all possible keys, which should be studied in depth in designing a cipher. The error function [16] is used here to determine the keyspace of the cipher. When $\mathbf{a}' \neq \mathbf{a}$, the decrypted plaintext, p'^j_n, can have deviated from the original one, p^j_n. The error function is defined as

$$e(j, \Delta a_i) = \frac{1}{T} \sum_{n=1}^{T} |p'^j_n - p^j_n|, \; j = 1, 2, ...L, \tag{14.14}$$

where $\Delta a_i = a'_i - a_i \; (i = 1, 2, ..., L), p'^j_n = \frac{P'^j_n}{2^{32}}, p^j_n = \frac{P^j_n}{2^{32}}$, T is the encryption time. The error function is plotted vs. Δa_1 with $T = 10^5$ in Fig. 14.10. It is shown that the error function does not equal zero but 0.25 even if Δa_1 is an extremely small value 2^{-50}. In other words, the key a'_1 is sensitive to any differences equal to or larger than 2^{-50}. In a similar way, it is shown that the keys $a'_i \; (i = 2, 3, ..., L)$ are also sensitive to any differences equal to or larger than 2^{-50}. Therefore, the keyspace is 2^{50L}.

14.5 Properties of the Cipher

To evaluate the performance of the cipher, its statistical properties, its security, and its efficiency are analysed in this section.

14.5.1 Statistical Properties of Keystreams

Since ciphertext is produced by using directly bitwise XOR between plaintext and keystream in the cipher, the statistical properties of the keystream have significant effects on the cipher's security. Some statistical properties, such

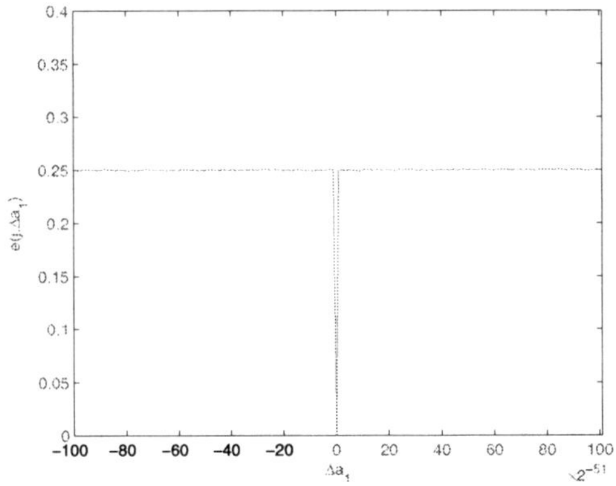

Fig. 14.10: Error function

as probability distribution, auto-correlation and run, of the keystream are investigated numerically.

The order-1 and order-2 probability distributions [16] of an arbitrary keystream in the cipher with arbitrary initial conditions and $\mathbf{a} = 0.5\mathbf{I}$ ($\mathbf{I} = [1, ..., 1]^T \in \Re$) are investigated.

The order-1 probability distribution of the keystream, $\rho(ks_n^j)(= \frac{\rho(K_n^j)}{2^{32}})$, is plotted in Fig. 14.11(a). Its order-2 probability distribution, $\rho(ks_n^j, ks_{n-1}^j)$ $(= \frac{\rho(K_n^j, K_{n-1}^j)}{2^{32}})$, is plotted in Fig. 14.11(b). The length of the keystream is 10^6. It is shown that the probability distributions are uniform.

A run of a binary sequence s is another postulate of randomness, and defined as a subsequence of s consisting of consecutive 0's or consecutive 1's which is neither preceded nor succeeded by the same symbol. The probabilities of 0/1 runs of length n ($n = 1, 2, ..., N$) in the keystream, denoted by $p_{0/1}(n)$ ($p_{0/1}(n) = \frac{R_{0/1}(n)}{R_{0/1}}$, $R_{0/1} = \sum_{n=1}^{N} R_{0/1}(n)$, $R_{0/1}(n)$ the number of 0/1 runs of length n), are investigated. $p_{0/1}(n)$ vs. n is plotted in Fig. 14.12. It is shown that $p_{0/1}(n)$ is in direct proportion to n, which is a characteristic of a truly random binary sequence of infinite length [16].

δ-like auto-correlation is also an cryptographic requirement for the keystream in a cipher. Auto-covariance, i.e., mean-removed auto-correlation, of the keystream is described as

$$\begin{aligned} &C_{ii}(\tau) = \hat{C}_{ii}(\tau)/\hat{C}_{ii}(0), \\ &\hat{C}_{ii}(\tau) = \frac{1}{T}\sum_{n=1}^{T}(K_n^i - \overline{K}_n^i)(K_{n+|\tau|}^i - \overline{K}_n^i), \\ &|\tau| = 0, 1, ..., T-1, \end{aligned} \tag{14.15}$$

where $\overline{K}_n^i = \frac{1}{T}\sum_{n=1}^{T} K_n^i$. Auto-covariance of the keystream with length $T = 10^6$ is plotted in Fig. 14.13. It is shown that the keystream has δ-like auto-correlation.

According to the analysis above, L keystreams have satisfactory random-like statistical properties.

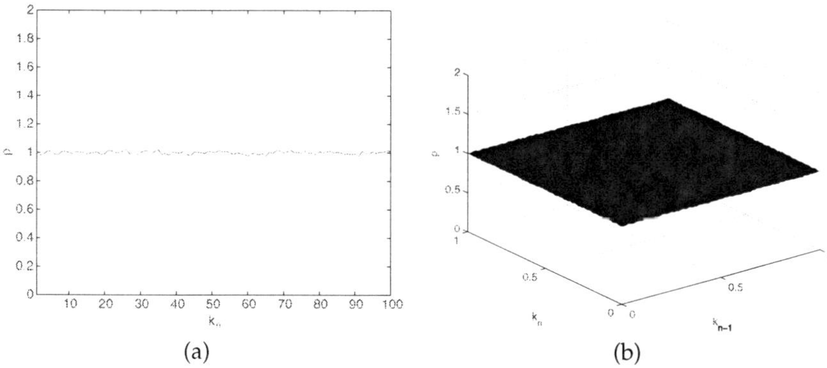

Fig. 14.11: Probability distribution of the keystream

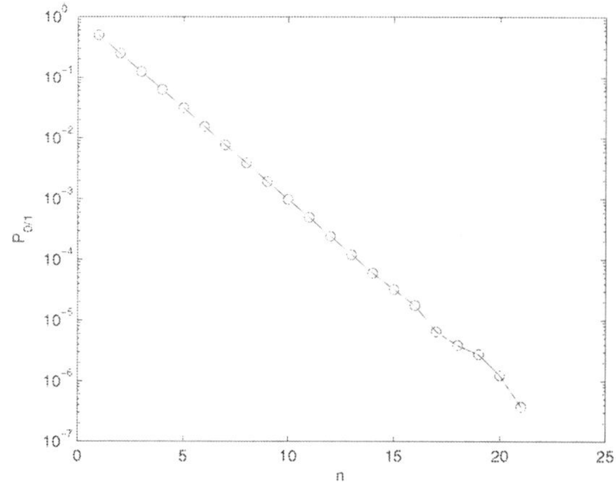

Fig. 14.12: Probability of the keystream's run

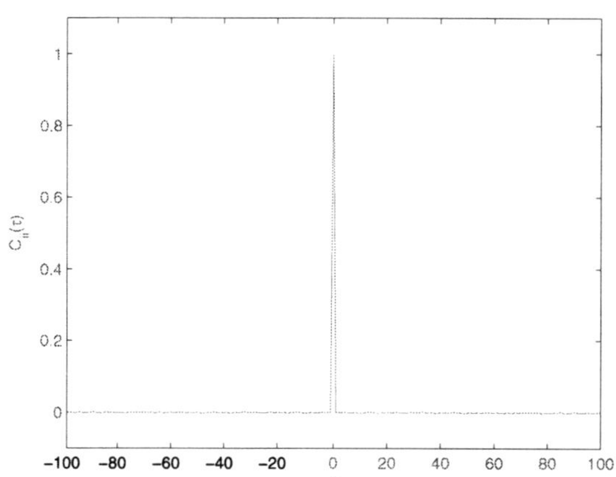

Fig. 14.13: Auto-correlation of the keystream

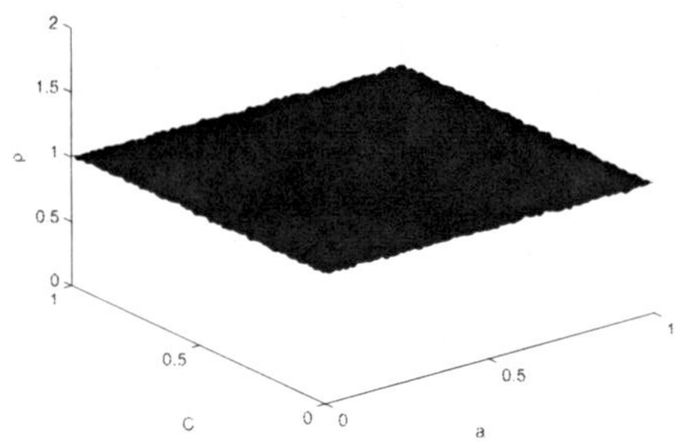

Fig. 14.14: Conditional probability distribution $\rho(c|a)$

14.5.2 Security Analysis

In this section, the security of the cipher is evaluated by investigating its confusion and diffusion properties and using various typical attacks, such as differential attacks, known-plaintext attack, brute-force attack, and chosen-plaintext/ciphertext attacks.

To resist common attacks, a cipher should have the following two basic cryptographic properties: confusion and diffusion. Confusion reflects the uniformity of all keys. Confusion of a cipher is guaranteed by the uniformity of the conditional probability distribution of the ciphertext $\rho(c|a)$ ($c = \frac{C}{2^{32}}$, $C = C_n(j; a)$, $\mathbf{a} = a\mathbf{I}$) for different keys, which is shown in Fig. 14.14.

Diffusion reflects strong sensitivity of a key to tiny changes. The sensitivity of the key to an extremely small change 2^{-50}, which is described in Section 14.4, approves the diffusion of the cipher.

Weaknesses in the differential relations between ciphertext and plaintext can be used to break a cipher by a differential attack. The conditional probability of the ciphertext $\rho(\Delta c|\Delta p)$ ($\Delta c = \frac{\Delta C}{2^{32}}$, $\Delta C = C_n(j; \hat{P}_n^j) - C_n(j; P_n^j)$) under the condition $\Delta p = \frac{\Delta P_n^j}{2^{32}}$ ($\Delta P_n^j = \hat{P}_n^j - P_n^j$), which is shown in Fig. 14.15, is uniform. Therefore, the cipher is immune to the differential attack.

With Kerckhoffs' assumption, one known-plaintext attack to the proposed cipher is via inverse analytical computation with known plaintext and accessible ciphertext, i.e., the keystream K_n^j. The cost of the known-plaintext attack to the cipher can be estimated as follows. To simplify the reasoning, $x_{n+1}^j = (1-\epsilon)f(x_n^j, a_n^j) + \epsilon f(x_n^{j-1}, a_n^{j-1})$ and $f(x_n^j, a_j) = (4+0.01a_j)x_n^j(1-x_n^j)$ in Eq. (14.12) are denoted by $x_{n+1}^j = G(x_n^j, x_n^{j-1}, a_j, a_{j-1})$. To obtain keys **a**, the following L equations are listed

$$\begin{aligned} x_{n+1}^1 &= G(x_n^1, x_n^L, a_1, a_L), \\ x_{n+1}^2 &= G(x_n^2, x_n^1, a_2, a_1), \\ &\cdots \\ x_{n+1}^L &= G(x_n^L, x_n^{L-1}, a_L, a_{L-1}), \end{aligned} \qquad (14.16)$$

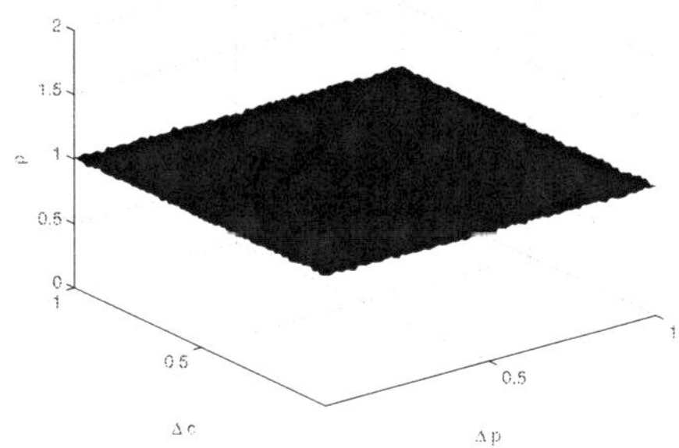

Fig. 14.15: Conditional differential probability $\rho(\Delta c|\Delta p)$

where $2L$ variables, x^j_{n+1}, x^j_n $(j = 1, 2, ..., L)$, should be known to solve $\mathbf{a}$. In addition, one x^j_n can be obtained from one K^j_n, however, one K^j_n corresponds to 2^{52-32} possible x^j_n. Consequently, the cost of the known-plaintext attack is not less than 2^{40L}.

A typical known-plaintext attack is a brute-force attack, where a cipher is attacked by trying every possible key one by one to decrypt plaintext with public ciphertext and checking whether the resulting plaintext is the original one. Since the keyspace is deduced as 2^{50L} above, the cost of the brute-force attack of the cipher equals 2^{50L}.

In applying chosen-plaintext and chosen-ciphertext attacks to a cipher, an attacker chooses some special plaintext and ciphertext to achieve certain keystreams if there exist some characteristic relations between keystream and key. The relation between keystream $k(= \frac{K^j_n}{2^{32}})$ and key $a(\mathbf{a} = a\mathbf{I})$ is investigated via the conditional probability distribution of k under the condition a, denoted as $\rho(k|a)$, which is plotted in Fig. 14.16. It is indicated that no characteristics of the key can be extracted from the keystream, consequently, no special plaintext or ciphertext can be chosen to break the key. In other words, the chosen-plaintext/ciphertext attack has the same efficiency as the known-plaintext attack to this cipher.

In summary, the known-plaintext attack is the most effective attack to the proposed cipher, whose cost is 2^{40L}. Moreover, the security of the cipher can be increased conveniently by adding one lattice to the CML in the cipher requiring only little more computation, which results in the cost for breaking the cipher to rise 2^{40} times.

14.5.3 Efficiency

Besides high security, the cipher is quite efficient. All the coupled maps in the cipher are used to encrypt plaintext simultaneously. A close-to-zero cross-correlation among L keystreams guarantees the efficiency of the parallel L encryptions/decryptions in the cipher. The cross-covariance, i.e., mean-

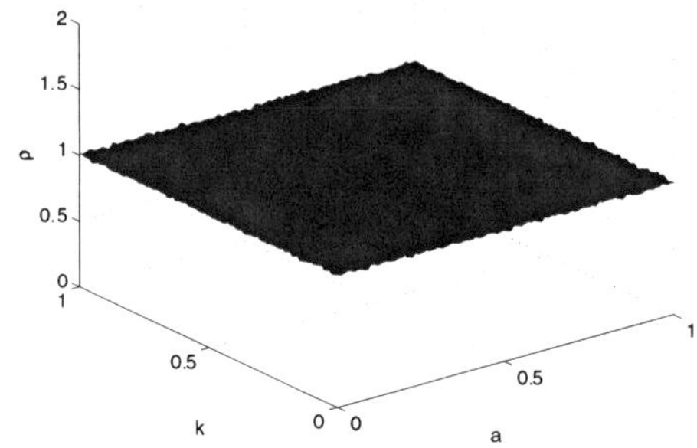

Fig. 14.16: Conditional probability distribution $\rho(k|a)$

moved cross-correlation, of two arbitrary keystreams with length 10^6, K_n^i and K_n^j, is described as

$$\begin{array}{l} C_{ij}(\tau) = \hat{C}_{ij}(\tau)/\sqrt{\hat{C}_{ii}(0)\hat{C}_{jj}(0)}, \\ \hat{C}_{ij}(\tau) = \frac{1}{T}\sum_{n=1}^{T}(K_n^i - \overline{K}_n^i)(K_{n+|\tau|}^j - \overline{K}_n^j), \\ \hat{C}_{ii}(\tau) = \frac{1}{T}\sum_{n=1}^{T}(K_n^i - \overline{K}_n^i)(K_{n+|\tau|}^i - \overline{K}_n^i), \\ |\tau| = 0, 1, ..., T-1 \end{array} \tag{14.17}$$

where $\overline{K}_n^i = \frac{1}{T}\sum_{n=1}^{T} K_n^i$. The results of computing the cross-covariance are plotted in Fig 14.17. It is shown that all keystreams are independent. Therefore, the parallel L keystreams can be used to encrypt plaintext at one time effectively.

Due to the parallel operation, around 350 Mbits plaintext can be encrypted per second in our computer with a 1.8 GHz CPU and 1.5 GB RAM. In addition, the encryption speeds of ciphers based on CMLs with various sizes are similar, which is indicated in the relation between the speeds and L as shown in Fig. 14.18. For comparison, the encryption speed of the cipher

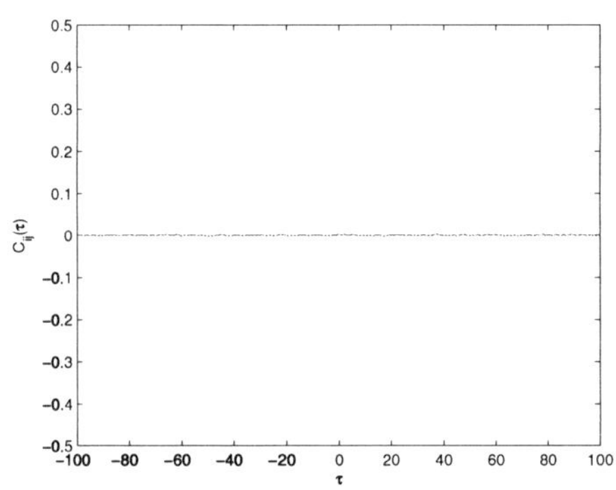

Fig. 14.17: Cross-correlation of any two keystreams

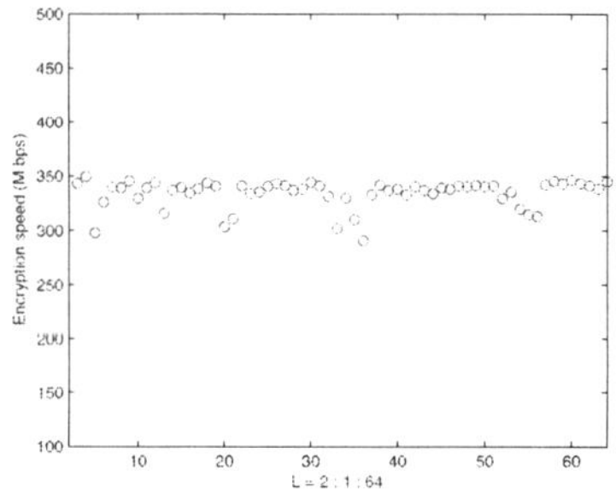

Fig. 14.18: Relation between the encryption speed and L

(a) Plain-image

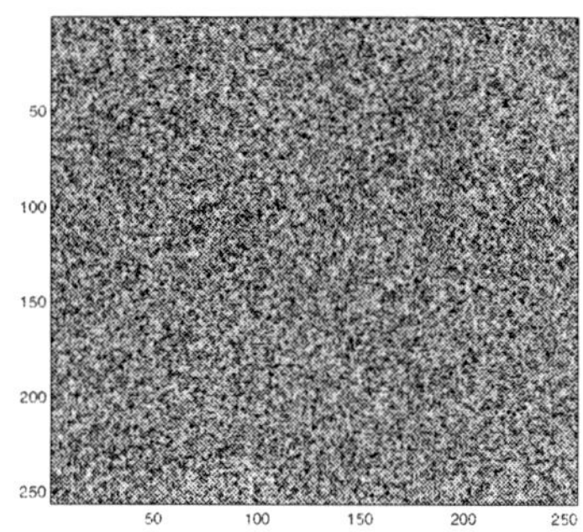

(b) Cipher-image

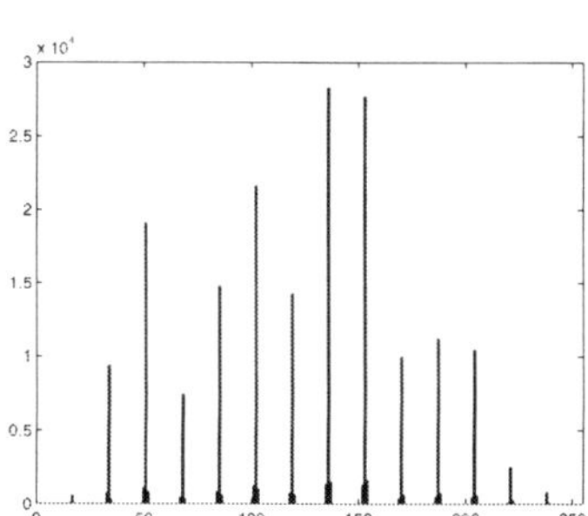

(c) Histogram of plain-image

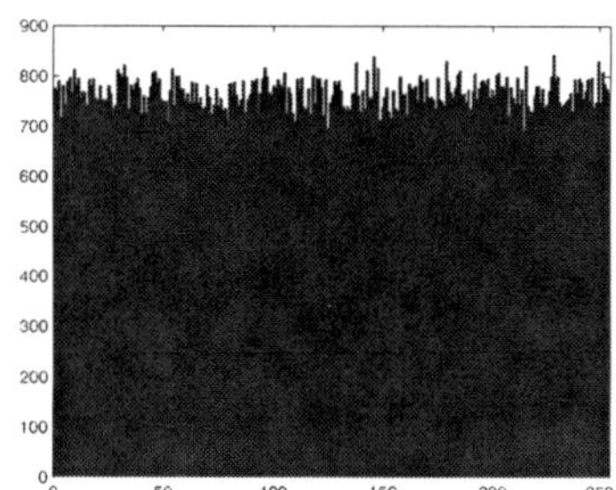

(d) Histogram of cipher-image

Fig. 14.19: An image encrypted with the cipher

[16] is computed as about 280 Mbits per second by our computer. However, its configuration is more complicated than that of the proposed cipher.

14.5.4 An Example: Image Encryption

A digital image is encrypted here to test the practical performance of the cipher. The plain-image, i.e., the well-known "Lena", and the cipher-image are shown in Figs. 14.19(a) and 14.19(b). Figs. 14.19(c) and 14.19(d) are the respective histograms. It is indicated that the plain-image is encrypted to the cipher-image with uniform histogram, which implies the perfect cryptographic properties of the cipher.

14.6 Conclusion

A multiple-output PRBG based on a spatiotemporal chaotic system was presented. A simple one-way CML based on r-adic maps was exploited as spatiotemporal chaotic system to yield a PRBG requiring low computational expense. The chaotic and cryptographic properties of the PRBG were investigated analytically and numerically, respectively, to determine the parameter ranges within which the PRBG is suitable for application in cryptography with satisfactory statistical properties. In addition, a stream cipher based on the PRBG was designed. Its cryptographic properties, security and efficiency were investigated. As an example, an image was effectively encrypted with the cipher. The cipher was approved to have high security and efficiency. Therefore, a spatiotemporal chaotic system is indeed a good choice for use in cryptography.

References

1. Baptista MS (1999) Cryptography with Chaos. Phys Lett A 240: 50–54
2. Baranovsky A, Daems D (1995) Design of One-dimensional Chaotic Maps with Prescribed Statistical Properties. Int J Bifurcation and Chaos 5 (6): 1585–1598
3. Batista AM, de S. Pinto SE, Viana RL, Lopes SR (2002) Lyapunov Spectrum and Synchronization of Piecewise Linear Map Lattices with Power-law Coupling. Physical Review E, 65:056209(9)
4. Garcia P, Parravano A, Cosenza MG, Jimenez J, Marcano A (2002) Coupled Map Networks as Communication Schemes. Physical Review E, 65:045201
5. Habutsu T, Nishio Y, Sasase I, Mori S (1991) A Secret Key Cryptosystem by Iterating a Chaotic Map. LNCS 547: 127–140
6. Kaneko K (ed) (2003) Theory and Application of Coupled Map Lattices, Ch. 1. John Wiley and Sons
7. Kocarev L (2001) Chaos-based Cryptography: A Brief Overview. IEEE Circuits and Systems Magazine 1 (3): 6–21
8. Kohda T, Tsuneda A (1993) Pseudonoise Sequence by Chaotic Nonlinear Maps and Their Correlation Properties. IEICE Trans Commun E76-B: 855–862
9. Lasota A, Mackey MC (1997) Chaos, Fractals and Noise: Stochastic Aspects of Dynamics. Springer, New York
10. Li P, Li Z, Halang WA, Chen GR (2006) A Multiple Pseudo-random-bit Generator based on a Spatiotemporal Chaotic Map. Phys Lett A 349: 467–473
11. Li P, Li Z, Halang WA, Chen GR (2006) Analysis of a Multiple Output Pseudo-random-bit Generator based on a Spatiotemporal Chaotic System. Int J Bifurcation and Chaos, 16 (10)
12. Li P, Li Z, Halang WA, Chen GR (2006) A Stream Cipher based on a Spatiotemporal Chaotic System. Chaos, Solitons and Fractals. To appear
13. Li SJ (2003) Analysis and New Designs of Digital Chaotic Ciphers. PhD thesis, School of Electronics and Information Engineering, Xi'an Jiaotong University. `http://www.hooklee.com/pub.html`
14. Li SJ, Mou XQ, Cai YL (2001) Pseudo-random Bit Generator based on Coupled Chaotic Systems and its Application in Stream-ciphers Cryptography. LNCS 2247: 316–329
15. Li SJ, Mou XQ, Cai YL, Ji Z, Zhang JH (2003) On the Security of a Chaotic Encryption Scheme: Problems with Computerized Chaos in Finite Computing Precision. Computer Physics Communications 153 (1): 52–58
16. Lu HP, Wang SH, Li XW, Tang GN, Kuang JY, Ye WP, Hu G (2004) A New Spatiotemporally Chaotic Cryptosystem and its Security and Performance Analyses. Chaos 14 (3): 617–629
17. Luekepohl H (1998) Handbook of Matrices. John Wiley and Sons
18. Masuda N, Aihara K (2002) Cryptosystems with Discretized Chaotic Maps. IEEE Trans Circuits and Systems-I 49 (1): 28–40
19. National Institute of Standards and Technology (2001) Security Requirements for Cryptographic Modules. FIPS PUB 140-2. `http://csrc.nist.gov/publications/fips/fips140-2/fips1402.pdf`
20. Protopopescu VA, Santoro RT, JS Tollover (1995) Fast and Secure Encryption/Decryption Method based on Chaotic Dynamics. US Patent No 5479513
21. Sang T, Wang R, Yan Y (1998) Perturbance-based Algorithm to Expand Cycle Length of Chaotic Key Stream. Electronics Letters 34: 873–874

22. Short KM (1997) Signal Extraction from Chaotic Communication. Int J Bifurcation and Chaos 7 (7): 1579–1597
23. Tang G, Wang S, Lü H, Hu G (2003) Chaos-based Cryptograph Incorporated with S-box Algebraic Operation. Phys Lett A 318: 388–398
24. Wang SH, Liu WR, Lu HP, Kuang JY, Hu G (2004) Periodicity of Chaotic Trajectories in Realizations of Finite Computer Precisions and its Implication in Chaos Communications. Int J of Modern Physics B 18 (17–19): 2617–2622
25. Wheeler DD (1989) Problems with Chaotic Cryptosystems. Cryptologia XIII (3): 243–250
26. Yang T (2004) A Survey of Chaotic Secure Communication Systems. Int J of Computational Cognition 2 (2): 81–130
27. Yang T, Yang LB, Yang CM (1998) Cryptanalyzing Chaotic Secure Communications Using Return Maps. Phys Lett A 245: pp 495–510
28. Zhang H, Wang H, Chen W (2002) Oversampled Chaotic Binary Sequences with Good Security. J Circuits, Systems and Computers 11: 173–185
29. Zhou H, Ling XT (1997) Problems with the Chaotic Inverse System Encryption Approach. IEEE Trans Circuits and Systems-I 44 (3): 268–271.

15

A Chip Performing Chaotic Stream Encryption

Yaobin Mao[1], Wenbo Liu[2], Zhong Li[3], Ping Li[3], Wolfgang A. Halang[3]

[1] Nanjing University of Science and Technology, 210094 Nanjing, China, maoyaobin@163.com
[2] Nanjing University of Aviation and Astronautics, 210016 Nanjing, China, wenboliu@nuaa.edu.cn
[3] Faculty of Electrical and Computer Engineering, FernUniversität in Hagen, 58084 Hagen, Germany {zhong.li|ping.li|wolfgang.halang}@fernuni-hagen.de

Summary. The design of chaotic pseudo-random number generators (PRNG) by use of non-linear feedback shift registers is shown. A secure stream encryption scheme, which uses M-sequence-like pseudo-random bit sequences generated by such a chaotic PRNG to mask plain-bits, is accomodated in a Field Programmable Gate Array (FPGA). It is shown that the proposed algorithm can easily be implemented in embedded devices such as IC cards or re-configurable chips. Finally, some experimental results are presented, and a cryptanalysis is reported as well.

Key words: Pseudo-Random Number Generator (PRNG), chaos, stream encryption, cryptanalysis, Field Programmable Gate Array (FPGA).

15.1 Introduction

The wide deployment of networks and the availability of digital consumer electronic devices for recording, storage, transmission, and playback have increased the demand for security of digital media. To guarantee confidentiality, integrity, and authenticity, which are crucial issues in many areas such as wireless communications, e-commerce, bank transactions, real-time video streaming, and remote medical applications, various new technologies like biological feature identification, digital watermarking, or chaos-based encryption have been developed, which are expected to act more efficiently than the traditional cryptographical algorithms.

Owing to the exceptional properties of chaotic non-linear maps such as sensitivity to initial conditions and parameters, which is also called butterfly-effect, mixing, pseudo-randomness, and broad spectrum [7, 4], chaos based encryption methods are very advantageous for cryptographical applications with respect to the aspects encryption speed and security [13]. Thus, since the idea of incorporating chaotic maps into cryptology was explicitly brought

Y. Mao et al.: *A Chip Performing Chaotic Stream Encryption*, Studies in Computational Intelligence (SCI) **42**, 307–332 (2007)
www.springerlink.com © Springer-Verlag Berlin Heidelberg 2007

about in 1989 [15], chaos-based encryption has attracted broad attention for more than a decade [13, 12, 8, 14, 17].

Just like traditional cryptology grounded on number theory, chaos-based cryptology can also be subdivided into block and stream ciphers. Unlike block ciphers, which operate on large blocks of data, stream ciphers typically operate on smaller units of text, usually short bit strings. Therefore, stream ciphers can be designed to be exceptionally fast, much faster than a typical block cipher. In his seminal paper [20], Shannon showed the latter to be unbreakable if plain-text is encrypted by bitwise antivalence with a key sequence of random bits only used once. Since securely distributing a key sequence (which should, theoretically, be longer than the plain-text) is difficult, the idea of this "one-time padding" has not been accepted for use in practice. There, from short keys long random numbers are algorithmically generated to be used as encryption pads. However, as stated by von Neumann, "anyone who considers arithmetical methods of producing random digits is, of course, in a state of sin", the algorithmic generation of random key-streams with good properties is a hard problem, to which a number of techniques has been proposed over the years, among which chaos-based methods show incomparable preponderance.

Theoretically, chaos-based Pseudo-Random Number Generators (PRNG) can be proven to exhibit good randomness and infinite periods, while their non-linear character significantly enhances the complexity of their structures. Furthermore, the wide variety of chaotic functions available provides for countless options of methods to generate pseudo-random numbers. Thus, many Chaotic Pseudo-Random Number Generators (CPRNG) have been proposed in the literature [1, 21, 22], but, to the best of our knowledge, hardware implementation, especially on a chip, was not considered, yet, although it increases the throughput significantly and enhances the security further. Moreover, a chip, as the one designed and presented in this chapter, can be embedded into devices such as mobile telephones, handheld computers, or Personal Data Assistants to en- and decrypt data sent over communication links of any kind and, thus, accomplish security of critical mobile applications such as bank transactions. However, due to limitations of power consumption, volume, and memory capacity, it is still a great challenge to tailor for chip realisation a CPRNG already well-proven on personal computers.

An investigation on generation methods for chaotic pseudo-random numbers by use of non-linear feedback shift registers, and on composition of a secure stream encryption scheme that can be accommodated in a Field Programmable Gate Array (FPGA) is described in this chapter. The fast chaotic stream cipher scheme uses M-sequence-like pseudo-random bit sequences that are generated by a chaotic PRNG to mask plain-bits. The structure of the generator is similar to that of one based on the Linear Feedback Shift Register (LFSR) method that is widely used in traditional cryptology, but which is different in its feedback functions: whereas the former adopts a non-linear chaotic map, the latter uses a linear map.

In following sections, after a brief review of the characteristics of chaos, first, we shall outline the proposed CPRNG algorithm as well as the corresponding encryption scheme and, then, detail the chip implementation of the algorithm. It will be shown that the proposed algorithm can easily be implemented in embedded devices such as IC cards or re-configurable chips. As an example, a design targeted at the Xilinx Spartan XC2S100 FPGA running with a 50 MHz clock is presented. Furthermore, some experimental results will be presented, and a cryptanalysis will be reported as well.

15.2 A Brief Review of Chaos

Chaos is an ubiquitous phenomenon existing in deterministic non-linear systems that exhibit extreme sensitivity to initial conditions and random-like behaviour. There are several definitions of chaos, which are similar, but actually not equivalent [5]. For brevity, only one textbook definition is quoted here [7].

For simplicity, one-dimensional maps are discussed here to explain the definition of chaos. Consider a discrete dynamical system in the general form

$$x_{k+1} = f(x_k), \quad f : I \longrightarrow I, \;\; x_0 \in I, \tag{15.1}$$

where f is a continuous map on the interval $I = [0, 1]$. This system is said to be *chaotic* if the following conditions are satisfied [7]:

1. Sensitivity to initial conditions:

$$\begin{aligned} &\exists \delta > 0 \; \forall x_0 \in I, \varepsilon > 0 \; \exists n \in \mathbf{N}, y_0 \in I : \\ &|x_0 - y_0| < \varepsilon \;\Rightarrow\; |f^n(x_0) - f^n(y_0)| > \delta. \end{aligned} \tag{15.2}$$

2. Topological transitivity:

$$\forall I_1, I_2 \subset I \; \exists x_0 \in I_1, n \in \mathbf{N} : f^n(x_0) \in I_2. \tag{15.3}$$

3. Density of periodic points in I:
   Let $P = \{p \in I | \exists n \in \mathbf{N} : f^n(p) = p\}$ be the set of periodic points of f. Then P, is dense in I: $\overline{P} = I$.

Many properties of chaotic maps are related to the features of pseudo-random number generators, among which the following three are prominent:

Sensitivity to initial conditions This property ensures that, if we take two different seeds in a chaos-based pseudo-random bit generator, then, with probability one, we obtain two different sequences of bits. In practice, due to chaos, and with some appropriate partitions, two different seeds lead to completely different sequences.

Ergodicity A Lebesgue measure μ is said to be *invariant* if and only if it satisfies

$$\forall A \in \sigma(X),\ \mu(A) = \mu(\phi(A)), \tag{15.4}$$

where $\sigma(X)$ is the σ-algebra of all measurable subsets in X, and $\phi(\cdot)$ is a map on X. Here, (X, ϕ) is called *ergodic* if and only if it has trivial invariant sets, only, i.e., $\phi(B) \subseteq B$ implies either $\mu(B) = 0$ or $\mu(B) = \mu(X)$. The ergodicity implies that the state space cannot non-trivially be divided into several subspaces. So, if some orbit starts from an arbitrary point x, then it will never be restricted within a small region. This property indicates that, if a chaotic map is used to compose a pseudo-random number generator, then the number space will not be restricted to a small subspace.

Mixing The aforementioned system is *mixing* if the following condition is satisfied (assume $\mu(X) = 1$):

$$\lim_{n\to\infty} \frac{\mu(\phi^n(B) \bigcap A)}{\mu(A)} = \frac{\mu(B)}{\mu(X)}. \tag{15.5}$$

This property implies that if one binary sequence was generated by a chaotic map by bipolarisation, the bits are asymptotically independent, i.e., for $n = 1, 2, ...$, the bits b_n, b_{n+k} (considered as random variables) generated by a given mixing dynamical system (X, ϕ) are asymptotically independent as k increases [11, 23].

The above-described features of chaos are the foundations of chaos-based pseudo-random number generators as well as stream ciphers. To design secure cryptographic algorithms, all these properties should be well utilised.

15.3 Obtaining Chaotic Random Numbers with Good Properties

A traditional PRNG uses the *Linear Congruential Generator* which is usually expressed by the following formula:

$$x_{n+1} = (a \cdot x_n + b) \bmod N$$

where N is a natural number, $x_0, a, b \in \{0, 1, ..., N-1\}$, and "mod" is the modulo arithmetic operator. The linear congruential generator is demonstrated to be periodic with the maximum period equal to N. The maximum period is reached if and only if the following conditions are satisfied [1, 10]:

- b is relatively prime to N;
- if N is divisible with the prime number p, then $a - 1$ must be a multiple of p;
- if N is a multiple number of 4, then $a - 1$ must be a multiple of 4.

Since a chaotic sequence is aperiodic, ergodic, and sensitive to parameters, it is possible to construct a random number generator with good performance based on chaotic systems. Although a sequence directly evolving from a chaotic system has many intrinsically good properties, it is still highly correlated between adjacent points and has non-uniform distribution [1]. Therefore, the main problem to be tackled is to eliminate the strong correlations between numbers in chaotic sequences, and to shape them into uniform distribution.

Our approach to construct a uniformly distributed uncorrelated chaotic random number sequence is based on the generation of chaotic pseudo-random bit sequences. So it is worthwhile to first illustrate the construction of Pseudo-Random Bit Sequences (PRBS).

15.3.1 Chaotic Pseudo-Random Bit Sequence Generator

Traditionally, linear feedback shift registers (LFSR) are popular generators of pseudo-random bit sequences like the M-sequence. In practice, the Gold sequence is also used frequently [19]. By using a chaotic map, a Chaotic PRNG (CPRNG) is easy to construct. As an example, the method introduced in [11] is employed here to describe the process. Assume that a dynamical system, denoted as (X, ϕ), has a normalised invariant measure μ. Divide the state space X into two disjointed parts, X_0 and X_1, such that $\mu(X_0) = \mu(X_1) = \frac{1}{2}$. Take an initial value $x_0 \in X$ as seed, and start to evolve the system governed by ϕ and x_0. Suppose that after n iterations, a value x_n is obtained. The nth bit b_n of the sequence is then determined by the following "coin-tossing" formula:

$$b_n = \begin{cases} 0 & if\ x_n \in X_0 \\ 1 & if\ x_n \in X_1 \end{cases}$$

Thus, one obtains a bit sequence, $\{b_1, b_2, \cdots, b_n, \cdots\}$. Owing to the intrinsic properties of chaos, like ergodicity and mixing, the CPRNG has many good features: unique dependence of the sequence on the seed, equiprobable occurrence of "0" and "1", and asymptotic statistical independence of bits.

The following dynamical system is used to demonstrate the feasibility of the method:

$$X_{n+1} = \sin^2\left(z \arcsin \sqrt{X_n}\right) \tag{15.6}$$

where $z > 1$ is an integer parameter. The test results, specified by the suite FIPS 140-2 (cp. Section 15.3.2) [16], are shown in Table 15.1, which demonstrate several good properties of the sequence [16].

Table 15.1: FIPS 140-2 tests on example sequence

Monobit test	Poker test	Run test						Long run test
10,025	13.3504	2453	1240	677	353	141	144	0

Apart from employing a map as described in formula (15.6), we could also use many other one-dimensional chaotic maps to generate chaotic bit sequences with the above mentioned method such as the logistic map (shown in formula (15.8)), the tent map (formula (15.7)), the sine map (formula (15.10)), the shift map (formula (15.9)), and a new sawtooth-like map defined in formula (15.11). All maps can obtain FIPS 140-2 specified bit sequences, that means, the way to generate chaotic pseudo-random bit sequence is universal. Formulae (15.7) to (15.11) are some one-dimensional discrete chaotic maps that can be used in pseudo-random number generation.

$$x_{n+1} = \begin{cases} ax_n & x_n < \frac{1}{2} \\ a(1-x_n) & x > \frac{1}{2} \end{cases}, \text{if } 0 \leq a \leq 2 \tag{15.7}$$

$$x_{n+1} = \lambda x_n(1-x_n), \text{if } 0 \leq \lambda \leq 4 \tag{15.8}$$

$$x_{n+1} = 2x_n \bmod 1 \tag{15.9}$$

$$x_{n+1} = \frac{a}{4} \sin \pi x1, \text{if } 0 \leq a \leq 4 \tag{15.10}$$

$$x_{n+1} = cx_n \bmod 1, \text{if } c > 2 \tag{15.11}$$

15.3.2 Assessment of Pseudo-Random Bit Sequence Generators

In calculating quantities over complete periods of periodic PRNGs, the following three conditions should be satisfied [9]:

1. the number of "0" bits should differ from the number of "1" bits by at most one;
2. among all runs, a half should be of length 1, a quarter should be of length 2, an eighth should be of length 3, and so on, and for each of these lengths there should be equally many runs of "0" bits and runs of "1" bits;
3. the value of the autocorrelation function is equivalent to the period of a generator when the offset is 0; otherwise, the value is equal to a certain constant integer.

A practical and widely used standard test is specified by the National Institute of Standards and Technology (NIST) in the United States, called FIPS 140-2 [16]. It consists of 4 tests on a total of 16 aspects. More specifically, a single stream of 20,000 consecutive bits should be subjected to the following 4 tests:

1. *Monobit test.* A monobit test first counts the number of "1" in the 20,000 bits long bitstream. Denote this quantity by X. If $9,725 < X < 10,275$, then the test is passed.
2. *Poker test.* The poker test first divides the stream of 20,000 bits into 5,000 consecutive 4-bit segments. Count and store the number of occurrences of the 16 possible 4-bit values. Denote $f(i)$ as the number of each 4-bit value i, where $0 \leq i \leq 15$. Evaluate the following expression:

$$X = \frac{16}{5000} \sum_{i=1}^{n} [f(i)]^2 - 5000$$

The test is passed if $2.16 < X < 46.17$.

3. *Runs test.* A run is defined as a maximum sequence of consecutive bits of either all "1" or all "0", which is part of the stream of 20,000 bits. The incidences of runs of all lengths in the bitstream are counted and stored. The test is passed if the runs that occurred are within the corresponding intervals specified in Table 15.2. Note that for the purpose of this test, runs greater than 6 are considered to be of length 6.

Table 15.2: Run test specification

Length of run	Required interval
1	2,315–2,685
2	1,114–1,386
3	527–723
4	240–384
5	103–209
6+	103–209

4. *Long run test.* A long run is defined to be a run of length exceeding 25, of either all "0" or all "1." On a sample of 20,000 bits, the test is passed if there are no long runs.

15.3.3 A Chaotic Pseudo-Random Number Generator

Since the sequence of bits, $\{b_n\}_{n \in Z}$, is very similar to the sequence obtained with the coin tossing method, it is infinite and aperiodic. It is well known that the binary representation of any integer number can be expressed as:

$$z = \sum_{k=0}^{M-1} b_k 2^k = (b_0, b_1, ..., b_{M-1}).$$

In order to obtain random numbers with uniform distribution, we can select at random each bit in the binary representation of an integer. Here we use the method described in Section 15.3.1 to obtain a chaotic pseudo-random bit sequence $\{Z_n\}_{n \in Z} = \{(b_0, b_1, ..., b_{M-1})_n\}_{n \in Z}$. As b_k is taken from set $\{0, 1\}$ with probability $\frac{1}{2}$, and an integer is obtained after M such independent events, the integer can take any value between 0 and $2^M - 1$ with the equivalent probability $\frac{1}{2^M}$.

15.3.4 A Naive Design Scheme for a CPRNG-Chip

Now we consider chip implementation of the CPRNG described above. As Fig. 15.1 shows, if we want to generate integers between 0 and $2^M - 1$, M sets of chaotic pseudo-random bit sequence generators have to be implemented. In each set, a chaotic sequence is first generated, and then fed into a comparator to obtain a binary sequence by one bit quantisation. Note that, although in Fig. 15.1 the logistic map is used as chaotic sequence generator, actually many other chaotic systems are available, depending on the feasibility of a practical implementation.

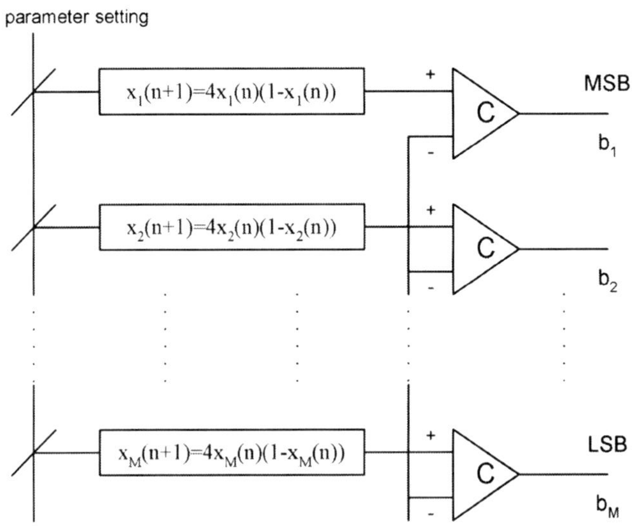

Fig. 15.1: Block scheme of CPRNG

15.4 Chaotic Pseudo-Random Number Generator Using Digitised Chaos

Since in practical applications chaotic pseudo-random numbers are generated by computers in which computation is performed in finite precision, performance degradation of a digital implementation is unavoidable. The most significant degradation is caused by the appearance of a period in the CPRN sequence. If we had a computer with infinite computational precision, we could generate aperiodic chaotic pseudo-random number sequences by the method mentioned in Section 15.3. But if we realise the method in finite precision, the non-linear map is no longer chaotic and, instead, it shows a kind of stabilisation which leads to a random cycle length in a CPRN sequence. Numerical simulation shows that the cycle length is dependent on computational precision, and heavily relies on initial values of the actual map [3]. In other words, suppose we digitise numbers in the map in N bits, and the upper boundary of the loop length is denoted by M, then the $M(N)$-dependence is scattered in an unknown manner obeying the relations $0 < M < (1/N)^{-\varepsilon}$, $\varepsilon = 0.68 \pm 0.05$.

Many people have encountered such a problem and wanted to solve it but, unfortunately, until now no systematic theory has been found. Only several engineering remedies have been proposed to deal with the problems such as using higher finite precision, cascading multiple chaotic systems, and the perturbation-based algorithm [12].

Here, we introduce another approach that uses a feedback mechanism to increase the cycle length of a digitised chaotic map. To reduce the numbers of multiplications, which is favourable for chip implementation, a sawtooth-like map is employed in our proposed CPRNG instead of a logistic map as described in the previous sections. First, we investigate the properties of a sawtooth-like map defined by $x_{n+1} = cx_n \bmod 1$.

15.4.1 Properties of Map $x_{n+1} = cx_n$ mod 1

Since the sawtooth-like map $x_{n+1} = cx_n \bmod 1$ is derived from a simple linear map, it can simplify the software and hardware design of a CPRBS generator. We shall demonstrate that above map has many good properties, and that it meets in full our requirements for a CPRBS generator.

First, we inspect the Lyapunov Exponent (LE) of map (15.11). According to the definition of the Lyapunov exponent of one-dimensional maps, the LE of the map, λ, can be calculated as:

$$\begin{aligned} \lambda &= \lim_{T\to\infty} \frac{1}{T} \sum_{i=0}^{T-1} \ln |f'(x_i)| \\ &= \lim_{T\to\infty} \frac{1}{T} \sum_{i=0}^{T-1} \ln |c| \\ &= \ln c \end{aligned} \tag{15.12}$$

So, when $c > 1$, the map is chaotic.

Furthermore, we are interested in the distribution of the sequence generated by the map. To analytically derive the distribution of a sequence generated by a map, the Frobenius-Perron equation has to be used. Suppose a one-dimensional map is denoted by $y = f(x)$, and the distributions of variables x and y are $v(x)$ and $v(y)$, respectively, according to [18, 6, 24], they should follow the Frobenius-Perron equation:

$$v(y) = \sum_{x\in f^{-1}(y)} \frac{v(x)}{|f'(x)|} \tag{15.13}$$

It is not an easy task to give an analytic formula that can describe the distribution of the sequence generated by map (15.11). But if the parameter c is restricted to be not less than two and is an integer, the distribution of $v(x)$ should be unified, i.e., $v(x) = 1$. This conclusion can be demonstrated by using the aforementioned Frobenius-Perron equation.

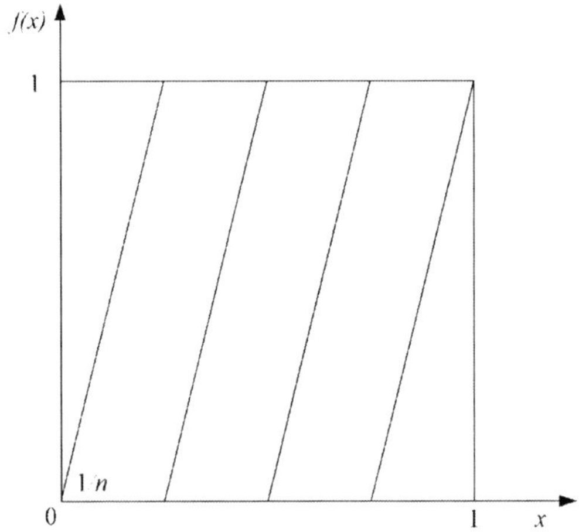

Fig. 15.2: Curve of map $f(x) = nx \bmod 1$

Proof: Without loss of generality, we set $c = n$, where n is an integer larger than 1. Then, map (15.11) can be expressed as:

$$x_{n+1} = \begin{cases} nx_n & \text{if } 0 \le x_n \le \frac{1}{n} \\ nx_n - 1 & \text{if } \frac{1}{n} < x_n \le \frac{2}{n} \\ \cdots & \cdots \\ nx_n - (n-1) & \text{if } \frac{n-1}{n} < x_n \le 1 \end{cases} \tag{15.14}$$

Its corresponding function, denoted by $f(x)$, can be described by formula (15.15).

$$f(x) = \begin{cases} nx & \text{if } 0 \le x \le \frac{1}{n} \\ nx - 1 & \text{if } \frac{1}{n} < x \le \frac{2}{n} \\ \cdots & \cdots \\ nx - (n-1) & \text{if } \frac{n-1}{n} < x \le 1 \end{cases} \tag{15.15}$$

The curve described by formula (15.15) is a piecewise zigzag line (as shown in Fig. 15.2), whose input-output relationship is linear, and has the same slope in each interval $(\frac{i}{n}, \frac{i+1}{n}]$. Thus, $|f'(x)| = n$, and

$$x = f^{-1}(y) = \begin{cases} \frac{y}{n} & \text{if } 0 \le x \le \frac{1}{n} \\ \frac{y+1}{n} & \text{if } \frac{1}{n} < x \le \frac{2}{n} \\ \cdots & \cdots \\ \frac{y+n-1}{n} & \text{if } \frac{n-1}{n} < x \le 1 \end{cases} \tag{15.16}$$

Therefore,

$$v(y) = \sum_{x \in f^{-1}(y)} \frac{v(x)}{|f'(x)|} = \sum_{x \in f^{-1}(y) = \frac{y}{n}} \frac{1}{n} + \ldots + \sum_{x \in f^{-1}(y) = \frac{y+n-1}{n}} \frac{1}{n} = n \times \frac{1}{n} = 1$$

which demonstrates that the distribution of the sequence generated by map (15.11) is even. In other words, to obtain FIPS 140-2 compatible chaotic pseudo-random bit sequences by using the method described in Subsection 15.3.1, the proper threshold should be set to 0.5.

15.4.2 A CPRBS Generator Scheme Based on Chaotic Non-linear Feedback Shift Registers

Recall the structure of a feedback shift register as shown in Fig. 15.3. It consists of two parts: an n-bit register used to right shift bits, and a Boolean feedback function $f(a_1, a_2, ..., a_n)$ that feeds a binary value determined by the current states of registers $a_1, a_2, ..., a_n$ back to the input. The Boolean feedback function f can be either linear or non-linear. To simplify design and analysis, a linear Boolean feedback function is usually employed. Such a structure is often called a linear feedback shift register, which has been investigated mathematically.

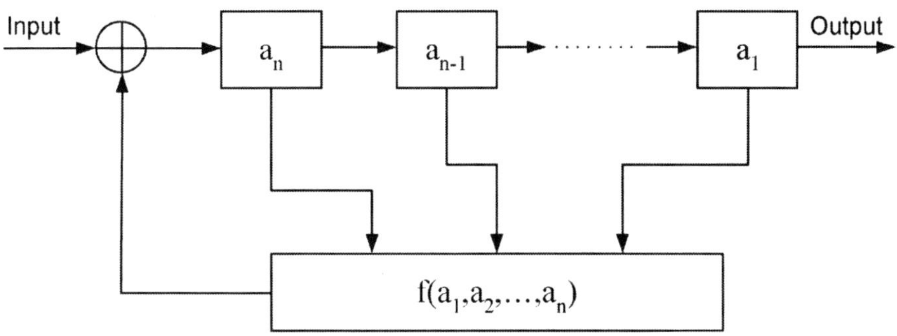

Fig. 15.3: Structure of feedback shift register

The structure of a chaotic non-linear feedback shift register (CNFSR) is similar to that of a linear feedback shift register, but only differs in the Boolean feedback function. The latter uses a non-linear Boolean feedback function that was derived from a one-dimensional chaotic map. Through experiments we can demonstrate that, by these means, the cycle length can be increased significantly.

A schematic drawing of a pseudo-random bit sequence generator is given in Fig. 15.4, where a digitised chaotic map, $x(k+1) = c(x(k)+1) \bmod 2^n$, followed by a comparator acts as a Boolean feedback function. Two parameters control the behaviour of the bit sequence, viz., c and initial value $x(0)$, both of whom are of length n bits. Suppose in the k-th step the integer input to the digitised chaotic map is $x(k) = x_{n-1}^k x_{n-2}^k ... x_0^k$, where x_i^k is the i-th bit in integer $x(k)$. After one round of iteration, we can obtain the output of the map, $x(k+1) = x_{n-1}^{k+1} x_{n-2}^{k+1} ... x_0^{k+1}$, which is further divided into two parts, viz., an even part $x_e(k+1) = x_{n-2}^{k+1} x_{n-4}^{k+1} ... x_0^{k+1}$, and an odd part $x_o(k+1) = x_{n-1}^{k+1} x_{n-3}^{k+1} ... x_1^{k+1}$. The ensemble output of the Boolean feedback function is determined by:

$$b = \begin{cases} 0 & if\ x_o(k+1) \leq x_e(k+1) \\ 1 & if\ x_o(k+1) > x_e(k+1) \end{cases}$$

The bipolarised value b is further exclusively ored with a plain-bit to obtain a feedback bit b^*. The bit b^* is appended at the end of $x(k)$, and the new $x(k)$ is left-shifted by one bit position to obtain a new integer $x^*(k+1) = x_{n-2}^k ... x_0^k b^*$ that is to be used in the next iteration round.

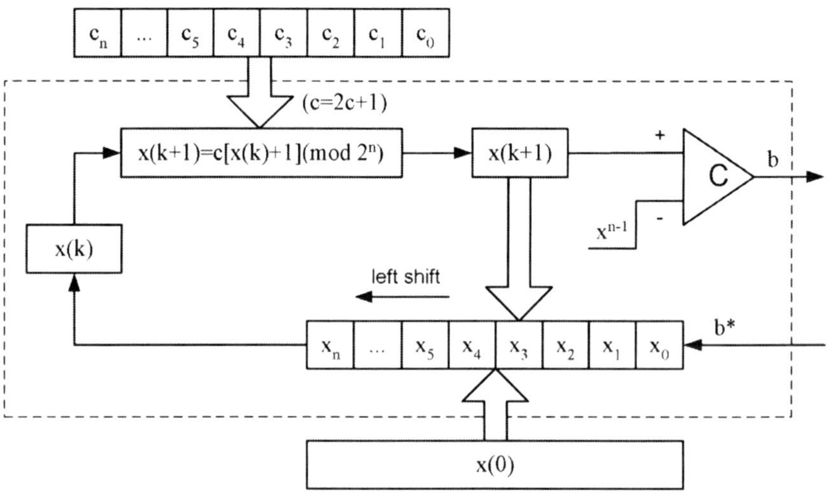

Fig. 15.4: Structure of chaotic pseudo-random bit sequence generator

There are three issues that should be noticed:

1. Simulation results show that parameter c should be an odd integer that is not less than 3, otherwise it is easy for the digitised map to fall into a stable state that leads to the appearance of short cycles. Thus, in Fig. 15.4, a left-shift and a plus-one-operation are designed to make all input parameters c be odd numbers and larger than 1.
2. The above mentioned non-linear feedback shift register scheme is a universal one. There are no special requirements for the digitised chaotic

map. Therefore, maps defined in formula (15.7) to (15.10) are all good candidates to be employed in the new non-linear feedback shift register scheme.

3. The non-linear feedback shift register scheme directly uses a digitised chaotic map to generate an integer sequence instead of maps operating on an infinite field, where a continuous chaotic map is first used to generate a real number followed by a digitising procedure. The former scheme is more accommodable to the implementation of hardware and has higher operating speed, since there are no floating-point operations.

15.5 A Stream Cipher Using Chaotic Non-linear Feedback Shift Registers

Using the above mentioned CNFSR scheme, a stream cipher can be constructed. Actually, multiple sets of CNFSRs have composed a pseudo-random number generator (PRNG), by which groups of pseudo-random numbers can be generated, and used to mask plain-texts. Every bit in encrypted byte streams is then fed back to each corresponding set of CNFSRs. So, the performance of the proposed stream cipher heavily depends on the structure of the CPRNG that consists of those CNFSRs.

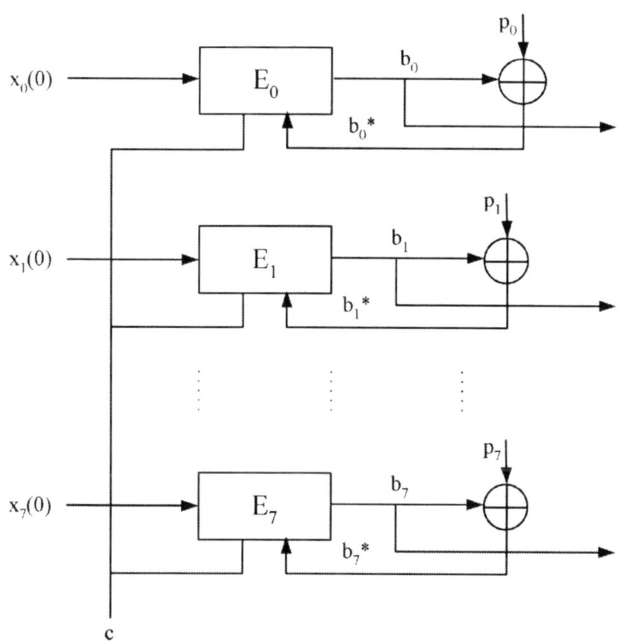

Fig. 15.5: CPRNG with uncoupled CNFSR sets

Since the pseudo-random bit sequence generated by a CNFSR is highly uncorrelated and exhibits long cycle length, the chaotic PRNG, which consists of sets of them, can easily be realised in hardware. Here, we shall show two types of possible CPRNGs. In the first one shown in Fig. 15.5, each set of CNFSR does not interfere with others, while in the second one, as shown in Fig. 15.6, each set of CNFSRs is coupled in cascade with its neighbouring set. Empirical results showed that the latter renders better performance.

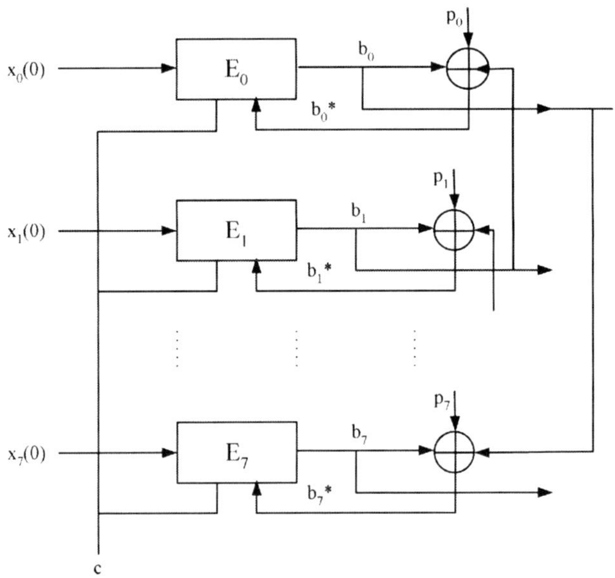

Fig. 15.6: CPRNG with coupled CNFSR sets

For convenience, we use CPRNGs with eight sets of CNFSRs as an example to investigate their structures in terms of coupled and uncoupled connections. Fig. 15.5 shows the CPRNG with uncoupled CNFSR sets. In this type of CPRNG, each set of bit sequence generator, namely, E_0, E_1, ..., E_7, is independent of any other, and all bits from eight channels form one byte. Here, $x_0(0)$ to $x_7(0)$ are initial values of chaotic maps, which together with parameter c act as cipher keys. Sets of plain-bits are denoted as p_0, p_1, ..., p_7, which are masked by each CNFSR's outputs b_0, b_1, ..., b_7 to form the cipher bits b_0^*, b_1^*, ..., b_7^*. All cipher bits are fed back to the CNFSRs. The feedback of each channel makes the statistical distribution of the chaotic pseudo-random numbers more uniform. However, because each set of CNFSR is independent, any disturbance of one channel will not interfere with any other, and the period caused by digitisation could not be completely diminished under the condition of zero input. That is, when $p_0 = p_1 = ...p_7 = 0$, the CPRNS shows certain kinds of periods.

We can select one channel output of the chaotic PRBS to demonstrate this assertion by drawing its auto-correlation plot, shown in Fig. 15.7, in which many peaks appear, meaning that the generated PRBS has many similar subsets. The coupling of CNFSRs could complicate the ensemble behaviour of the CPRNG by diffusing and confusing each CPRBS. Fig. 15.8 demonstrates the excellent performance obtained with the use of coupling among CNFSRs, since only one peak is found in the auto correlation plot. Another FIPS 140-2 test also shows the improved performance of coupled CPRNGs over uncoupled ones. Table 15.3 provides the results. Both of the tests of coupled and uncoupled CPRNGs are performed on 256 groups of data with initial value $x_0 = 31$, and c varying from 3 to 513 with an interval of 2. Since the coupled CPRNG scheme prevails over the uncoupled one, in the following section, we shall discuss and make performance tests on the former scheme, only.

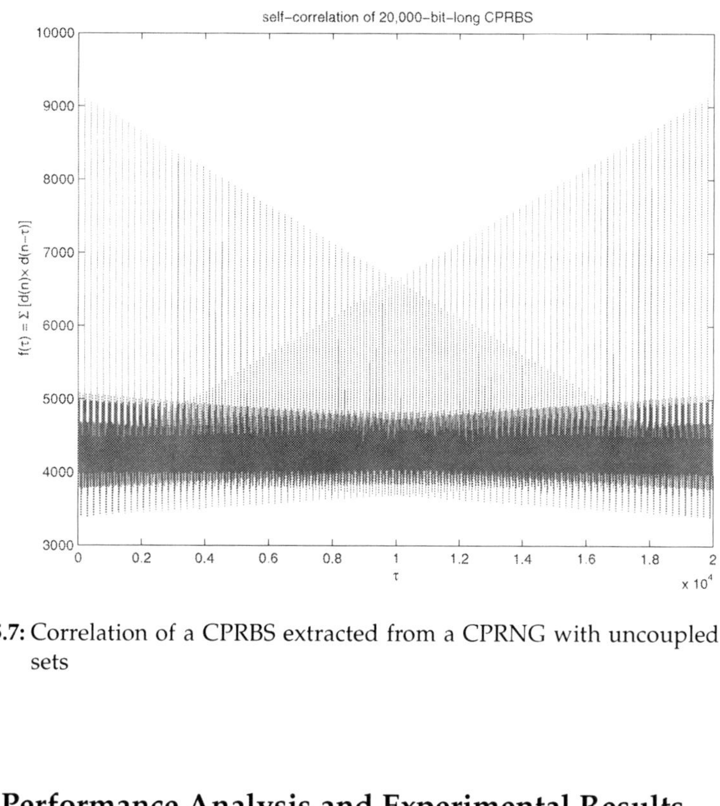

Fig. 15.7: Correlation of a CPRBS extracted from a CPRNG with uncoupled CNFSR sets

15.6 Performance Analysis and Experimental Results

By use of the proposed chaotic pseudo-random number generator, we can encrypt digitised texts, speech, or images byte by byte. In this section, we shall take image encryption as an example to further analyse the security of the proposed scheme.

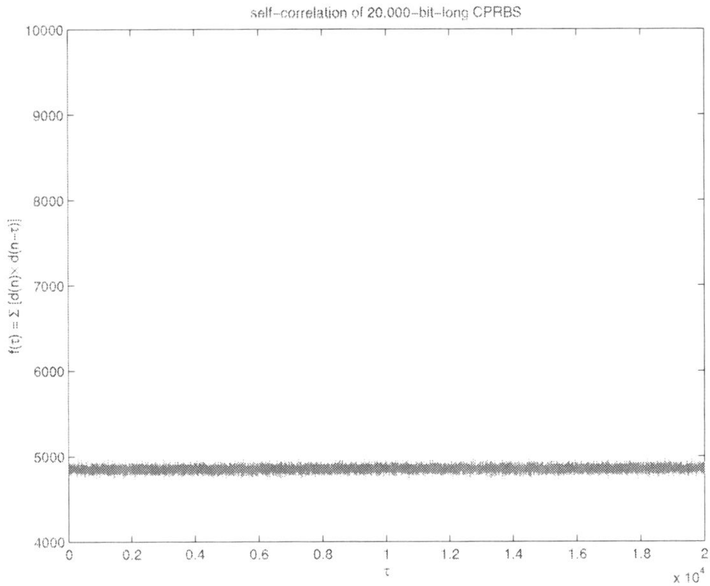

Fig. 15.8: Correlation of a CPRBS extracted from a CPRNG with coupled CNFSR sets

Table 15.3: Results of FIPS 140-2 tests on bit sequences in terms of zero-input

	Uncoupled	Coupled
Monobit test	10,041	10,009
Poker test	851.7	15.239
Run length = 1	2,546.5	2,499.5
Run length = 2	1,225.2	1,248.3
Run length = 3	618.28	623.95
Run length = 4	295.82	311.71
Run length = 5	146.81	157.89
Run length ≥ 6	159.7	154.73
Long run test	0	0

15.6.1 Key Space Analysis

In this encryption scheme, the size of the key space depends on the precision of the computation. As shown in Section 15.4, each non-linear feedback shift register (NFSR) has two control parameters, namely, c and the initial value x_i, that determine the behaviour of the bit sequence generated. If we set the computational precision to $n = 20$ bits, each set of NFSRs needs 40 bits to manipulate the generation of a PRBS. To simplify hardware design, we use the same c for all m sets of NFSRs. Since one byte has eight bits, we need 8 sets of NFSRs to accomplish the task of simultaneously encrypting one byte at a time. Thus, the key length of our scheme is about $20 + 20 \times 8 = 180$ bits. In this regard, the key space is really huge, and it is absolutely impossible to break the cryptosystem by exhaustive search.

15.6.2 Key Sensitivity Test

Thanks to the employment of a chaotic map, our encryption scheme is very sensitive to the encryption key. We have performed two kinds of tests on images and bit sequences.

- The first test is performed on a bit sequence generated by a NFSR. At first, we set $c = 37$ and $x(0) = 100$ to generate a bit sequence denoted by $s_{37,100}$. Then, we change c to 36 and keep $x(0)$ unchanged, and generate another bit sequence denoted by $s_{36,100}$. Finally, we compare these two sequences to find out the percentage of bit differences. Then, in the next round, we fix $c = 37$ and slightly change $x(0)$ to check the percentage of bit differences. The results of these two tests are 50.31% and 51.11%, respectively, which means in both scenarios about half the quantity of bits has been changed as a result of a slight modification of keys, only.
- Next, we use a pair of keys which differ in just one bit to encrypt the same image twice, and find out the differences between the two encrypted images. The two pairs of keys are $X(0) = [332\ 32\ 124\ 83\ 251\ 72\ 172\ 12], c = 37$ and $X(1) = [331\ 32\ 124\ 83\ 251\ 72\ 172\ 12], c = 37$ respectively, and the difference of the two encrypted images is about 99.56%, which demonstrates the high key sensitivity of the proposed encryption scheme.

15.6.3 Performance of the Encrypted Bit Sequence

The encrypted bit sequence in each set of NFSRs also exhibits good statistical properties. Here, we still use the FIPS 140-2 test to inspect the performance of an encrypted bit sequence. Table 15.4 presents the test results in which the testee is a bit sequence extracted form a cipher image that is encrypted by XOR-ing its counterpart plain-image with a pseudo-random number generated by the aforementioned CPRNG.

Table 15.4: Results of FIPS 140-2 tests on a bit sequence after being XOR-ed with plain-text

Monobit test	Poker test	Run test						Long run test
10,013	16.851	2,530	1,252.5	616.19	305.55	150.25	146.16	0

15.6.4 Statistical Tests

The histograms of several size 256×256 grey-scale images were analysed with different contents and natures. Test results show that histograms of encipher-images are very uniform, which makes statistical attacks difficult. Fig. 15.9 shows one typical result.

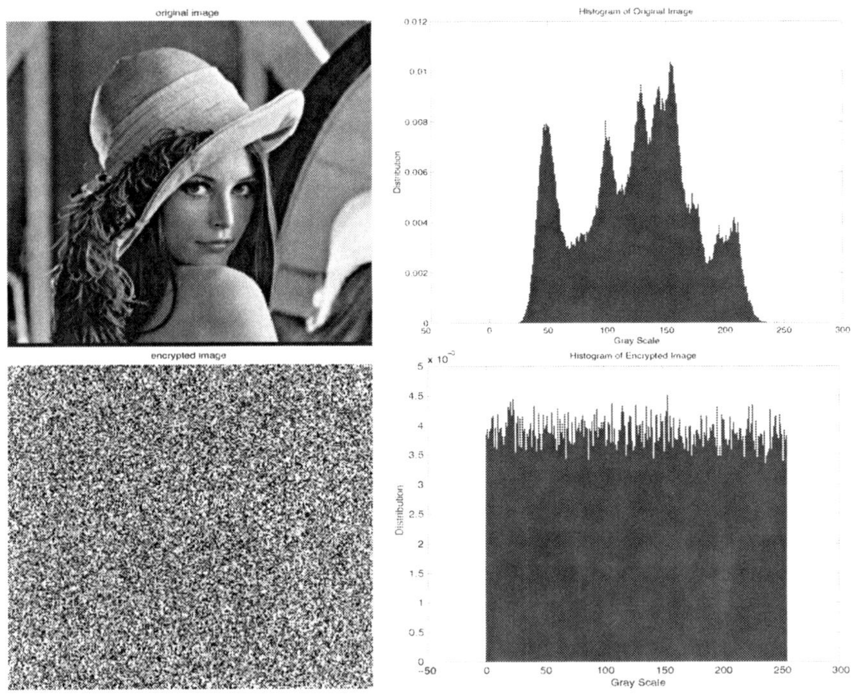

Fig. 15.9: Histograms of plain-image and encrypted image XOR-ed with a sequence of CPRNs

15.6.5 Sensitivity to Slight Changes in Plain-text

A secure encryption scheme should resist known plain-text or chosen plain-text attacks, each of which can be used by opponents to guess cipher keys so that the cryptosystem is broken. If we have two pieces of plain-text, namely P and P^*, such that $\|P - P^*\| << \varepsilon$, and if their corresponding cipher-texts $C = E(P)$ and $C^* = E(P^*)$ are significantly different, i.e., $\|C - C^*\| >> M$, we say the encryption scheme E is sensitive to plain-text. If an encryption scheme is so, it can resist known plain-text or chosen plain-text attacks. Since our encryption scheme uses a mechanism to feed back cipher-bits, the encrypted output is highly correlated to its corresponding plain-text. Thus, the scheme can resist known plain-text or chosen plain-text attacks.

We use two images to demonstrate this conclusion. These two images are different in only one byte at their top-left regions (The result of this test heavily depends on the position of the byte difference in plain-bit streams; therefore, we choose the position of byte difference in the top-left corner of both images to make the test reasonable). The experimental results are shown in Fig. 15.10, where we find that almost all pixels have been changed after one pixel in the top-left corner was modified. Actually, the ensemble pixel differences of the two encrypted images is about 92.58%.

Fig. 15.10: Sensitivity to slight changes in plain-text

15.7 Chip Implementation

The encryption scheme proposed above is highly suitable for hardware implementation, as there is no need for floating-point operations. To map our algorithm onto a chip, just some registers, fixed-point multipliers, comparators, and some other logic circuits are required. The most important observation is that the algorithm is essentially a parallel one. In Fig. 15.11 a diagrammatic sketch of one set of NFSRs is presented. An encryption chip consists of n sets of such modules (denoted by E_i) that can encrypt n data bits simultaneously, as shown in Fig. 15.12. Besides the aforementioned merits, another advantage of the design is the flexibility in setting the parallel operating window-length, n, because n can arbitrarily be set to meet the requirements of different applications. For instance, one can set $n = 7$ for ASCII text encryption, $n = 8$ for 256 grey-scale imagea, and $n = 24$ for 24-bit true colour images, respectively.

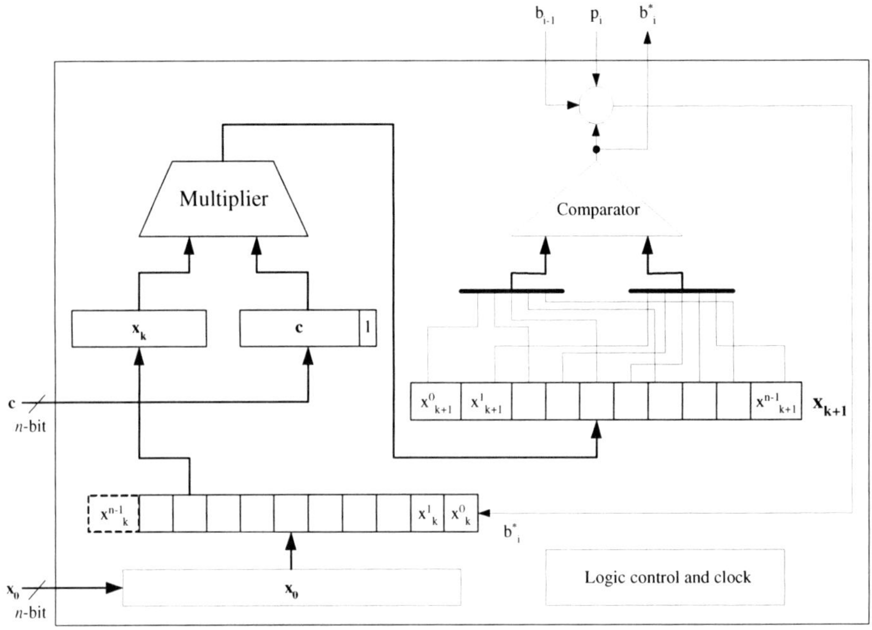

Fig. 15.11: Chip design of a set of NFSRs

To verify the design realising our stream cipher algorithm on a chip, an FPGA chip along with some accompanying circuits and some Electronic Design Automation (EDA) tools like VHDL, Xilinx ISE (Integrated System Engineering) are needed. A Field Programmable Gate Array (FPGA) is a type of programmable device. The computing core of an FPGA consists of a highly complex re-programmable matrix of logic integrated circuits, registers, random access memory, and routing resources. These can be used to perform logic and arithmetic operations, storage, and data transfer between different parts of the system. Moreover, as no CPU controls the entire chip, and no sequential instructions have to be processed, thousands of operations can be performed in parallel on an FPGA during every clock cycle [2]. With the VHSIC Hardware Description Language (VHDL) digital integrated circuits are described. It enables the execution of concurrent statements in contrast to sequential statements. Therefore, it is widely used in the design of real-time systems, especially FPGA based ones [2]. Here we use a Xilinx Spartan XC2S100 FPGA chip and Xilinx ISE 5.2 to accomplish our chip design and algorithm verification work. The whole realisation procedure of our algorithm went along as follows: the algorithm was first described in VHDL,

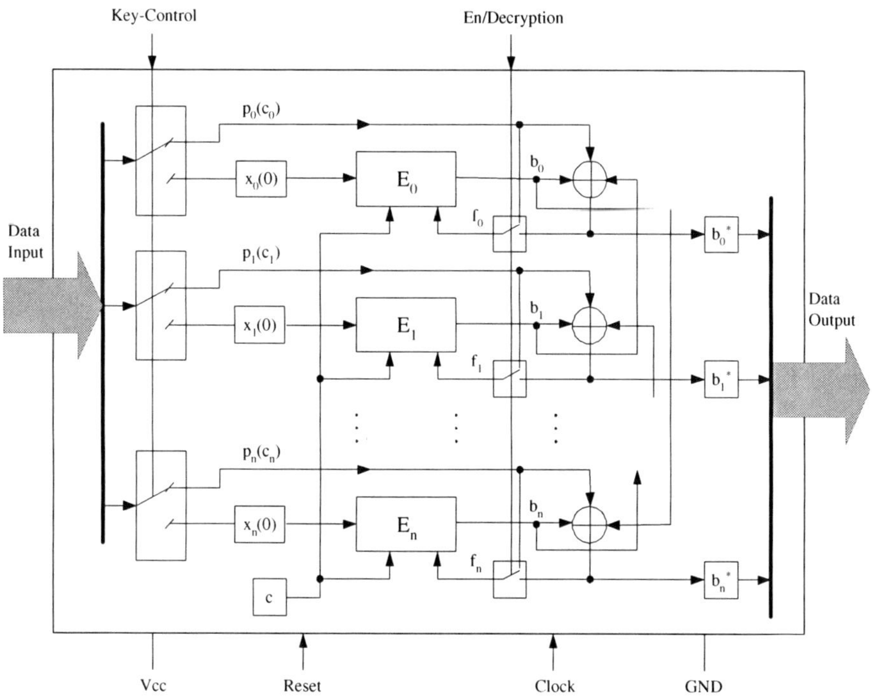

Fig. 15.12: Block diagram of a chip for chaotic stream encryption

then, the VHDL program was synthesised and translated into code that an FPGA chip can recognise by means of the Xilinx FPGA/CPLD design platform, ISE. Finally, the code was loaded into the chip to make it function as desired.

The flow-chart of the algorithm proposed in Section 15.5 is shown in Fig. 15.13, based on which the VHDL routine was programmed. We can observe that the algorithm is implemented with successive passes through the data path. In Fig. 15.13, the arrows indicate the direction of data flowing. In fact, the units *cell-2*, *cell-3*, and *cell-4* are implemented with combinational logic circuits, while the units *cell-5*, *cell-6*, and *cell-7* are operated as sequential logic circuits. The clock cycle used in the sequential logic circuit affects the encryption speed.

Fig. 15.14 shows the interface of the encryption chip. The interface circuit is very simple, and the chip is also not complex, which makes it suitable to be integrated into handheld or mobile devices. As shown in Fig. 15.14, besides the power supply part (pins Vcc and GND), only two types of pins are present, viz., there are five input pins for control signals, and two sets of pins act as buses for in- and out-going data. In detail, the function of each pin on the chip is:

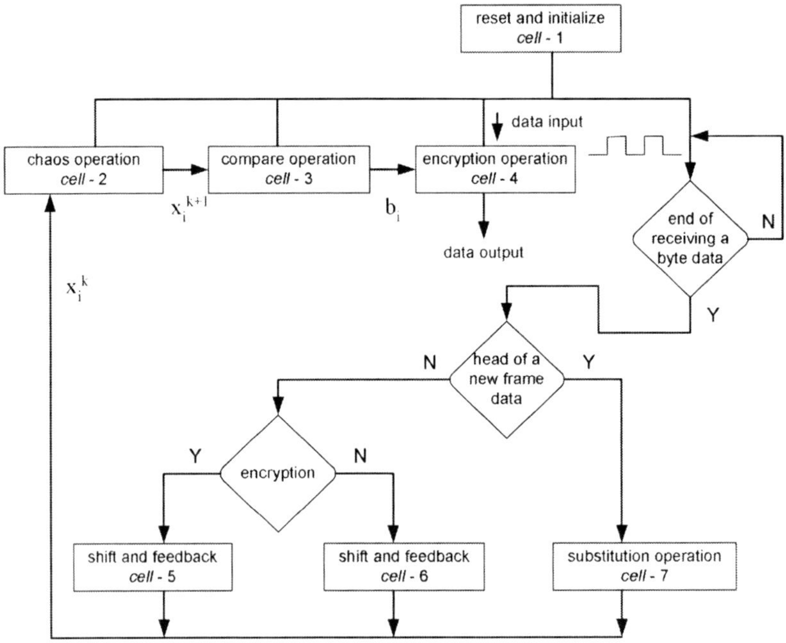

Fig. 15.13: Flow-chart of the proposed algorithm

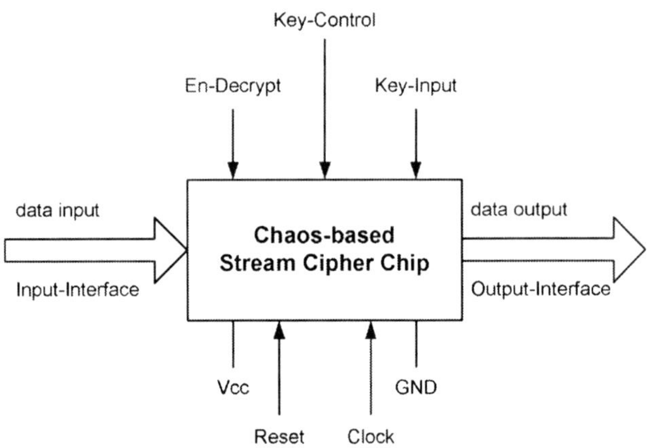

Fig. 15.14: Interface of the chip for chaotic encryption

- The two databuses constitute a bidirectional path for data input and output, which links the chip with its peripherals such as a recorder, a camera, or some other electric devices. Since the peripheral devices can vary, the communication protocols between peripheral devices and the chip are different. Two commonly used communication standards are supported in our design, which are the RS-232 serial communication protocol, and the standard parallel-port transmission, respectively. Other databus standards like USB, IEEE 1394, I^2C, and IEEE 488 are under development.

- As usual, the RESET pin, that most chips should have, is used to restore the chip's initial states.
- The CLOCK pin is used to provide clock-pulses that ensure the chip to work normally. In our design, the chip can reach its maximum working frequency of 100 MHz.
- The EN-DECRYPT pin is an on-off switch that controls the chip to switch between encryption and decryption. Since all encryption and decryption programs are loaded into the chip at the time when it is reset, there must be a signal indicating whether the chip is to act as encrypter or decrypter. When the EN-DECRYPT signal is set to logical one (high electrical level), the chip performs encryption, otherwise, decryption.
- The KEY-CONTROL pin acts as a selector. When it is set to logical zero, a fixed key preset in the chip will be used. Otherwise a user-specified key that has previously been transmitted from a peripheral device (say, a computer linked through RS-232) is to be used.
- The KEY-INPUT pin controls the transmission of the user-specified key as mentioned above. When logical-zero signal is applied to the pin, the chip receives and interprets data input from a peripheral device as key instead of plain-text, whereas, when the electrical level on the pin is set to high, all input data are subject to encryption or decryption, depending on the EN-DECRYPT signal.

The FPGA design implemented together with a testing device are shown in Fig. 15.15. Some tests were made, and one typical result each for text and image encryption is shown in Fig. 15.16. Here, the number of CNFSRs was set to 7 for text encryption, and 8 for image encryption, and the computational precision was set to 20 bits. Further experimental results show that the encryption operation consumes 2.5 clock-cycles, only, i.e., if a block of data is subject to byte-wise encryption given a 50 MHz clock frequency, the throughput is 20 Mbytes per second. Due to the symmetry of the encryption and decryption algorithms, the time consumed for decryption is the same as for encryption. As the operation speed scales linearly with the clock frequency, implementations as application specific integrated circuits promise much higher performance.

15.8 Conclusion

This research focused on developing a simple and practically useful data encryption algorithm, which can be accommodated in an FPGA. Although methods based on software implementation have many merits like ease of use, ease to upgrade, or portability, an implementation on a chip, like an

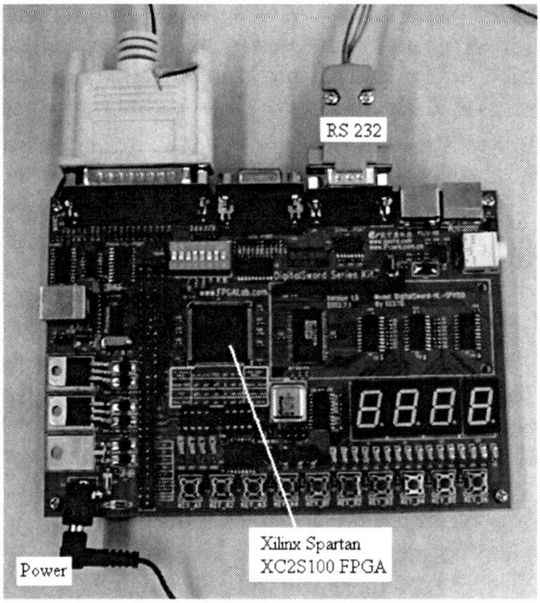

Fig. 15.15: FPGA chip and device used for testing

FPGA, is even more attractive when considering a cryptosystem. On one hand, a hardware implementation is faster than software. On the other, cryptographic algorithms implemented in hardware are physically more secure as they cannot easily be read or modified by an outside eavesdropper.

Driven by the above considerations, in this chapter a chaotic stream encryption scheme was constructed that is suitable for chip implementation. The design philosophy was borrowed from classical LFSR design, which is broadly accepted as a mature technology, and is employed in the generation of widely-used M-sequences. To complicate the structure of the pseudo-random number generator and make its cryptanalysis difficult, a one-dimensional chaotic map was introduced. Unlike other known chaotic pseudo-random number generators, this scheme uses a digitised map instead of a continuous one, which has simplified the chip design and has facilitated its implementation in hardware. Security analysis as well as experimental tests have demonstrated the feasibility of the design.

The proposed algorithm is easy to implement on any re-configurable chip such as an FPGA or a Complex Programmable Logic Device (CPLD). As a demonstrator, a Xilinx Spartan XC2S100 FPGA chip was used for an experimental implementation, on which the algorithm has been realised successfully. Experimental results show that the proposed algorithm runs very fast on the FPGA chip, since the implementation is essentially parallel. The encryption speed is as fast as up to 20 Mbytes per second under a 50 MHz clock

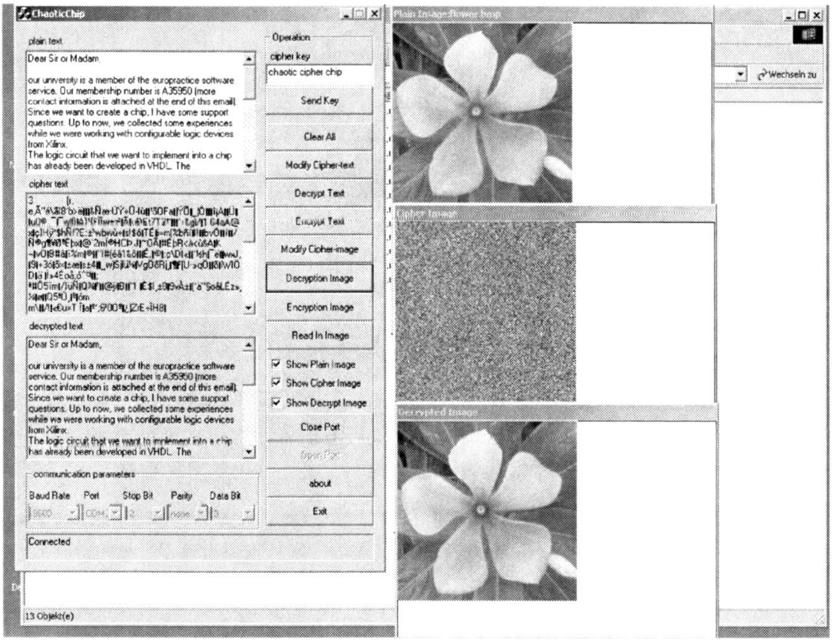

Fig. 15.16: Text and image encryption using the chip designed

frequency. Due to the symmetry of the encryption and decryption algorithm, the time consumed for decryption is the same as for encryption. The operation speed scales linearly with the clock frequency, allowing for continuous increases in performance.

Acknowledgment

The first author was partially supported by the Natural Science Foundation of Jiangsu Province in China under grant BK2004421.

References

1. Andrecut M (1998) Logistic Map as a Random Number Generator. Int Journal of Modern Physics B 12 (9): 921–930
2. Ashenden JP (1997) The Designer's Guide to VHDL, 2nd edn. Morgan Kaufmann Publishers
3. Černák J (1996) Digital generators of chaos. Phys Lett A 214: 151–160
4. Chen G, Dong X (1998) From Chaos to Order: Methodologies, Perspectives and Applications. World Scientific, Singapore

5. Chen, G (2003) Chaotification via Feedback: the Discrete Case. In: Chen G, Yu X (eds) Chaos Control: Theory and Applications. Springer, Berlin Heidelberg New York, pp 159–177
6. Collet P, Eckmann J-P (1980) Iterated Maps of the Interval as Dynamical Systems. Birkhäuser, Basel
7. Devaney RL (1989) An Introduction to Chaotic Dynamical Systems, 2nd edn. Addison-Wesley, Reading, MA
8. Fridrich J (1998) Symmetric Ciphers based on Two-dimensional Chaotic Maps. Int J Bifurcation and Chaos 8(6): pp 1259–1284
9. Golomb SW (1967) Shift Register Sequences. Holden-Day, San Francisco
10. Knuth DE (1998) The Art of Computer Programming, vol 2: Seminumerical Algorithms. Addison-Wesley
11. Kotulski Z, Szczepański J (2000) On Constructive Approach to Chaotic Pseudorandom Number Generator. In: Proc Regional Conf on Military Communication and Information Systems, CIS Solutions for an Enlarged NATO, pp 191–203
12. Li S, Mou X, Cai Y (2006) Chaotic Crypography in Digital World: State-of-the-art, Problems and Solutions. `http://www.hooklee.com`
13. Mao Y, Chen G (2005) Chaos-based Image Encryption. In: Bayro E (ed) Handbook of Computational Geometry for Pattern Recognition, Computer Vision, Neurocomputing and Robotics, Springer, pp 231–265
14. Masuda N, Aihara K (2002) Cryptosystems with Discretized Chaotic Maps. IEEE Trans Circuits and Systems-I 49(1): 28–40
15. Matthews R (1989) On the Derivation of a 'Chaotic' Encryption Algorithm. Cryptologia, XIII (1): 29–41
16. National Institute of Standards and Technology (2001) Security Requirements for Cryptographic Modules. FIPS PUB 140-2. `http://csrc.nist.gov/publications/fips/fips140-2/fips1402.pdf`
17. Pichler F, Scharinger J (1995) Ciphering by Bernoulli-shifts in Finite Abelian groups. In: Kaiser HK, Muller WB, Pilz GF (eds) Contributions to General Algebra 9: 249–256
18. Schmitz R (2001) Use of Chaotic Dynamical Systems in Cryptography. J Franklin Institute 338: 429–441
19. Schneier B (1995) Applied Cryptography: Protocols, Algorithms, and Source Code in C, 2nd edn. Wiley, New York
20. Shannon CE (1949) Communication Theory of Secrecy System. Bell System Technical Journal 28: 656 – 715
21. Stojanovski T, Kocarev L (2001) Chaos-based Random Number Generators — Part I: Analysis. IEEE Trans Circuits and Systems I: Fundamental Theory and Applications 48 (3): 281–288
22. Stojanovski T, Pihl J, Kocarev L (2001) Chaos-based Random Number Generators — Part II: Practical Realization. IEEE Trans Circuits and Systems I: Fundamental Theory and Applications 48 (3): 382–385
23. Szczepanski J, Kotulski Z (2000) Chaotic Pseudorandom Numbers Generators based on Chaotic Dynamical Systems. Open Sys And Information Dyn 7: 1–10
24. Walters P (1982) An Introduction to Ergodic Theory. Springer

Index